THE CHICAGO
PUBLIC LIBRARY

DESIGN OF CONSTRUCTION AND PROCESS OPERATIONS

Design of Construction and Process Operations

DANIEL W. HALPIN
Head, Construction Program
Georgia Institute of Technology
Atlanta, Georgia

RONALD W. WOODHEAD
Head, Department of Engineering
Construction and Management
University of New South Wales
Sydney, Australia

JOHN WILEY & SONS
NEW YORK   SANTA BARBARA   LONDON   SYDNEY   TORONTO

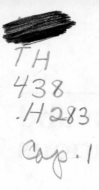

Library of Congress Cataloging in Publication Data:

Halpin, Daniel W.
  Design of construction and process operations.

  Bibliography: p.
  Including indexes.
  1. Construction industry—Management—Mathematical
models.   2.   Building—Data processing.
I. Woodhead, Ronald W., joint author.   II.   Title.

TH438.H283   624'.01'84   76-9784
ISBN 0-471-34565-2

Printed in the United States of America

10 9 8 7 6 5 4 3 2 1

Preface

The past 15 to 20 years have seen the proliferation of a large number of management methods, many of which are concerned with project control and management. Emphasis has been on methods designed to give managers, at and above the project level, a means of quantifying progress to date and projecting future expected progress. These control methods have been based on the concept of breaking projects into management subcomponents or activities. Activity definition allows the manager to view the project as a network of logically linked work packages. Such network representations provide a framework within which progress can be measured and in terms of which upper management can formulate procedures for controlling the project.

Difficulties have been encountered in adapting these activity-oriented management methods to the needs of the site manager. The site manager's problem relates to the continuous and dynamic commitment of resources to the various construction operations, and his objective focuses on keeping the project on schedule. Delays and failures to meet critical milestones are detected by the activity-oriented control models. However, they seldom provide insights regarding recommitment of resources or changes in technology required to get things back on schedule. In order to *analyze* and *solve* such problems, a method closely related to the actual processes of construction or assembly must be used. Such process level investigation is needed by the site manager to determine the *how* and *why* of delays. Analysis at the process level provides a basis for identifying and eliminating imbalances that lead to deviation from scheduled progress.

Central to the analysis of construction processes is the idea of a work task. A *work task* is the basic or elemental descriptive unit in construction practice. If a work task is further broken down into components, these involve human factor considerations of detailed equipment motions and assume microanalysis of motions and actions. Work tasks should be readily identifiable components of a construction process or operations. Their description must be such that any member of a construction crew can readily grasp and visualize what is involved and required by the work task. Work tasks are therefore the basic building blocks of processes and operations.

This text presents a method for modeling and analyzing operations at the process level in terms of work tasks and idle states. The technique is designed to be responsive to the needs of a manager charged with the commitment of resources on the job site as well as the instructor in the classroom faced with the task of introducing the whole subject of construction operations. The modeling concept is called the CYCLONE modeling method and, when used in conjunction with the CYCLONE problem-oriented language computer processor, establishes a system analysis methodology for construction and process operations. The CYCLONE methodology enables operations to be described, modeled, analyzed, and designed in whatever level of detail is relevant to the needs of the engineer, head office planner, or field agent.

The modeling concepts of the CYCLONE graphical format are simple and versatile and enable the ready portrayal of work sequences, construction or process technology, and conditional interrelationships between the various work tasks and resources involved in the operational pattern to be studied. When properly developed and labeled, CYCLONE models are easily understood by field personnel and provide a vehicle for communication as well as identification of crew interaction and individual work assignments for crew members. The method can be quickly grasped by personnel familiar with or using time scheduling methods. The graphical nature of the technique provides virtually unlimited opportunity for development of paper and pencil laboratory and homework assignments, both in theory and practice. For this reason, emphasis has been placed on the CYCLONE approach as one that can be applied to think a construction or operational process through to its logical conclusion.

The book can be used in teaching the topic of construction and process design at several levels. The basic modeling technique can be introduced in sophomore level courses and engineering technology programs. Details relating to statistical analysis and computer processing can be presented in senior elective and graduate level programs. The book can be used both as a text and as a source book to generate original ideas for term projects involving the modeling of major real-world situations. In addition, rich use is made of figures and tables with the intent of making study material accessible to the professional reader interested in using the book for self-study. The numerous examples presented are designed to stimulate the reader's interest in experimenting with the technique in his own area of application and expertise.

The opening chapters describe the existing state of the art as it relates to construction and process modeling techniques. Chapters One to Five introduce the CYCLONE modeling concept and the basic elements and procedures used in model building and design. Chapter Six discusses transit time concepts, and Chapter Seven describes hand simulation procedures. Chapters Eight and Nine look in greater detail at some control structures and mechanisms that are continuously encountered in real-world process modeling. Closing chapters present statistics collections, computer implementation, problem-oriented computer definition, and application of the technique to real-world problems (e.g., construction of the Peachtree Center Plaza Hotel).

We would like to thank the numerous individuals who have contributed to the development of the CYCLONE concept and provided critical and constructive comment throughout the writing of this text, especially Boyd C. Paulson of Stanford University for his frank and thoughtful comments included in prepublication review of the manuscript. In addition, we would like to acknowledge the contributions of V. Summersby, I. McIntyre, G. Birdsall, M. S. Huang, and A. M. Burger, who read the manuscript and furnished many helpful suggestions for the improvement of the presentation. We also thank the many students at Georgia Institute of Technology and the University of New South Wales who generated process modeling ideas and examples while participating in construction management courses at these two institutions. Finally we are deeply indebted to E. Caterson, whose dedicated typing of the final manuscript and management of correspondence between Australia and the United States made the book a reality.

Daniel W. Halpin

Ronald W. Woodhead

Contents

CHAPTER FIVE   Model Formulation   111

CHAPTER SIX   Work Task Transit Times   145

CHAPTER SEVEN   Hand Simulation   185

DESIGN OF CONSTRUCTION AND PROCESS OPERATIONS

CHAPTER ONE

The Nature of Construction Operations

Construction operations are work processes which are basic to the accomplishment of the physical components of a project. They are also fundamental to the performance of support services on construction sites. Typical examples are excavation, earthmoving, dewatering, formwork erection, concreting, and bricklaying. The complexity of modern construction dictates the use of many different construction processes that bring together sophisticated resources and technologies.

Each construction operation has an associated technology that prescribes and constrains the basic work processes and operates in a characteristic material and work site environment. The operation itself requires the use of labor and equipment resources, and its proper planning and execution require the know-how and skill of operators and managers at all levels from the workman up. Often, for a general contractor, the availability of know-how and skills required for an operation are the limited resources that decide whether the operation will be handled in-house or subcontracted. In highly competitive situations where a single operation has a critical and dominant influence, limited resources may dictate rejection of the project.

In construction operations a sequence of work tasks or work processes can often be identified as an intrinsic property of the operation. While specific project activities using a common construction operation may produce different end products, an underlying similarity will exist in the technologies, work task sequences, and processes at the various work sites. For example, masonry operations have a number of basic processes that can be identified in any masonry job, such as bricklaying, brick and mortar supply, storage and access, and scaffold erection. The specifics of any particular site and masonry project activity, however, will dictate the relative magnitude and mix of these basic work processes and the nature of the task assignments to crew members.

A construction operation can be characterized by its technology, its resource use, and its breakdown into work tasks and sequences. Therefore, concrete delivery processes can be performed manually using wheel barrows, mechanically with buggies, by a crane and bucket, by pumping, belt conveyors or chuting, or even by pressure spraying. Although the selected technological format may vary, the definition of the basic processes associated with the operation remains the same. In each concreting operation, for example, a sequence of work tasks associated with the concrete manufacture, delivery to work site, temporary storage, transfer to work face location, placement, and subsequent treatment can be identified. A full technical description of the operation would then require that each of these work task sequences be defined within the framework of the selected technology.

The selection of a particular technology and material handling process for a construction operation will depend on several factors, including magnitude of job, site conditions, and availability of equipment and skilled labor. The specific crew mix and work assignments for crew members will similarly affect the on-site physical working of the operation and its productivity and cost. On large jobs, the volume of work may permit the use of large crews and hence the use of sole purpose work assignments (e.g., a mason's laborer on mortar supply only); on smaller jobs, a single workman may be required to handle many work tasks (e.g., a general purpose mason's laborer handling brick and mortar supply, scaffold erection and adjustment, and cleanup).

Some construction operations are closely related to trade classifications and skills and are formally taught in apprenticeship schools in conjunction with continuous on-the-job work site experience. Many operations, however, are performed by men who transfer their skills from other areas to the specific requirements of the current operation. If construction volume permits, specialized crews emerge who acquire experience relating to efficient methods, crew sizes and mixes, balanced work assignments, work rates, and production. Tunneling is a specific example of this situation; the determination of an optimum shift size and day servicing crew for a given site, tunneling method, and rock conditions may take several weeks or months and may change depending on ground conditions and haulage distances. Only in exceptional cases is the construction operation considered as a system problem of process design requiring a methods study or crew balance study.

A basic problem for the construction engineer is the selection of the best technology available to him and the definition of the relevant construction method for each work situation he meets in the field. He must understand the nature of the technology and the physical environment at the work face, and thus design the construction operation and work plan. He must describe the nature and working of the construction operation in such a way that superintendents, foremen, and crews may perform the field operations properly and efficiently. Finally, he must monitor, manage, react to changed conditions, and thereby gain experience during the performance of the field operation.

This environment creates the need for the description, modeling, analysis, and design of construction operations. In the field it is necessary to record field data relating to the influence on productivity of different resource and crew mixes. This need establishes a fundamental modeling problem for construction operations. A methodology is required that provides a venue and format for modeling construction operations that are relevant and acceptable to field agents.

This chapter introduces basic concepts and definitions relating to the nature of construction operations and thereby outlines the composition of this text.

1.1 CONSTRUCTION OPERATIONS

Modern society is becoming more technological in nature. Its needs are being met by an industry that is becoming more technical at a time when the delay lag between scientific discovery and technical implementation of the fruits of scientific research is diminishing. Today we are being deluged with and are absorbing the fruits of science and technology.

The demands, inventiveness, and consumption of the industrial society react heavily with a service industry such as construction. Consequently, a bewildering array of construction projects, activities, and operations exists today; this array calls for the continuous development of new construction methods and new material handling and placement requirements as basic project component characteristics (size, quality, sensitivity) change from past practice. In addition, equipment manufacturers are producing more and more general and special function equipment. Each new demand and situation requires the design and specification of new construction operations or the adaptation of existing methods to the specifics of the new problem.

Projects that require construction are owner- and user-developed; they rarely consider construction difficulties in achieving the desired end product. The construction industry must develop construction methods and operations that efficiently, economically, and safely achieve the desired physical product.

The construction industry, however, has some unique features and structures that influence the current approach and practice to construction operations.

While the construction industry is the largest industry in the United States, it is highly fragmented; most new construction is performed, on contract, by firms engaged primarily in construction work for others. The work is carried out in a project format.

Intense competition exists for projects so that for any one firm, continuity of construction volume in specific areas of construction is hard to achieve. In almost all cases, construction firms operate with a dynamic portfolio of projects; projects are continually being sought, added to the portfolio, and removed by completion. The projects are usually of short duration, and little scope is offered to the contractor for a controlled construction environment or economies of scale by mass production. The structure of the construction industry has been forced to adapt to the requirements placed on it by the construction volume demands and nature of the technological society.

Most construction companies and subcontractors concentrate their activities within limited areas in order to master the skills, processes, and technologies involved in their specialization and to reduce unit placement costs to such an extent that they can stay in business. Many companies hesitate to broaden their activities or enter new areas of construction because the lack of knowledge, skill, or resources in these areas compounds the risk associated with the construction process. In practice the in-depth design or analysis of a construction operation is rarely formally considered. It is either implicit in the adoption and modification of past methods or it is realistically solved by the practitioner in the field.

Construction operations may be grouped into different classifications, depending on which features are selected for the classification rationale. Various rationales have been suggested: by finished product or structure (e.g., high-rise building operations or dam construction operations); by material (e.g., excavation, pumping operations, and hand or equipment operations); or by function (e.g., haulage or demolition operations). Thus, for example, construction operations can be classified as follows.

By structure: pile driving operations
pile capping operations
pile-cap formwork operations   that is, a focus on
pile-cap concreting operations, etc.   the finished product.

By material: rock drilling operations
rock blasting operations
rock ripping operations
rock haulage operations   that is, a focus on
rock crushing operations   the specific material.
rock grouting operations, etc.

By function: material hoisting operations
                man hoist operations
                winching operations
                crane operations   that is, a focus on
                cableway operations   the specific function
                belt conveyor operations   of haulage.
                gantry operations
                boom operations, etc.

These operations all require labor and machine resources in various proportions to accomplish a productive activity that contributes to the completion of the construction. Usually several technologies are available that describe the methodology and manner in which these resources will be committed. The selected technology depends on many factors such as site conditions, quantities involved, equipment availability, crew skills, and available time, and each technology leads to variation in the manner of task accomplishment.

For instance, as mentioned previously, concrete can be poured down a chute, be placed by buggies or by crane and bucket, or be pumped. Each operation is subject to constraints that aid in selecting the proper technology or method and complement of resources. For example, if site access is constrained, quantities are large, and money is available, pumping the concrete may be the most economical method. In this case the delivery process may require a small crew, but the actual placement crew at the work face may require a larger crew (i.e., vibrators and cement finishers) because of the higher rate of delivery achieved by pumping. Therefore crew size and composition depend on the technological basis of the construction operation.

It is possible to characterize operations as labor intensive or equipment intensive based on the pattern of resource commitment dictated by the technology and the controlling environmental constraints. Bricklaying and slab preparation are traditionally labor intensive operations; earthmoving is now generally an equipment heavy operation. In some cases, a mixed team equally composed of labor and machine elements may be appropriate to the working of the operation.

A construction operation as defined in this book (see Section 1.2) is a collection of work tasks and processes embodying the commitment of resources within a technological format or methodology that leads to the placement of construction. The technological format is defined in terms of a recognizable sequence of work tasks that provides a framework within which productive units interact to accomplish a physical placement of construction.

For example, if the physical object considered is a brick wall, then the relevant construction operation is bricklaying. In order to describe fully the bricklaying construction operation, a comprehensive list of basic work tasks must be identified. An initial list of bricklaying work tasks is as follows:

stack brick pallets	erect scaffold	lay out wall
break open pallets	adjust scaffold	set corners
carry bricks to face	mix mortar	establish line
moisten bricks	carry mortar to mason	lay bricks
	dismantle scaffold	strike joints
		wash walls, etc.

These work tasks must be related in sequences to suit the technological requirements of the construction operation. This development is considered in detail in later chapters. The further description of the operation requires the enumeration of the resources required and their involvement in the various work tasks. Therefore the resources involved in bricklaying may be:

Labor resources	mason foreman
	masons
	mudmen (mason laborers)
Equipment resources	scaffold
	fork lift trucks
	wheel barrows
	shovels, trowels, etc.
Material resources	brick pallets
	bricks
	mortar
	cement, sand, water, etc.

The relevant technology in this case is the hand laying of individual bricks using adjustable scaffolding.

The full description of construction operations requires the identification of the construction technologies involved, the enumeration and sequencing of the various work tasks that make up the various processes, and the enumeration and allocation of the required resources for the work tasks and processes of the operation. Later chapters develop modeling, analysis, and design concepts that are relevant to the description of construction operations.

1.2 DEFINITIONS

It is essential in establishing a unified system for analyzing construction operations that a clear and readily understandable terminology be defined. Any methodology for the consideration of the operations and processes of construction must immediately consider the hierarchical nature of construction.

Furthermore, relationships between existing and commonly used terms must be established.

In construction management, four levels of hierarchy are definable.

1. *Organizational.* The organizational level is concerned with the legal and business structure of a firm, the various functional areas of management, and the interaction between head office and field agents performing these management functions. The organizational level is not considered in this book, since its terminology does not conflict with that relating to construction operations.

2. *Project.* Project level vocabulary is dominated by terms relating to the breakout of the project for the purpose of time and cost control (e.g., *the project activity* and *the project cost account*). Also, the concept of resources is defined and related to the activity as either an added descriptive attribute of the activity or for resource scheduling purposes.

3. *Operation (and Process).* The construction operation and process level is concerned with the technology and details of how construction is performed. It focuses on work at the field level. Usually a construction operation is so complex that it encompasses several distinct processes, each having its own technology and work task sequences. However, for simple situations involving a single process, the terms are synonymous.

4. *Task.* The task level is concerned with the identification and assignment of elemental portions of work to field agents.

The relative hierarchical breakout and description of these levels in construction management are shown in Table 1.1. It is clear that the organizational, project, and activity levels have a basic project and top management focus, while the operation, process, and work task levels have a basic work focus. In order to develop definitions that distinguish between terms used at these levels, it is important to start at the work task level.

A *work task* is the basic descriptive unit in construction practice. If a work task is broken down into components, human factor considerations or detailed equipment motions are involved that assume microanalysis of motions and actions.

A work task should be a readily identifiable component of a construction process or operation. Its description must be so clear that any member of a construction crew can readily grasp and visualize what is involved in and required of the work task. Work tasks are therefore the basic building blocks of processes and operations.

A *work assignment* is the collection of work tasks specifically assigned to a crew member for performance. Work assignments usually involve sequences of work tasks appropriate to a certain trade and skill level of the worker and, therefore, they may define a construction process.

Table 1.1   HIERARCHICAL LEVELS IN CONSTRUCTION MAN-
AGEMENT

Hierarchical Level	Description and Basic Focus
Organizational	Company structure and business focus. Head office and field functions. Portfolio of projects. Gross project attributes: total cost, duration, profit, cash flow, percent complete.
Project	Project definition, contract, drawings, specifications. Product definition and breakdown into project activities. Cost, time, and resource control focus.
Activity	Attainment of physical segment of project equated to time and cost control. Current cost, time, resource use. Status focus.
Operation	Construction method focus. Means of achieving construction. Complete itemized resource list. Synthesis of work processes.
Process	Basic technological sequence focus. Logical collection of work tasks. Individual and mixed trade actions. Recognizable portion of construction operation.
Work task	Fundamental field action and work unit focus. Intrinsic knowledge and skill at crew member level. Basis of work assignment to labor.

Focus on project attributes and physical component items

Focus on field action and technological processes

A *construction process* is defined as a unique collection of work tasks related to each other through a technological structure and sequence. A construction process therefore represents a technological or readily identifiable segment of a construction operation. It may represent the impact on an operation of a member or trade component of a mixed crew. It may, however, represent one of many processes that a single worker can perform because of his trade training and skill. In this sense a work assignment to a crew member may involve him in different construction processes at different times or at different sequences in a construction operation. Many construction processes are highly repetitive: an operator continuously cycles through work tasks processing or using resources, thereby achieving progress in construction.

A *construction operation* results in the placement of a definable piece of construction and has implicit in it some technological processes and work assignment structure.

The construction operation is closely related to the means of achieving an end product and can be repetitive in nature. It is therefore an expression of and linked to a construction method and only indirectly with a physical segment of the completed product. It is, in fact, a synthesis of construction processes. In simple operations the operation is identical with its single construction process.

The time frame duration of the basic cycle of the construction operation is measured in hours or days.

An *activity* is a time and resource consuming element of a project normally defined for the purpose of time and cost control by a planner, estimator, scheduler, or cost engineer.

An activity is usually related to the production of a physical segment of the required finished product. It may refer to an actual item of work listed in the itemized bill of quantities or to a portion of the project defined by contract drawings. In some cases an activity refers to a servicing function such as the procurement of materials, inspection, or formwork preparation where the time impact of the servicing function on the project is to be considered.

Activities are the aggregation of operations or processes that contribute to the completion of a physical component of the structure or the performance of a support service. In this sense an activity is unique and must be completed once, although its completion may require the repeated performance of a number of operations or processes, some of which may be unique to the activity. Construction operations or processes are often common to many activities and assume a unique magnitude and significance for a specific activity. In this book, activities represent a significant portion of the project and have a duration of days, weeks, or months.

The distinction between an activity and an operation is firmly tied to the duration of the function and whether it is primarily concerned with a physical segment unique to the time cost control of the project (i.e., as it is for an activity) or to the technological process or method required to achieve an end product (i.e., as it is for an operation). The operation is therefore more fundamental for an understanding of field methods.

As mentioned above, further breakouts of construction functions are possible at lower microlevels or higher macrolevels. Sometimes a microanalysis of a construction operation or process is useful and is undertaken in the field when a highly repetitive process is required for a work item of large magnitude. Although all construction projects are unique and few are of sufficient magnitude to warrant special studies, contractors often fail to realize that many operations and processes are common to every project and, therefore, they require special study.

To illustrate the above definitions, consider a masonry subcontract on a building project. A breakout of typical items under each definition is given in Table 1.2. Work task assignments would depend on the specific character of the job site and foreman decision making. For example, one mason's laborer may break open brick pallets, wet the bricks down, and load and deliver wheelbarrows to hoist; another mason's laborer only mixes mortar. Similarly, one mason may lay out the wall, lay corner bricks, and establish lines and levels; other masons only lay brick.

It is possible to describe construction operations through a careful selection, description, and assignment of work tasks. Work tasks then become the elemental model-building component.

Table 1.2   MASONRY EXAMPLE OF HIERARCHICAL TERMS

PROJECT	All masonry work including face brick walls with back-up block, and partition walls.
ACTIVITY	Lay first floor face brick and backup block. $80,000, 20 days, 3 crews, work force 24 men.
OPERATION	Bricklaying operation using hand mixer, Morgan scaffolding, and hoist. 3 foremen, 12 masons, 8 laborers, 1 equipment operator.
PROCESS	Brick delivery process by mason laborer. Mortar delivery process by mason laborer. Erect and adjust scaffold process. Wall layout process. Cut and lay brick, and strike joints process.
WORK TASK	Break open brick pallets. Wet down bricks. Load wheelbarrow with bricks and deliver to hoist. Select, cut, and fit brick. Strike joints.

1.3 CONSTRUCTION MANAGEMENT

The basic characteristics of a project and its impact on the selection of construction methods and equipment requirements are considered at the planning and estimating stages of the building process. These initial decisions establish the framework for field construction operations. Management then expects field agents to work within these constraints and to supply the skill and know-how required for each construction operation.

This attitude explains why existing methods of construction management focus on the time and cost scheduling nature of project control.

Time scheduling methods (e.g., Critical Path Method) are particularly interesting to project-level managers and higher-level managers, since construction deadlines and gross resource commitment are emphasized. These concepts are relevant at the upper-management level and help to determine whether or not a project is on schedule. Furthermore, they aid in deciding whether or not planned resource commitments (i.e., dollars, man-hours) are consistent with the estimate.

Time scheduling methods help to give the site manager goals and objectives toward which to work, but they provide no assistance in defining the means of accomplishing assigned tasks. Moreover, they tend to establish goals that constrain the site manager without providing a clear picture as to how he can achieve these goals. Therefore, time scheduling at the site level has met with some resistance.

These methods are characterized by the breaking down of a project into elements called activities or cost accounts, depending on the type of control being considered. The obvious interest of management is to insure that the project is on schedule or within budget constraints. Clearly these elements focus on the project at the project level.

The day-to-day management of the project, however, must be focused at the level of construction processes, since site management is concerned with the method of accomplishment required to meet goals imposed by higher management. Although the goals are always in the background, the immediate concern of the site manager is the commitment of resources, the method or technological format of an operation, and the elimination, as much as possible, of delays and idleness of resources that come from poor planning. Therefore, site managers are interested in a much more detailed breakout of the construction activity as a means of revealing potential delays and imbalances affecting the rate of placement of construction.

Difficulties have been encountered in modifying activity-oriented management methods to the needs of the site manager. Essentially, the site manager's problem relates to the continuous and dynamic commitment of resources to the various construction operations and in response to sudden site emergencies; the modification of project focused activity-oriented methods to cope with the dynamic use of resources has not been insight producing at the field level. At best, they have been descriptive in nature, showing projected resource commitment over a planning horizon. However, the interference of site events on the preplanned smooth interplay of resources and the resulting imbalance and loss of productivity at the construction process or operational level is not addressed by such methods. Resource conflicts at a higher project level on a weekly or monthly basis can be considered and resolved or used as a basis for establishing the overall scale of resources to be made available for the project as a whole.

The site manager's problem, however, relates to the day-to-day conflicts that must be reconciled immediately; this leads to the partial invalidation of methods operational over a longer time. Activity-oriented methods, therefore, do not solve problems at the site manager's level, since they detect long-term problems but provide no immediate solution mechanism.

Existing management systems are postulated on the concept of "management by exception." The project is defined in terms of scheduled activities and cost accounts that provide an accepted profile or flight plan for project completion. The project is set in motion, and monitoring systems are set up to detect departures from the plan or estimate. "Exceptions" to the plan are detected, and efforts are launched to determine why these exceptions have taken place. Two reasons or sources of exception can be defined. Is the plan wrong? Has the flight plan been developed based on incorrect assumptions? Or is the plan valid, and inefficiencies in the implementation have led to "exceptions" and potential loss?

Such detected "exceptions" can lead to an investigation of construction processes that reveals inefficiency because of improper resource commitment and results in interference and delays. However, the investigation of such situations is normally carried out by cost analysis methods that may or may not detect the inefficient commitment of resources.

It is typical of this approach that no action is taken if there is no departure from the estimated cost. If, however, there is a departure from the estimated cost, then action is taken to remedy the (financial) situation.

The concept of "management by exception" emphasizes the breakout of a project for the purpose of defining an acceptable project progress profile both with regard to time (duration) and cost. This profile relies on definition of the construction project activities in terms of the physical components of the facility or structure to be completed; it is therefore referred to as an activity-oriented method.

In general, construction processes as defined above are not directly analyzed unless (1) the process is new, unique, and central to the overall project completion, or (2) problems develop relative to the process as the project proceeds. It is generally accepted that "normal" processes (e.g., concreting and pile driving) will take care of themselves. That is, the job manager will have enough experience and intuitive understanding of the site adaptation required to achieve a tolerable efficiency. However, the management training environment by means of which the site superintendent is expected to acquire these skills is extremely uneven and tied to tradition and historical precedents. Very limited systems or techniques are available to the site manager within which he can organize and sort out his approach to process design and implementation. Moreover, the site superintendent is forced to come up with his own "ad hoc" system for organizing construction operations. This system is a function of his training, his exposure, and company tradition and policy. Only limited models now exist that aid the site manager in sizing up an operation. No comprehensive system for viewing construction processes in a general modeling format has been developed and incorporated into manager training systems.

Existing estimating systems that set up the cost "flight plan" for the project tend to integrate existing field practice into the estimate without looking directly at the procedure. This implicit inclosure of existing field practice tacitly accepts the level of professional operation in the field and consequently does not motivate any close look at the efficiency of ad hoc systems that contribute to the current cost basis. Therefore, estimators use recent field cost information that includes all the good and bad aspects of the ad hoc systems and cost guidelines.

No general system for analyzing construction operations is generally available to capture the expertise and improvement in field operational techniques and design that are continually occurring. Therefore, the improvements are captured only by the ad hoc system of the practical experience and know-how of the construction agents involved. A system for communicating and systematizing information developing at the field level is needed. This system should provide a vehicle for compiling the insights available in ad hoc models developed by a wide range of managers. It should also provide a basis for standardizing the method by which construction operations are viewed and analyzed.

1.4 THE NEED FOR OPERATIONAL ANALYSIS

Activity-oriented models are prevalent in construction management, but they do not solve or address the site manager's daily problems regarding methods and resource commitment. Activity-oriented models that are expanded to handle resource leveling and allocation aspects of project management are projected at an upper-management level and across a time horizon of weeks or months. They are not responsive to the day-to-day problems of site managers, since they are unable to focus on or supply solutions to these problems.

Site managers need a method for modeling, analyzing, and establishing the correct design of construction operations that determines the proper quantity and sequencing of labor and equipment resources within the context of a selected field construction technology. This method must allow examination of the interaction of the committed resources to determine imbalance and poor utilization of resources. A conceptual modeling format is required within which the site manager can "tinker" with these interactions until a smooth and productive process is achieved. This system capability will allow determination of system sensitivity to various policies adopted by the manager.

One problem that the superintendent and the estimator both face is the transferring of past experience and knowledge of operations into a new site environment. This is usually done by intuition. The properly designed methodology provides a vehicle or modeling environment within which the manager can experiment with past procedures and arrive at an initial design.

Furthermore, such an environment provides for system structure within which operations can be analyzed during construction. Departures from expected productivity can be traced to specific process tasks, and the source of inefficiency can be identified.

Once the operations are designed, many criteria for evaluating performance become available. Standards of delay, utilization, and idleness can be established.

Traditionally, construction operation design has received little if any attention in the construction industry. It has been generally accepted that construction operations and processes are unique and must be solved on the spot, using experience and engineering judgment. The concept of designing and analyzing processes before the actual construction operation commences has not gained much support. Industrial engineers confronted with more repetitive situations have given more consideration to the study of process design.

Because of this attitude, the young engineer is expected to learn as he goes on the job. And yet there is no medium, no notation, no closed form or set of equations in which he can organize his observations about the various situations with which he is confronted. He is in the position of the musician who is forced to learn to play music by hearing it and playing it back "by ear," since no written form of notation is available. As most construction managers can attest, this can lead to some strange music, particularly in the early stages of an engineering graduate's career. The engineering graduate himself, if he is fortunate, may have a good "ear" and learn the construction operation practice quickly. However, since there is no notation available, he is at a disadvantage in passing what he has learned on to others and in transferring things learned on one process to another. In addition, because of the lack of a medium for documenting them, many good solutions to various construction problems have been developed over the years and then lost after being handed down from one job to the next, since no method of documenting them was available. It may turn out that just a slight variation in work assignments or in the order in which a task in a process is handled leads to a higher degree of efficiency.

What is needed is a notation or format in terms of which construction operations and processes can be defined, documented, designed, analyzed, and improved; this format must allow for knowledge gained on one site operation to be made available to another. This notation should be simple and graphical so that it can be understood at all levels, allowing a wide use and input from the level of the tradesman or journeyman up to the level of top management.

A modeling rationale for the description of construction operations is introduced in Chapter Three and developed in the later chapters into a complete modeling, analysis, design, and decision tool.

PROBLEMS

1. Develop a list of construction operations involved in a construction project of your choice and reclassify them according to one or more of the following rationales.
   (a) By major equipment use or by trades.
   (b) As work face or support operations.
   (c) As preparatory, main structure, or finishing operations.
   (d) As listed in a construction handbook.

2. Illustrate the hierarchical nature of the terms and definitions introduced in this chapter by describing several construction project work areas using the format of Table 1.2.

3. Identify the various work tasks and processes involved in the following material handling situations.
   (a) Readimix concrete delivery to a high-rise city building material hoist skip with floor hoppers and buggy distribution at floor level.
   (b) The delivery of rebars to a ground site area with subsequent manufacture of column cages and insertion into column forms on the working floor of a high-rise building.
   (c) Spoil removal from a tunneling project.
   (d) A quarry and crushing plant operation.

4. Develop a list of work tasks and work assignments for a crew that places the concrete for a typical building floor slab.

5. Identify the major equipment and labor resources involved in a tunneling project. Assume that the tunnel face is advanced by drilling, mucking is mechanized, and tunnel lining is required. List the various construction operations involved and indicate the resource mixes required for each.

6. An earthmoving operation uses scrapers for haulage and a pusher dozer for loading operations. Develop a detailed list of work tasks for the operation.

7. Identify a construction situation where an operational analysis study might be found useful. List the construction agents involved in the current situation and those that would be involved in the decision processes for an operational analysis and subsequent field implementation of any improvements. What arguments would you develop to support the need for the operational analysis?

CHAPTER TWO

Existing Modeling Concepts

Historically, the construction industry has adopted a multilevel structure as an expedient in organizing its activities. From the construction company's viewpoint, this hierarchy begins at the foreman or superintendent level and ascends to the president of the company.

Management viewpoints depend on the decision level and functional areas of responsibility of the manager involved in the problem under consideration. Many management viewpoints are possible even when they basically consider the same situation. For instance, the foreman may be interested in the efficient use of crew and equipment for a given operation, while the project management is interested in leveling resources across the job and organizing supporting activities such as procurement and payroll. By contrast, the company president may be interested in labor and equipment utilization factors, cash flows, and capital investment ratios.

Top management is interested in broad project statements and gross time-cost profiles that allow comparison between actual and estimated progress of each project in the company's portfolio. At lower levels, interest focuses on gross determination of equipment availability, suitability, and use for feasibility or efficiency analyses of project, activity, and operation technologies. At the field level, attention is on the physical, human, and work task sequences of the construction operation itself.

The fact that construction management is hierarchical in nature, focuses on different problems, and requires that varying types of models and levels of information must be emphasized. The modeling viewpoint is, therefore, a function of the hierarchial level of the manager, the decision process and management function being served by the model, and the project time horizon of the agent involved.

Typical management time horizons for agents involved in the construction area are shown in Figure 2.1. Top management is interested in a time horizon encompassing this, other, and potential future projects and may extend several years. As the management hierarchy is descended, the management focus is oriented to the construction operation at the work face, its supply, and the problems associated with labor, crews, and the actual performance of the operational technology.

Therefore, any consideration of a modeling concept must consider the modeling rationale, the problem areas and decision processes it purports to assist, and its relevance and acceptance by field decision agents. This chapter introduces a number of modeling techniques that are available or are in use in the construction management area; it discusses their relevance for the analysis and management of construction operations and processes.

2.1 MODELING VIEWPOINTS

A number of modeling tools have been specifically developed or proposed for the construction management area. These include the networking techniques of the critical path method (CPM) with its many variants and computer-based information systems. Certain modeling tools have been borrowed from other management fields and adapted to the construction environment. Typical examples are the GANTT or bar chart models originally developed for the industrial process area and now commonly used in construction field and head offices.

Time Horizon Decision Agent	Days (1–10)	Weeks (1–8)	Months (1–N)	Years (1–5)
Top management		XXXXXXXXXXXXXXXXXXXXXXX		
Project manager	XXXXXXXXXXXXXXXXXXXXXX			
Job superintendent	XXXXXXXXXXXXX			
Foreman	XXXXXXX			

FIGURE 2.1   Management time horizons.

Some of the modeling techniques have a widespread relevance and validity and, therefore, are used in various construction decision processes and field situations. Others are special purpose techniques relevant only to certain types of building processes, and their use depends on opportunity, awareness by management of the modeling techniques, and the belief that benefits will result from the application of the modeling techniques. In some cases contract terms may specifically require the development, use, updating, and reporting of project status based on specific modeling techniques such as the networking techniques; in others the contractual relationship may render unnecessary the use of any but the simplest modeling concepts such as that of actual and expected cost curve representations of a project.

The nature, purpose, and use of a model is highly influenced by the management function addressed by the model and the project life stage involved in its use. Thus a planning model at the preplan stage may not be relevant as a planning model at the project initiation stage, or even as a planning model on the site at the project implementation. Planning models may not be good reporting models, although estimated cost and deviation statements do provide a good format for cost and duration reporting models.

It is possible to develop unique modeling viewpoints on a management function and project life basis. Figure 2.2 indicates such an approach using the management function breakdown into planning, scheduling, directing, and reporting functions against a project life breakdown of definition, initiation, implementation, and completion. As an example, a preplan technology process model (such as the Multiple Activity Chart) may be relevant in a head office project definition stage of a project, but it may not be relevant for field use if field management and labor attitudes are unfavorable. During actual construction (project implementation phase), field scheduling techniques may focus on the crane and hoist schedules, material delivery, inventory problems, and the determination of immediate work sequences. If a specific model is capable of use at more than one project life stage and serves management over a number of functions, it is more attractive and liable to be adopted than a special purpose model. The framework of Figure 2.2 provides a basis for classifying and comparing the different models available for construction managers, as well as for the identification of areas for which useful models may be developed.

The actual specifics of a construction operation are not fully tested until the field management level of the construction hierarchy is reached and the project implementation phase is under way. In some cases a preplanning group may consider field operations and construction technology when they establish the nature and magnitude of resources that will be made available to a project; in this sense their decisions influence the technology and productivity of an entire project.

If the preplanning group is both experienced and sophisticated and if project magnitude warrants it, they may develop a detailed Multiple Activity

Project Life Stage ↓ \ Management Function →	Planning	Scheduling	Directing (Actual Field Operations)	Reporting
Project definition	Preplanning technology models Estimating models Multiple activity charts Budget models	Resource availability Resource-use-time models	Site investigations Labor availability and attitude models	Bidding models Rise and fall models
Project initiation	Equipment allocation models Site layout	Procurement Inventory	Expediting access models Priority models	Reassessment of adequacy of site resources, access and constraints
Project implementation	Work order models Crew compositions Productivity models	Crane and hoist use Schedule charts Delivery models Inventory models	Work order Crew assignments Work face layout models Labor relations	Status reports Time and cost reports Work sampling Delivery
Project completion	Contract time Project duration Risk models	Change order and variation impact models	Strategy models for settlement of claims and disputes	Predicted completion date Final budget Project summary Historical data

FIGURE 2.2 Modeling viewpoints.

Chart model of an operation that will become the basis for the development of a field schedule. Nevertheless, their primary focus is on feasibility and the desire to control time and cost instead of on the actual performance of the construction operation itself.

The estimator, for example, may contemplate the actual construction process, but only in sufficient depth to establish resources and duration estimates so that he can develop cost data for cost projections. Similarly, cost engineers focus on a construction operation or process whenever field costs significantly exceed estimated costs and rates. In this way they operate on a management-by-exception basis, and only when cost overruns signify trouble do they investigate in detail a construction operation with a cost-saving focus.

The most commonly available and implemented models are bar charts and activity-oriented models that break the project out according to the physical configuration of the facility. These models basically are designed to act as planning, scheduling, and monitoring systems for managers at the project manager level or above. These models do not provide solutions for field managers at the superintendent or foreman levels. The disenchantment of field managers with these methods indicates that the models, from their viewpoint, constitute a harassment instead of an aid for problem solution and decision making. Information obtained from these models may indicate that something is not going as planned, but no information is given on how to proceed. In order to proceed the field manager must search inside his professional "bag of tricks" (i.e., ad hoc models and experience). Furthermore, the field manager often questions the rationality of the time-cost profile against which he is forced to perform, since the project performance models often are compiled at higher management levels without the benefit of his practical knowledge of the situation.

Existing activity-oriented models have an abstract nature that does not take into account actual interrelationships and environment at the project level. The field situation is extremely dynamic and complex, involving managerial skill, labor inclination, materials delivery, quality of finished product, and many continually varying environmental conditions. Existing activity-based models abstract this complexity down to a bar or arrow that neglects the dynamic interaction of these field factors. These models do perform a useful function in establishing progress profiles and goals that motivate the site manager to achieve a certain degree of time and cost efficiency.

Before proceeding with the presentation and discussion of useful construction management-oriented models, it is helpful to consider the nature of various modeling forms.

2.2 CONCEPTUAL MODELS

A model is a representation of a real-world situation and usually provides a framework within which an investigation and analysis can be made of the given situation. Models contain or portray data about a situation that, when interpreted according to certain rules or conventions, provide information about the situation relevant to pertinent decision processes.

Models may be physical or conceptual. A physical model is a mock-up or scale model of the prototype. Physical models are often used in the preplan analysis of industrial projects; if they are built out of components that equate to individual physical units to be delivered and connected, they enable a

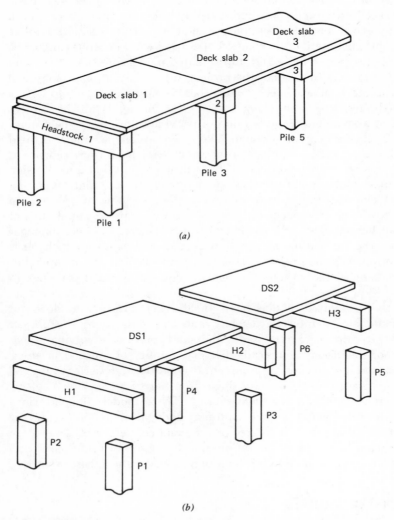

FIGURE 2.3   Simple schematic models. (*a*) Schematic view of pier. (*b*) Exploded view of pier.

mock-up of the construction or erection sequence to be performed. Conceptual models are abstractions of reality and are not intuitive to the uninstructed observer. Conceptual models are developed on a set of modeling and interpretive rules. Network models and bar charts, for example, are conceptual models that have their own individual modeling and interpretive rules. Schematic models are representations that, to some extent, portray a physical situation, so that a physical modeling reaction or perception is induced in the user through a conceptualizing of the situation. Exploded drawings of a physical facility can be considered to be schematic models.

As an illustration of modeling concepts, consider a simple pier made up of two lines of piles with connecting head stocks and simply supported deck slabs. A schematic view of a portion of the pier is shown in Figure 2.3a. The various physical components of the pier have been identified and labeled. An exploded view of the pier is shown in Figure 2.3b, which shows each physical component individually separated but in the same relative positions. Notice that abbreviated labels have now been introduced. Clearly, these figures are schematic models (i.e., not physical models), but they have rather simple conceptual rules so that most construction men would immediately capture the physical situation.

Now suppose that the conceptual modeling rule is adopted of letting a labeled circle (or node) portray a physical component. Figure 2.4 gives a "plan" view of such a model for the pier components shown in Figure 2.3. Such a model can be used as the basis for portraying information about the physical makeup of the pier or about the order in which the physical components will actually appear on the site.

For example, an indication of the adjacency of physical components or the relational contact of physical components may be required. A model to portray these properties requires a modeling element (say a line) to indicate that the property exists. Assuming the modeling rationale of Figure 2.5a, the various nodes of Figure 2.4 can be joined by a series of lines to develop a graph structure portraying the physical component adjacency or contact nature of the pier. If the idea of contact is expanded to indicate the order in which ele-

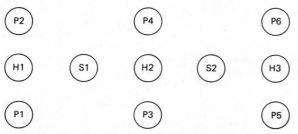

FIGURE 2.4   Conceptual model of pier components.

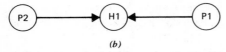

FIGURE 2.5   Logical modeling rationales. (*a*) Adjacency or contact model-ing. (*b*) Physical structure order modeling. (*c*) Physical construction order modeling.

ments appear and physical contact is established, a directed arrowed line modeling rationale may be used, as shown in Figure 2.5*b*. Using this concep-tual modeling rule, Figure 2.6 can be developed. This figure shows, for ex-ample, that headstock 1(H1) can only appear (i.e., be built) after piles 1 and 2 (i.e., P1, P2) appear; in fact, headstock 1 is built around, on top of, and therefore in contact with piles 1 and 2. Finally, if the order of appearance of physical elements is to be modeled for all elements, whether or not in contact, a modeling rationale such as that shown in Figure 2.5*c* may be necessary.

   As an example of the above modeling techniques, consider the pier pile driving operation. A number of possible pile driving sequences are shown in Figures 2.7*a*, 2.7*b*, and 2.7*c*. In Figure 2.7*a* it is assumed that the pile driving rig is swung by its mooring cables to drive the piles alternatively from one line to the next (i.e., P1, P2) before being relocated for the next set of piles (P3, P4), and so forth. In Figure 2.7*b* the pile driving sequence is along one line first (i.e., piles P1, P3, P5) and then along the other line (P2, P4, P6, etc.). Figure 2.7*c* shows a situation that may result if field events interrupt the

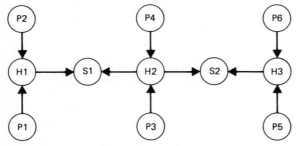

FIGURE 2.6   Conceptual model of pier component relationships.

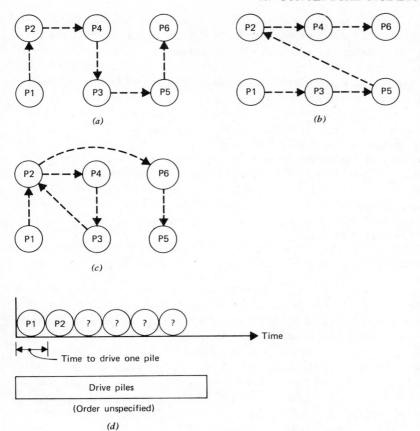

FIGURE 2.7   Construction sequence and activity modeling. (*a*) Alternate row pile driving. (*b*) Sequential row pile driving. (*c*) Field mishap alteration to pile driving sequence. (*d*) Bar chart model of pile driving operation.

planned sequence. In this case the figure indicates a situation where, for example, pile 2 (P2) is broken or lost during pile driving operations, so that to conserve time the pile driving rig moves on to drive piles P4 and P3 and then returns to redrive a new pile P2 before resuming normal pile driving sequences. This situation is common in practice and indicates the major difficulty with planning models in relation to what actually happens in the field. Generally, planning models portray a fixed field construction logic that has been decided off the site by head office planners. These models are not receptive to field conditions and vagaries, and their rigorous use (and interpretation) only constrains the field manager.

Figure 2.7*d* indicates the basic modeling rationale of bar charts wherein specific identification with individual piles is hidden. It implies first the concept that each pile requires a certain time to appear in its driven position on site; second, it implies that the actual sequence of driving piles on the site is not absolutely fixed or essential to field management. The full consideration of bar charts is given in Section 2.3.

The above examples illustrate the gradual development of the modeling rationale for a particular conceptual modeling methodology. The practicing construction engineer can readily construct a number of modeling rationales to suit a variety of construction situations or sequences for the pier project outlined above. Whether the modeling technique is relevant or not to his field problems depends on project details, construction methods, management philosophies, and attitudes beyond the scope of this book. The following sections of this chapter consider in detail a number of tested modeling techniques available to construction managers.

2.3 BAR CHARTS

The bar (or GANTT) chart has been an accepted method of portraying a project plan and work progress since its original development early in this century. Originally developed for the industrial production management area, bar charts have been widely used in construction management.

The basic modeling concept of the bar chart is the representation of a project work item by a time scaled bar whose length represents the planned duration of the project work item. Figure 2.8*a* indicates a bar chart for a work item requiring four project time units (i.e., weeks or months, depending on the time scale adopted). The bar chart can be located in calendar time to

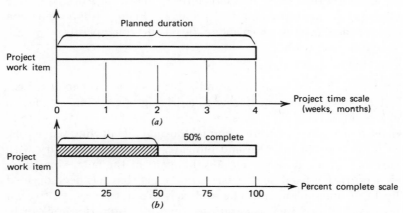

FIGURE 2.8   The bar chart model. (*a*) Plan focus. (*b*) Work focus.

indicate the schedule for planned starting, execution, and completion of the project work item it models.

In practice the scaled length of the bar is also used as a graphical base on which to plot actual performance toward completion of the project work item (see Figure 2.8b). In this way the bar chart model acts both as a planning-scheduling model and as a reporting-control model. In this use of the bar chart model, the bar chart length has two different connotations.

1. The physical bar chart length represents to a given scale the planned duration of the project work item.

2. It also provides a proportionally scaled base line on which to plot at successive intervals of time the current percentage complete status of the project work item.

Figure 2.8b indicates a bar chart for a project work item when it has been half completed. In a situation where the work rate is constant and field conditions permit, this would occur in half the planned duration. If, however, actual work rates vary from time to time according to resource use and field conditions, then the work will be half completed sooner or later than half the planned duration. In this modeling concept actual work progress is modeled independently of the actual time scale of the bar chart.

Project bar chart models are developed by breaking down the project into a number of components. In practice the breakdown rarely exceeds 50 bars and generally focuses on physical components of the project. If a project time frame is established the relative juxtaposition of the project work item bars indicates the planned project schedule and a gross project logic. Figure 2.9a, for example, shows a schedule for a project comprising three work items A, B, and C. Work item A is to be carried out in the first four months, item B in the last four months, and work item C in the third month. Actual progress in the project can be plotted from time to time on these bars, as shown in Figure 2.9b. This figure shows that at the end of the second month, work item A was 40% complete and B was 15% complete. Consequently, bar A is shaded over 40% of the length and bar B over 15% of its length. In this way project status contours can be superimposed on the bar chart model as an aid to management control of a project. By using different shading patterns, the bar chart can indicate monthly progress toward physical completion of the work items.

The information content of the bar chart model can be increased by incor-porating planned and actual rates of progress of working the project item onto the bar model itself. Figure 2.10a shows a graph of planned rate of completing the work items (10% in the first month, a total of 30% by the end of the second month, etc.). This information can be added as shown to the bar chart model. Similarly, the actual rate of progress experienced in the field can be

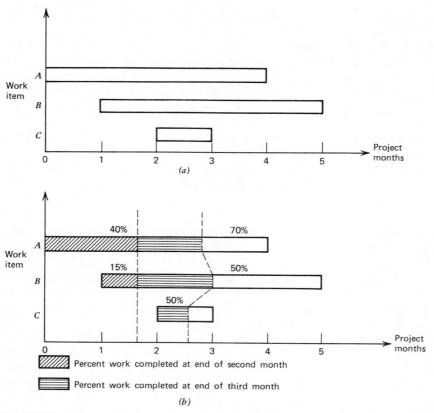

FIGURE 2.9   Bar chart project models. (a) Bar chart schedule (plan focus). (b) Bar chart updating (control focus).

assessed and plotted to give the project curve shown in Figure 2.10b. This information can also be readily incorporated into the bar chart model.

A further modeling development is shown in Figure 2.11, whereby a symbolic plotting of daily activity on the project work item is incorporated into the bar chart model. Assuming a 20-workday month, the figure shows that in the first month there was a five-day delay in starting work and a one-day strike on the eleventh day. Thus, at the end of the first month, the work item was 5% complete in 14 workdays against a planned 10% complete with 20 workdays. Similarly, at the end of the second month, the work item was 25% complete with 31 workdays out of a potential total of 40 workdays. This level of modeling bar charts provides management with meaningful information about the overall scheduling, use of resources, and productivity. The bar

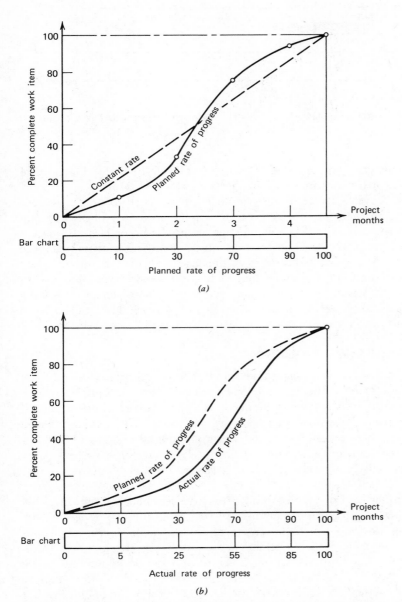

FIGURE 2.10 Bar chart planning and control models using percentage complete attributes. (*a*) Planned rate of progress. (*b*) Actual rate of progress.

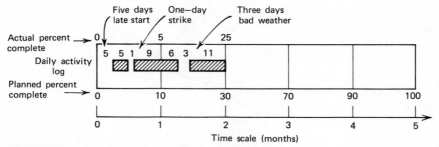

FIGURE 2.11   Bar chart model of work item progress (status at end of second month).

chart model, however, gives no clue as to the details of the technology and of the construction operations involved in the project work item.

Construction projects are commonly modeled in network form (i.e., CPM; see section 2.6) and then transformed into bar charts for ready understanding by field personnel. These bar charts can then be used to plot actual progress against planned progress and thus provide management with criteria for the dynamic allocation of project resources. When used in conjunction with percentage completion curves, bar chart models become an effective field project control technique. A number of planning, field, and control techniques have their origin in the bar chart modeling rationale. These include line-of-balance models and multiactivity charts and are treated in the following sections.

2.4 LINE-OF-BALANCE MODELS

In certain types of projects repetitive sequences of activities develop from the nature of the project and the construction technology adopted. In such cases the individual activities must be properly synchronized in duration and productivity if gross delays and misuse of resources are to be avoided. Often mishaps and delays in working the activities compound the instability between the sequence of activities, even in properly balanced situations, with consequent loss of overall productivity and the generation of chain reaction management problem areas.

Typical examples of projects involving repetitive sequences of activities are high-rise buildings, tunnels, and pipelines. In high-rise buildings repetitive activity sequences develop from the nature of floor-to-floor operations. Thus formwork erection, steel rebar placement, and concreting interact with formwork dismantling if a limited formwork situation exists and influences the following "close-in" trades working on exterior walls and window panels. Similarly, in tunneling, the drill, blast, muck, and tunnel lining operations are often built around a shift basis so that delays affect the entire tunnel cycle and the work force. Again, in pipeline construction, the individual crews of a properly balanced spread move forward at the same pace, minimizing the overall spread distance along the pipeline; however, if one or more crews are delayed, progressive individual crew downtimes occur, with uneconomic spreading along the length of the pipeline.

The balanced progression of the activities, trades, and crews is desirable to project management and is the basis for scheduling crew sizes. However, variations resulting in smaller than scheduled work forces and delays of all types tend to perturb the normal synchronized progression. This construction and management environment enables line-of-balance concepts developed for industrial situations to be applied to the management of construction operations, work tasks, and the daily available labor resources.

The line-of-balance model is a refinement of bar charts. It focuses on monitoring the current status of an activity relative to its scheduled status, incorporates updating as a function of production progress, and enables management decisions to be made to maintain or retrieve the balanced progression relationships between project activities by the way it reallocates labor resources. The line-of-balance approach is to determine the technological time and resource relationship between the sequential set of activities involved and, by means of this line of balance, monitor the relative departures from the desirable balanced position and thus initiate corrective remedial action by the dynamic allocation of resources. The line-of-balance model can be considered a means for determining priorities for labor allocation of "dragging" activities or for the slowing down or stoppage of "crowding" activities and diversion of their labor resources or crews to other activities.

Figure 2.12 shows a number of activities involved in a concrete frame building with a brick wall and window frame facade. The technological line of progress is shown based on the following assumptions: four floors of formwork and each floor divided into four stages; the set windows activity is one floor below the bricklaying activity, which is planned to be half a floor behind the dismantle forms activity; the dismantle forms activity is four floors below the erect form activity so that as one stage of forms is dismantled, it can be cleaned and reerected the following day. Steel rebar cages are one floor stage behind form erection, with concreting operations one stage further behind. In this way, with constant crew size and productivity, a four-day balanced building construction cycle exists.

In the normal balanced situation each activity crew is so matched with its work volume and relative location in the building floor sequence that as it completes one floor stage (i.e., the A or B or C or D areas of a floor) and moves to the next, each of the other activity crews moves on to its next work stage. This matched performance and technological sequence produces the so-called "line of balance" in the total work sequence and is shown in Figure 2.12 as the technological line of progress. In a current status bar chart model it appears as a vertical line, as shown in Figure 2.13.

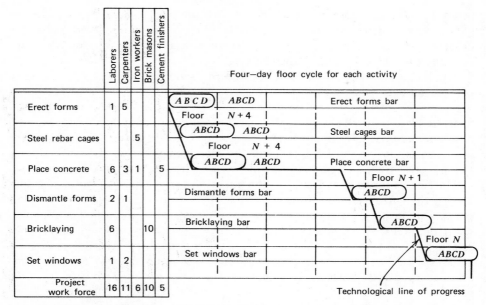

FIGURE 2.12 Building project bar chart, crews, and work force.

The development of line-of-balance concepts is illustrated in Figure 2.14. Each activity has associated with it a time-dependent reservoir of potential work that is technologically available to it for completion. As a unit of work is completed in one activity, it generates different units of potential work in successive related activities and thus updates their reservoirs of potential work as it decrements its own reservoir.

For example, assuming formwork for four floors is available once these have been erected, the activity "erect forms" reservoir is empty and cannot be replenished to permit the erection of forms until activity "dismantle forms" operates to create a pool of unused forms. Similarly, once forms have been newly erected, it is possible to place steel rebar cages and the required concrete. The erection of forms thus generates the need for a certain quantity of steel work that, in turn once completed, generates the need for a concrete pour. In this way the reservoirs of potential work correspond to logical permits for work, which can then be started. The cyclic nature of the activities involved in the building plus the constraints imposed by the limited number of form sets and the technological constraints implied in the sequence of activities on any floor provide a management environment in which the rates of working the various activities are interrelated.

The number of units of work that can be completed in a given time on an activity depend on the potential work reservoir available, the crew size allo-

Erect forms bar	D_{N+4} →	A_{N+5} →	B_{N+5}
Steel cages bar	C_{N+4} →	D_{N+4} →	A_{N+5}
Place concrete bar	B_{N+4} →	C_{N+4} →	D_{N+4}
Dismantle forms bar	A_{N+1} →	B_{N+1} →	C_{N+1}
Bricklaying bar	C_N →	D_N →	A_{N+1}
Set windows bar	C_{N-1} →	D_{N-1} →	A_N

Technological line of balance given by floor numbers and activity stage (A, B, C, D) ↗

Time now bar chart status ↖

Bar chart status in one day with balanced progress ↖

FIGURE 2.13 Normal line of balance progress.

cated, and its resulting efficiency. Different crew sizes produce different rates of progress. The basic problem faced by the superintendent is how to allocate labor resources across activities constrained by both the availability of labor and the reservoirs of potential work.

Figure 2.15 illustrates a situation where the bricklaying crews follow hard behind (i.e., two floor stages) the dismantling crews. In Figure 2.15a the normal balanced situation is shown where, after completing the brick work on

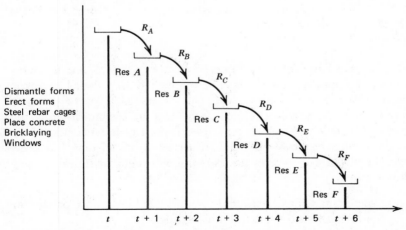

Dismantle forms
Erect forms
Steel rebar cages
Place concrete
Bricklaying
Windows

FIGURE 2.14 Cascaded potential work reservoirs that constrain individual activity work rates.

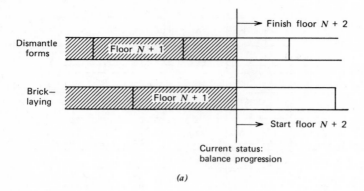

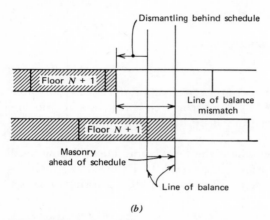

FIGURE 2.15   Current status on line of balance plot. (*a*) Normal balanced situation. (*b*) Out of balance situation.

floor $N + 1$, the masonry crews can proceed to floor $N + 2$ and find clear access, since the dismantling crews have half finished this floor. In Figure 2.15*b*, however, the masonry crews are ahead of schedule (either through better productivity or higher attendance), whereas the dismantle crews are behind schedule. In this situation the activities are out of balance, and either the masonry crew must be slowed down or diverted or the dismantling crews speeded up. If the dismantling crews are understaffed, then the superintendent faces the choice of:

1.   Working the dismantling crew longer hours to achieve more production so that the masonry crew can move up on schedule.

2.   Bringing up the strength of the crew by borrowing labor from the erection crews, which retards the form erection activity.

3. Accepting the delay to the masonry crew and also as a consequence on the next day a delay in form erection on the leading floor and hence on the following days delays on the steel cage and concreting crews, and so on.

The superintendent's decision will depend on the magnitude of the departures from the line-of-balance status as well as on labor availability and attitudes to the breaking up of crews and the giving of new work assignments.

The standard line-of-balance chart is shown in Figure 2.16, together with illustrative entries relating to the status and productivity progress. The project manager updates the chart daily by entering the values of actual production on each activity. These are then used to update pictorially the bar charts to indicate current project status in relation to the planned status and thus to determine the extent of departures (a', b', etc.) as a means of deciding activity priorities for project resources.

Square feet	Plan	585	585	585	1070												
	Actual	552	7572														
Dismantle forms	7020				a'												
Square feet	Plan	585	585	585	1070												
	Actual	575	7010														
Erect forms	6435				b'												
Tons	Plan	1.83	1.83	1.83	3.66												
	Actual	1.42	19.72														
Steel rebar cages	18.30				c'												
Cubic feet	Plan	20	20														
	Actual	20	160														
Place concrete	140				d'												
Bricks	Plan	3160	3160														
	Actual	3260	3741														
Bricklaying	9480				e'												
Windows	Plan	4	4	4	8	4	12										
	Actual	5	13														
Windows	8																

f'

a' Dismantle forms behind 6% (units working day)
b' Erect forms behind 2%
c' Steel behind 22%
d' Place concrete on schedule
e' Bricklaying 3% ahead
f' Windows 30% ahead

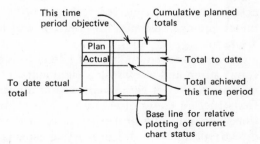

FIGURE 2.16 Line-of-balance chart.

Line-of-balance concepts are intuitively followed by most building construction superintendents. The technique is sometimes used formally by builders on large multibuilding projects, where crews can be supplemented or borrowed from neighboring building sites.

2.5 MULTIPLE ACTIVITY AND CREW BALANCE CHARTS

In many situations, and especially in high-rise building construction using precast units on congested sites, there is a need to identify and coordinate many activities with crew allocations and to schedule delivery and material handling systems. In this way it is possible to analyze and establish the basic building frame cycle duration. A technique in common use in the preplan phase is the Multiple Activity or Multiactivity Chart. It is also often applicable and used in the field.

The multiactivity chart is a modified bar chart consisting of a series of equal length vertical bars. A vertical bar is used for each crew, hoist, or crane unit involved. The scheduled sequence of tasks or activities for each item is then defined, and estimates are made for their individual durations. In this way individual crew or equipment cycle times result. These are then all simultaneously analyzed for compatibility, so that two crews are not simultaneously in the same area or do not require the same crane. Finally, the bars are adjusted to a common length building cycle duration by equating durations to practical, readily usable part days, hours, and the like. The multiactivity chart thus portrays the planned sequence and schedule for all activities on a building site relevant to the basic frame erection cycle. The individual vertical bars portray the work assignments and schedule for each individual crew and major piece of equipment, and they can be used as a reference in the field to measure current status and as a statement of the next work assignment.

A typical multiactivity chart for the example introduced in the line of balance section is shown in Figure 2.17. A four-day cycle is laid out for each crew in line with Figure 2.17. Additional bars are added for the crane and hoist schedules and as a means of avoiding simultaneous use clashes by different crews. Sufficient detail should be included to identify each different activity involved, but there should not be so much detail that breakdowns under one or two hours for crews or half hours for crane and hoist operators are used.

A more detailed analysis of site activities requires an examination of crew composition and work task assignments. Most construction operations require the interaction of equipment with a number of workers arranged into crews. Construction crews often spread over a number of trade and skill classifications and must be properly established to maintain a balance between the directly productive and supporting aspects of the operation. In addition, crew sizing and individual work assignments need to be defined so as not to over-

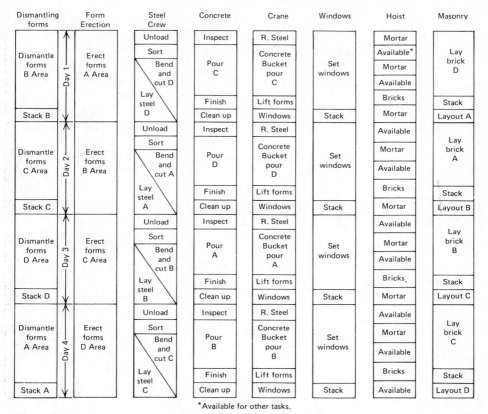

Dismantling forms	Form Erection	Steel Crew	Concrete	Crane	Windows	Hoist	Masonry
Dismantle forms B Area	Erect forms A Area	Unload	Inspect	R. Steel	Set windows	Mortar	Lay brick D
		Sort		Concrete		Available*	
		Bend and cut D	Pour C	Bucket pour C		Mortar	
						Available	
		Lay steel D	Finish	Lift forms		Bricks	Stack
Stack B			Clean up	Windows	Stack	Mortar	Layout A
Dismantle forms C Area	Erect forms B Area	Unload	Inspect	R. Steel	Set windows	Available	Lay brick A
		Sort		Concrete		Mortar	
		Bend and cut A	Pour D	Bucket pour D		Available	
		Lay steel A	Finish	Lift forms		Bricks	Stack
Stack C			Clean up	Windows	Stack	Mortar	Layout B
Dismantle forms D Area	Erect forms C Area	Unload	Inspect	R. Steel	Set windows	Available	Lay brick B
		Sort		Concrete		Mortar	
		Bend and cut B	Pour A	Bucket pour A		Available	
		Lay steel B	Finish	Lift forms		Bricks	Stack
Stack D			Clean up	Windows	Stack	Mortar	Layout C
Dismantle forms A Area	Erect forms D Area	Unload	Inspect	R. Steel	Set windows	Available	Lay brick C
		Sort		Concrete		Mortar	
		Bend and cut C	Pour B	Bucket pour B		Available	
		Lay steel C	Finish	Lift forms		Bricks	Stack
Stack A			Clean up	Windows	Stack	Available	Layout D

*Available for other tasks.

FIGURE 2.17 Multiple activity chart for small building construction cycle.

load individual workers unnecessarily in relation to other crew members and their relative trade skills. Finally, in efficient operations, there must be a balance between the productive time profile use of equipment and the labor-intensive contribution to the operation.

Crew balance charts are similar to multiactivity charts except that the model focus is internally on a crew's operations, and the individual bars are used to model individual crew member work task assignments and sequences. Consequently, crew balance charts have a work improvement usage and are especially useful for situations that involve highly repetitive work such as tunneling, pipelaying, pile driving, and jetty construction.

Crew balance charts are useful as a field technique for monitoring crew composition and work assignments, especially in situations that demand changing crew sizes over time. Typical examples are crew build-up situations in masonry and in tunneling where, first, it may be necessary to match masonry output to parallel the steel erection rate of a building and second, where the crew composition changes as the driven tunnel length increases.

Improperly balanced crews will mean that some crew trades or individuals will have low work rate requirements and frequent idle time periods while awaiting the outcome of more heavily loaded worker operations. Operational efficiency and costs are thus adversely affected directly by idle nonproductive labor hours and more subtly by a lowering of crew morale and labor motivation.

Properly balanced crews will match work levels with skill requirements and ensure reasonable recuperative idle times to offset physical fatigue and the corresponding drop in quality of work and attitude to work. Whenever the nature of the construction operation is such that only vague work assignments can be specified, the individual crew members will adopt and regularly interchange work roles so that general work levels are made compatible with the capacity of individuals. If, however, the work roles are defined by trade classifications and the construction operation has a repetitive nature, considerable attention must be paid to work assignment, crew composition, and the monitoring of work levels.

A typical crew balance chart is shown in Figure 2.18 for a simple pipe loading operation involving the loading by crane of five pipes onto a truck. The pipe loading crew is composed of five men: two operating on top of the pipe stack [pipe stack man far end (PSF); and pipe stack man near end (PSN)]; two men operating on the ground [ground man far end (GMF); and ground man near end (GMN)]; and a truck loading man (TLM) who supervises the actual pipe loading of the truck.

The crew balance chart indicates the individual basic work tasks for each crew member and crane together with their logical interactions whenever joint work tasks are involved, as well as indicating the individual idle times associated with the need to wait until permissive conditions permit resumption of active work tasks. The crew balance chart enables management to establish efficient crew sizes, analyze pipe loading cycle costs, and compare different competing technological methods. If sufficient detail is incorporated into the chart activities, it also is an instructional device for foremen and for capturing field experience.

2.6 THE CRITICAL PATH METHOD (CPM)

The critical path method is a new and powerful tool for the planning and management of all types of projects. Essentially it is the representation of a project plan by a schematic diagram or network that depicts the sequence and interrelation of all the component parts of the project, and the logical analysis

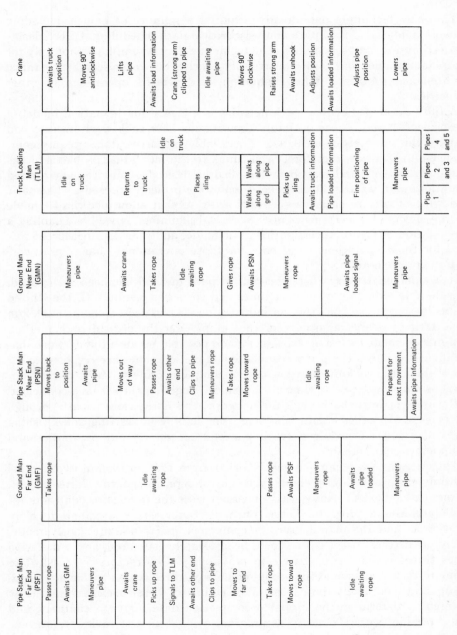

FIGURE 2.18   Crew balance chart for pipe loading operation.

and manipulation of this network in determining the best overall program of operation.

This new technique has already acquired a number of names, including network analysis, critical path analysis, critical path scheduling, PERT, least-cost estimating, and scheduling. The technique may be employed not only for the planning and control of construction works, but for research programs, maintenance problems, sales promotion, and related operations in other industries.

A fully developed critical path network is a logical mathematical model of the project, based on the optimum time required for each work process and making the most economical use of available resources (labor, equipment, finance, etc.). It has, therefore, been tuned to the individual problems of the particular project, and may be as detailed as desired to suit the anticipated conditions and hazards. During execution of the project, it permits systematic reviewing of current situations as they arise, so that allowance can be made for the effects of uncertainties in the original planning, as well as enabling a reevaluation of future uncertainties to be made, and remedial measures initiated for those operations—and only for those operations—that require correction or acceleration.

It is significant that, where the critical path method has been introduced, considerable reduction of project times and costs has resulted. In the United States its use in the construction industry has led to decreases up to 20% in project times over similar projects not employing the critical path method as a management tool. This has been made possible because the network diagram clearly shows the processes whose completion times are responsible for establishing the overall duration of the complete project; these critical operations must be kept "on time all the time." Together they form a connected pathway of operations through the network; this is the critical path through the project. All other operations have some leeway in starting and finishing dates and may be arranged (within limits) to smooth out manpower and equipment requirements.

The critical path technique had its origin from 1956 to 1958 in two parallel but different problems of planning and control in projects in the United States. In one case, the U.S. Navy was concerned with the control of contracts for its Polaris Missile program. These contracts comprised research and development work and the manufacture of component parts not previously made. Therefore neither cost nor time could be accurately estimated, and completion times therefore had to be based on probability. Contractors were asked to estimate their operational time requirements on three bases: optimistic, pessimistic, and most likely durations. These estimates were then mathematically assessed to determine the probable completion date for each contract, and this procedure was referred to as program evaluation and review technique, abbreviated to PERT. It did not consider cost as a variable. It is therefore important to understand that the PERT systems involve a "probability approach" to the problems of planning and control of projects and are best suited to reporting on works in which major uncertainties exist. Relative uncertainty of performance times was reflected in statistical output that allowed prediction of the probability of project completion or the achievement of any other intermediate project event by a specified date.

In the other case, the E. I. du Pont de Nemours Company was constructing major chemical plants in America. These projects required that time and cost be accurately estimated. The method of planning and control that was developed was originally called project planning and scheduling (PPS) and covered the design, construction, and maintenance work required for several large and complex jobs. PPS requires realistic estimates of cost and time and is thus a more definitive approach than PERT. This approach has since been developed into the critical path method (CPM), which is finding increasing use in the construction industry. Although there are some uncertainties in any construction project, the cost and time required for each operation involved can be reasonably estimated, and all operations may then be reviewed by CPM in accordance with the anticipated conditions and hazards that may be encountered on the site.

Both CPM and PERT were introduced at approximately the same time and, despite their separate origins, they were very similar. Both required that a project be broken down into component activities that could be presented in the form of a network diagram showing their sequential relationships to one another. Both used the "arrow diagram" in which network lines are arrows that represent activities and network nodes are events that represent the point in time at which activities may be commenced or are completed. Both required time estimates for each activity that are used in very routine calculations to determine project duration and scheduling data for each activity. These calculations are used to determine which activities must be kept on schedule if the calculated project duration is to be realized ("critical" activities) and which activities have extra time ("float" time in CPM and "slack" time in PERT) available for their performance.

Critical path method is essentially a graphical process. The network diagram is a graph of the project activities that portrays the logic of the construction plan. The project activities are collected and synthesized into a connected whole, using a graphical process to show the construction logic. The network model uses labeling techniques to assign a variety of attributes to the model in such a way that simple computational algorithms directed toward a wide spectrum of project management questions become graphical problems.

The modeling focus on "connectivity" enables graphical models to provide simple tools and concepts for model construction. It is the area of concentration, and its severe simplicity, that makes network graphs capable of serv-

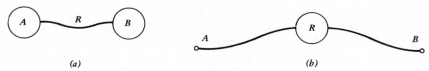

(a)   (b)

FIGURE 2.19   Different graphical representations for A and B are related; that is ARB.

ing a wide variety of uses. Connectivity interpretations can be developed for organization, logical, and combination concepts that enable graphical models to be developed. Thus, for example, if A and B are related to each other in some way, graphical methods enable this fact to be modeled in two ways (see Figure 2.19a). In this figure the objects of discussions A and B are modeled by "nodes," and the relationship R between A and B is portrayed by the connecting line between the nodes. In Figure 2.19b, A and B are modeled by "lines," and the relationship R is portrayed by the node connecting the two lines together. Both representations of the relationship between A and B are valid in linear graph theory and are easily recognized as the basis for the circle and arrow network models of CPM.

In an activity-oriented network or arrow diagram each line or arrow represents one activity, and the relation between activities is represented by the relation of one arrow to the others; each circle (or node) represents an event (see Figure 2.19b). Diagrams of this type are shown in Figure 2.20. The length of the arrow has no significance; it merely represents the passage of time in the direction of the arrowhead. Each individual activity is represented by a separate line (or arrow), and the start of all activities leaving a node depends on the completion of all the activities entering that node.

The network diagrams in Figure 2.20 illustrate some of the logical procedures adopted by CPM. In part (1), it is obvious that A must precede B, and B must precede C. In (2), A must precede both B and C. In (3), A and B must precede C. In (4), A must precede C, and B must precede D. In (5), A must precede C and D, and B must precede D; this necessitates using a connecting arrow (called a dummy) to maintain the logical sequence of events between A and D. Dummy activities have zero time; they are shown by broken arrows. Dummies may also be required to maintain specific activity identification between events, as shown in (6), where A must precede B and C, and B and C must precede D. Events and activities should be labeled, and they are usually numbered for computer identification of the network.

There is another type of network diagram in which the nodes represent activities, and the lines or arrows represent relationships between activities. These networks, called event-oriented networks, precedence diagrams, and circle diagrams are constructed in a manner similar to activity-oriented networks. Like these, the length of the arrows has no significance, since they

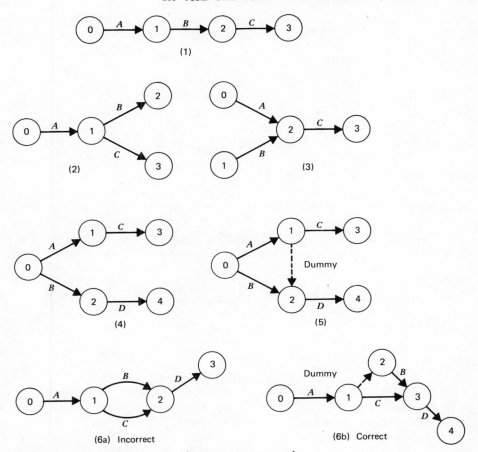

FIGURE 2.20   Elements of an arrow network.

merely point in the direction of the increasing time; circle diagrams, however, eliminate the need for dummy activities, identify each activity with a single reference number, and can be readily adapted to changes in logical relationships between activities. Figure 2.21 shows the elements of a circle network, corresponding precisely to the situations presented in Figure 2.20 for an arrow network.

The first step in the preparation of a network is the division of the project into its activities. No specific order of precedence is necessary, but systematic listing by trades, skills, location, or plant requirements is often helpful.

The next step is to formulate the construction logic, or the specific ordering of the activities. This involves a precise statement of the relationships between the activities as a means of formulating the construction technology and pre-

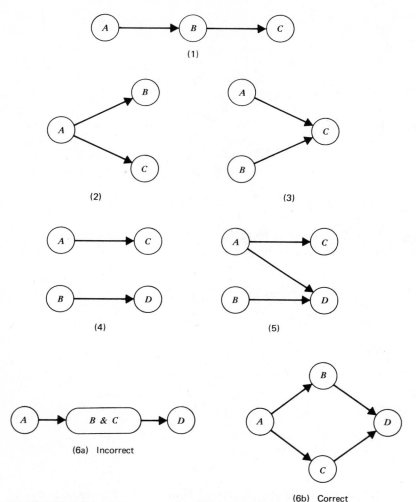

FIGURE 2.21   Elements of a circle network.

scribing management options. A general ordering of activities within the project is not difficult, since their description often implies a relative location within the job; the specific ordering, however, is more difficult, and requires very careful consideration.

A good approach to specific ordering is first to determine the obvious physical and safety constraints, then the crew and other resource constraints, and finally the management constraints. The physical constraints initially lead to chains of activities, simply determined and coupled. The consideration of other constraints and the detailed determination of physical requirements usually lead to the branching and intermingling of the chains into networks. It is often helpful to tabulate the activities systematically in order to note those that must precede each activity, those that must follow each activity, and those that may be carried out simultaneously. The network layout is then determined by trial and error, first satisfying some of the conditions and then refining the portions of the network that violate the remainder.

Designing a network that satisfies all the constraints requires a great deal of skill. In some cases, managerial decisions are extremely difficult to formulate in a diagram. However, it is reasonably simple to test a given network. Consequently, to determine improvements to the diagram, it is easier to start with a rough network (incorporating finer details successively) than to attempt a detailed diagram at the outset.

For example, consider the simple construction of concrete footings, which involves earth excavation, reinforcement, formwork, and concreting. A preliminary listing of activities might be:

A.   Lay out foundations.
B.   Dig foundations.
C.   Place formwork.
D.   Place concrete.
E.   Obtain steel reinforcement.
F.   Cut and bend steel reinforcement.
G.   Place steel reinforcement.
H.   Obtain concrete.

Examination of the list of activities shows that some grouping is obvious. Thus, considering physical constraints only, the following physical chains are developed.

1.   From a consideration of the actual footings: *A*, *B*, *C*, *G*, *D*.
2.   From a consideration of the steel reinforcement: *E*, *F*, *G*, *D*.
3.   From a consideration of the concrete only: *H*, *D*.

When the project is seen from these different viewpoints, individual chains of activities emerge; but, on viewing the job as a whole, it is obvious that interrelationships exist. For example, it is useless to pour concrete before the steel reinforcement is placed and the formwork is installed. Therefore, all the chains must merge before pouring the concrete. And if steps are to be taken to obtain the steel and the concrete immediately when work begins (this would be a management decision or constraint), then the chains all start at the same point or event with the laying out of the foundations.

The development of a preliminary network for the project is possible at this stage because first, a list of activities has been defined and second, a rough construction logic has emerged.

The actual representation and appearance of the network depend on the modeling form adopted and on the spatial locations of the symbols as drawn. As mentioned previously, there are two basic ways in which activities can be

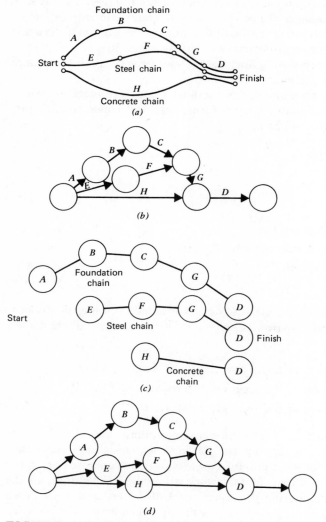

FIGURE 2.22 Preliminary network diagram. (a) Initial sketch, arrow network. (b) First draft. (c) Initial sketch—circle network. (d) First draft.

modeled: (1) when the activities are represented by "arrows" in an activity-oriented network and (2) when the activities are represented by "nodes."

In Figure 2.22 a preliminary network is developed, in both arrow and circle forms, from the above information.

2.7 QUEUEING MODELS

Many situations in which units are processed can be viewed as queueing or waiting line problems. Systems in which two units, designated as processor and calling unit, interact with one another can be modeled as queueing situations. Systems in which more than two units are required are beyond the capability of conventional queueing techniques. Commonly, units that are processed move directly to the processor or are delayed because the processing unit is busy with another unit. Since the arriving unit must wait pending availability of the processor, a waiting line or queue develops from time to time, depending on the processor rate and the arrival rate of units to be processed. If the idea of a permanent queue associated with the processor is established, the queue at any time is either empty or contains a number of units that are delayed pending availability of the processor.

The arrival of trucks at a shovel to be loaded with earth is a classical example of a queueing situation in construction. In this case, the processor (the loader or shovel) maintains its position and the trucks or haulers cycle in and out of the system. The queueing system is defined as shown in Figure 2.23. It consists of the processor activity (loading) and the delay position (the truck queue). Units enter the system from outside its borders, are processed, and exit. In the earth loading system, the truck exits, travels to a dump (fill) location, releases the material, and returns to the system to be reloaded. The system is not, however, directly concerned with activities that occur outside its boundaries. The "back cycle" activities of the truck (i.e., travel to dump, dump, return to load) are characterized for the model in terms of the arrival rate of trucks at its input side. The basis (i.e., back cycle activities) for this arrival rate is not of interest to the model. Only the result in the form of the arrival rate is important.

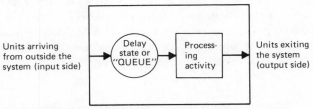

FIGURE 2.23   Simple queueing model.

In construction, the processor sometimes moves from point to point to serve the units to be processed. For instance, a laborer supplying bricks to masons moves from mason to mason. In this case, the masons can be thought of as the processed units and the laborer as the processor performing the processing activity of resupplying bricks. The arrival times of the masons to be processed is the rate at which they run through the packet or pallet of bricks supplied by the laborer. The processor time is the time required by the laborer to complete one resupply cycle.

The rate of units arriving at the input side of a queueing system can be described in terms of random or constant (deterministic) time intervals. Mathematical solutions of the basic queueing model normally assume exponentially distributed interarrival times and service (processing) times.* The assumption of exponentially distributed arrival times simplifies the mathematical development of the model. When using the exponential arrival rate assumption, the model input is defined by the parameter λ where:

$$\frac{1}{\lambda} = \int_0^\infty tf(t)\, dt$$

is simply the mathematical expectation of an interarrival value t. It is not difficult to establish that the probability of N arrivals in the period $(0, t)$ is given as

$$P_N(t) = \frac{(\lambda t)^n e^{-\lambda t}}{n!}$$

This is a discrete probability distribution called the Poisson distribution. Therefore, systems in which the time interval between arriving units is defined in terms of an exponential distribution are said to have a Poisson input process.

A second factor distinguishing the input process in queueing models relates to the size of the population from which the arriving units enter the system. The population is defined as infinite or finite. In the infinite input situation, the units enter the system from an infinitely large pool, transit the system, and then exit. There is no reentry on the part of previously processed units. That is, units that have exited the system are not returned to the entry population to be recycled through the system. A typical example of a construction queueing system in which units enter from an infinite population is the operation of a hoist being used to place concrete on a high-rise building. The "lift loads" of concrete arrive at the hoist, are potentially delayed (i.e., in the transit mix truck), are lifted, and exit the system. No recycling of lifted units occurs.

*For a more detailed discussion of the mathematical development of queueing theory, see Appendix A.

A finite input population implies that, following their exit from the system, processed units may reenter the system at a later time (established by inter-arrival rate). The most common example of this is a population of machines (e.g., fleet of trucks or graders) that periodically break down and are repaired by a maintenance crew. Following repair by the crew, the machines exit the system and operate until they break down again, at which time they reenter the system and await repair. Since this type of model processes a finite number of units, it is referred to as a finite queueing model. The shovel-truck model described above is a finite queueing model, since a finite number of trucks, M, exit and reenter the loader station.

Another important characteristic of queueing models is the processor or server rate. Although this rate can be either deterministic or probabilistic to include any number of distributions, the most common assumption regarding the server rate is that it is also exponential. The server rate, μ, is defined as

$$\frac{1}{\mu} = \int_0^\infty tf(t)\,dt$$

where $f(t)$ is the probability density function defining the randomness of the processing times; $1/\mu$ is then the expected mean processing time.

Therefore, if the value of $1/\mu = \frac{1}{4}$ hour, the mean duration of processing is 15 minutes and the rate of processing, μ, is four units per hour. Similarly, if the mean arrival time of units is 20 minutes, $1/\lambda = \frac{1}{3}$ and the interarrival rate, λ, is three units per hour.

The final distinguishing characteristic of queueing models is the manner in which units are sequenced while delayed in the waiting line. This is referred to as the *queue discipline*. Most construction processes are best modeled as first-in-first-out (FIFO) systems. This means simply that units are processed and exit the system in the order of their arrival. This discipline precludes the possibility of a unit arriving late getting ahead of a unit already in the queue. A wide variety of disciplines is possible. Completely random sequencing of units in the queue is sometimes used when appropriate. A last-in-first-out (LIFO) policy is characteristic of some systems encountered in practice. In general, however, construction processes at the job site level are FIFO processes.

It is possible to solve for the productivity of a finite queueing model such as the shovel-truck system by determining the probability that no units are in the system, P_0. Having determined P_0, the probability that units are in the system is $(1 - P_0)$, and this establishes the expected percent of the time the system is busy (i.e., productive). The production of the system is defined as:

$$\text{Prod} = L(1 - P_0)\mu C = L(\text{P.I.})\mu C$$

where

> μ = the processor rate (i.e., loads per hour)
> C = capacity of the unit loaded
> L = period of time considered
> P.I. = $(1 - P_0)$ = productivity index (i.e., the percent of the time the system contains units that are loading.)

If for instance, the P.I. is 0.65, the μ value is 30 loads per hour, the L value is 1.5 hours, and the hauler capacity is 15 cubic yards, the production value becomes

$$\text{Prod} = 1.5 \ (0.65) \ 30 \ (15) \cong 439 \ \text{cubic yards}$$

The value of P_0 can be determined by writing the equations of state for the system and solving for the values of P_i, $(i = 0, M)$.* The value of P_0 can be reduced to nomograph format so that, given the values of λ and μ, the production index (P.I. = $1 - P_0$) can be read directly from the chart. These nomographs for some typical queueing systems are given in Appendix D. Queueing solutions to production problems involving random arrival and processing rates are of interest, since they allow evaluation of the productivity loss caused by bunching of units as they arrive at the processor. Analysis of this effect is not possible using deterministic methods, since they consider only the interference between units caused by imbalances in the rates of interacting resources (e.g., truck and loader). This loss of productivity because of random transit times and its impact will be discussed in detail in Chapter Six.

The queueing formulation of certain construction processes has been extended to include the concept of a storage or "hopper" in the system. This extension allows the server (i.e., loader in an earthmoving system) to store up productive effort in a buffer or storage during periods when units are not available for processing. In the context of the shovel-truck production system, this allows the loader to load into a storage hopper during periods when no trucks are available to be serviced directly.

2.8 SIMULATION METHODS

Because of the complexity of interaction among units on the job site and in the construction environment, queueing models can be applied to only a limited number of special cases. As noted in the section on line of balance, the output from one operation tends to be the input to following operations. This leads to the development of chains of extremely complex queues as well as situations in which many units are delayed at processors pending arrival

*See Appendix A.

FIGURE 2.24   Two-link system.

of a required resource. Such chained or linked situations are too complex to be modeled using queueing models. Simulation techniques offer the only general methodology that affords a means of modeling such situations.

One class of simulation models that has been used to model chains of queues in which units interact is referred to as the "link-node" modeling format. The link-node model was developed originally to investigate problems associated with cargo handling activities at marine ports. The method gets its name from the chainlike or linked appearance of the graphical representation of the model. Nodes are located at the end of each link. Figure 2.24 is a diagram of a two-link system with the arrows indicating the direction of entity flow. Production units cycle between the nodes at either end of each link. These nodes are transfer points and normally represent operations such as loading or off-loading. Units remain in assigned links and at one end of the link can be thought of as the calling or served unit and at the other end as the server. Times or durations are associated with each node and with the half-link elements connecting each node. In the construction adaptation the nodes are designated as load and dump activities, with the upper link element being the haul activity while the lower element represents the return travel. The nodes are constrained by the requirement for only two units (one from each link). Therefore, the ingredience requirements for the nodes in the link-node model are the same as those of the classical queueing model—a calling unit and a server. The difference is that the entity acting as the server may be delayed at the other node in its cycle awaiting realization of the ingredience requirements there. Therefore the server is not always available, and delays can occur because of the lack of a required unit.

The simulation formulation of the "link node" has been applied to the equipment fleet selection problem. A study by Teicholz (1963) utilizes a two-link simulation model to investigate both shovel-truck and pusher tractor-scraper systems. In each case, units representing the server (i.e., loader or pushers) cycle in link 1 and units representing the processed units (i.e., trucks or tractor scrapers) cycle in link 2. The types of delays considered in the study are:

1. Balancing delays.
2. External delays.
3. Delays caused by passing disciplines on the haul roads.

In Teicholz's construction adaptation of the link-node model it is possible to simulate external delays such as maintenance-and-repair activities. These delays are incorporated into the haul-and-return elements of the appropriate cycle.

This modeling format can also be used to model material distribution systems such as those encountered in asphalt paving operations. A typical model consisting of three links is shown in Figure 2.25. The three links represent:

1. The asphalt plant production cycle.
2. The movement of asphalt from plant to job.
3. The placement cycle.

Units representing truck load size batches of asphalt are generated in link 1, transferred into trucks cycling in link 2, and placed into the paver(s), which cycle in link 3 [Sprague, (1972)].

Systems simulating both direct transfer from link to link (e.g., plant to truck and truck to paver) as well as hopper storage effects at the transfer nodes (e.g., surge capacity at the plant and immediate placement as windrows at the job) can be developed using this model. Use of storages has a decided effect on the operation of such systems because of the reduction in interdependency among links. The use of a storage or "hopper" system will be discussed in Chapter Five in conjunction with a simple masonry system.

A variety of simulation program languages is available for the modeling of processes in which discrete units cycle or flow through systems of active and idle states. These simulation "languages" allow the investigation of complex queueing networks that cannot be treated using the mathematical methods of queueing theory. Among the more popular of these languages are the General Purpose Simulation System (GPSS), SIMSCRIPT, and General All-purpose Simulation Program (GASP). A detailed description of these simulation systems can be found in various references [see, for instance, Gordon (1969) or Naylor et al. (1966)].

To illustrate the nature of such a language, consider a GPSS model developed by Gaarslev (1969) to model a two-link system (link-node format) with several independent servers. A GPSS block diagram for a link-node system containing a variable number of haul units (e.g., trucks) in the processed unit link and three loaders in the server link is shown in Figure 2.26. This model

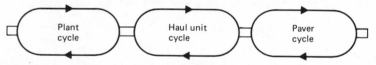

FIGURE 2.25   Asphalt distribution system model.

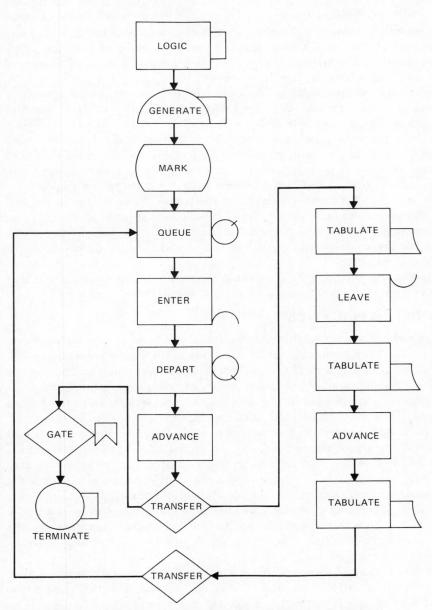

FIGURE 2.26   GPSS model of multiprocessor two-link system.

illustrates the set of symbols used in GPSS to develop a flow diagram of the process being modeled. The symbols or blocks represent functions that must be performed in the simulation of the system under consideration. Each function required is tied to a subprogram of GPSS. Therefore, the modeling focuses on programming function and sequence instead of on the physical layout and flow unit interaction of the process being modeled. Some of the more commonly used GPSS block symbols are shown in Figure 2.27. A set of numerical parameters is associated with each modeling block.

Since the use of the modeling blocks is oriented toward the programming functions involved in simulating the system, the modeler must be familiar with programming concepts in order to use GPSS. Each of the blocks corresponds to a program subroutine that performs the required programming function. Therefore, the configuration of the model emphasizes the functions to be performed from a computer standpoint without providing insight into the interaction of units within the process from the manager's viewpoint.

In the model shown, units representing the trucks are generated in the system at element 2 and activate a series of functions that provides for delaying and processing them as well as the collection of statistics. Information is developed by the model relating to delays and travel times observed in the operation of the system.

The model is presented here to indicate the nature of graphical models that are presently used in association with process simulation.

2.9 MODEL APPLICATION

The models presented in this chapter indicate the wide range of decision problems that have been addressed by existing methods. In general, application of these models in a rigid fashion to obtain decimal point accuracy is not relevant to the management of construction. Instead, these models and techniques provide a framework in terms of which various decision problems can be viewed. They must be adapted to the management situation. In this sense, they are analogous to the recipes used by a chef. They provide the guidelines for approaching a problem and identify the ingredients that must be blended or considered. However, too rigid an application of the recipe or model may yield a very poor result. The degree and amount of application of the model is a management decision, just as the blending of ingredients to obtain a good sauce relies on the chef's intuition. A chef must take into consideration the quality and source of the ingredients, and the manager must consider the management environment to determine the degree to which a model is relevant and can be applied.

An understanding of these and other management models, however, is very helpful in arriving at an approach to a given problem. For this reason, the study of such models is justified and aids the manager in sizing up given situations and gaining insight into the relevant aspects of his problems.

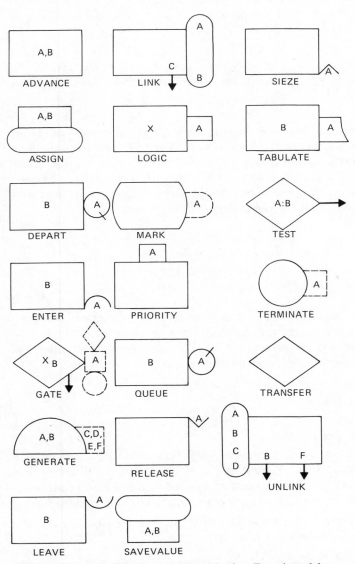

FIGURE 2.27 GPSS modeling blocks. Reprinted by permission of Prentice-Hall, Inc. from Gordon (1969).

PROBLEMS

1. Prepare a list of work tasks and rough duration estimates for the layout, excavation, and construction of a small reinforced concrete footing. Then develop a bar chart model for the footing construction that portrays the basic construction logic and sequence of the construction by the use of properly scaled and located bars.

2. Develop an activity-oriented CPM model for the footing construction project of Problem 1. Then identify the chain of work tasks (the critical path) whose summed duration establishes the minimum project duration. Correlate the CPM critical path chain with the bar chart model of Problem 1.

3. Sketch the layout of a construction site, indicating the location of special equipment (such as cranes and hoists) and of special use areas (such as site access, material handling areas, storage, and rebar and block manufacture areas). Indicate on the sketch a typical material flow from delivery to final inclusion in the permanent project structure. Use the following symbols (and attributes) for the material flow process.

        ○   operation (description)
        ⇨   transport (distance)
        △   storage (time period, quantity)
        □   inspection (detail, quality)

4. Develop a crew balance chart for a particular construction operation identifying individual crew members and major equipment items. Use approximate times for the various work tasks, an average cycle time, and cross-link concurrent and dependent work tasks.

5. Develop a multiple activity chart for a typical:
   (a) City building frame construction.
   (b) Finish trade activity on a city building.
   (c) Tunneling project.

6. Develop a model for a crane or hoist operation on a city building site. Indicate approximate loading, hoist and unloading times for different floor levels and locations. Can the above data be used as a basis for establishing daily crane and hoist schedules or is work face activity a dominant factor?

7. Identify a problem area facing a field construction decision agent. How would you go about developing a model:
   (a) To describe the situation?
   (b) That could be useful in solving the problem?
   (c) That would be acceptable to the decision agent?

CHAPTER THREE

Modeling Concepts

Construction operations contain the basic work processes in construction. Their definition requires a knowledge of the construction technology involved, a breakdown of the processes into elemental work tasks, the identification of equipment resources, and the work assignment to the labor force involved. The description of a construction operation must indicate what is to be done and how (i.e., a technology and process focus) and who is to do it with what (i.e., a resource use focus). Practical descriptions for the performance of the construction operation must also indicate the conditions under which the various processes and work tasks can be initiated, interrupted, or terminated. The planning and management of an efficient construction operation also requires information relating to the impact on productivity and resource use of different spreads of equipment for different crew composition and sizes.

Any modeling methodology for construction operations must be capable of meeting most (if not all) of the above requirements. Construction operation models can be developed at several levels, depending on whether the model purpose is to describe, analyze, or assist the user in a decision process relating to the operation. Descriptive models require simple modeling concepts if field and head office agents are to find the models useful. Analysis models require the development of solution processes that operate on relevant descriptive data. Decision models must focus on decision variables pertinent to the construction operation itself. The decision variables must be available to the head office and field agents for manipulation in the design and management of the construction operation. A modeling methodology that is capable of integrating a construction operation model through all these levels is preferable to a number of independent and fragmented modeling methodologies.

This chapter introduces modeling concepts concerned with the description of construction operations and leaves for later chapters the development of modeling concepts for the analysis and decision processes associated with the planning, design, and management of construction operations.

3.1 A MODELING RATIONALE FOR CONSTRUCTION OPERATIONS

Construction operations can be considered and defined in terms of specific collections of work tasks where the work task is a basic or elemental component of work. The work task (see definition Section 1.2, page 7) is a readily identifiable component of a construction process or operation whose description to a crew member implies what is involved in and required of the work task. The various work tasks are logically related according to the technology of the construction process and the work plan. The work plan prescribes the order in which the resources made available to the construction operation carry out the various work tasks. The nature of and the relationships between the work tasks, including the type of equipment and material used by the work force in a construction operation, define the construction technology.

A meaningful description of a construction operation therefore requires the definition of the basic work tasks and the manner in which the available resource entities perform (or sequentially process through) the work tasks. In this sense individual resources can be said to traverse or flow through work tasks. The sequential, relative, and logical relationships between the various work tasks as defined by the various resource flows portray the plan or static structure of the construction operation and technology. The actual working of the operation can then be described by locating and monitoring, from time to time, the various resource entities as they dynamically traverse the static structure of the operation.

At the construction level, the available resources are labor, equipment, material, and the skills shared by labor and management in their knowledge of the operation and the construction technology involved. Construction operations at the work face level are carried out purposely by explicit work assignment allocations to particular tradesmen, crews, and equipment. Furthermore, the orderly progression of the work and the technology of the construction operation will dictate the sequence in which a specific resource (i.e., equipment piece or tradesman) will carry out each assigned work task.

An initial step in the development of a construction operation model requires the enumeration of the various resource units involved in the operation. Once a resource unit has been defined, it is easy, for a person with a knowledge of the construction operation, to identify the specific work tasks in the operation with which the resource entity is involved, and thereby to determine the sequential ordering of the work tasks.

At any one time, resource units can be considered to be either in an active state (i.e., working at, or traversing, a work state) or to be in idle state (i.e., idle awaiting the opportunity to become active in a succeeding work task). In this way flow units can be considered as passing from state to state whenever the transfer becomes possible. The need for distinguishing between active and passive states is caused by the requirement for the modeling and monitoring of idle and unproductive resources in construction operations as measures of the productivity, or lack of productivity, of the operation.

The identification of the individual resource units associated with a construction operation, the elemental work tasks, and the resource unit flow routes through the work tasks can be made the basic rationale for the modeling of construction operations. The requirement for modeling the conditions that must be met for a work task or process to become active can be satisfied in terms of conditional logic relating passive resource states with active work tasks and the use of special symbols distinguishing between unconstrained and constrained work tasks.

3.2 BASIC MODELING ELEMENTS

A graphical modeling format for work states and entity flow modeling can be developed using three basic modeling shapes representing:

1.   The active state square node model of a work task.

2.   The idle state circle model of a resource entity.

3.   The directional flow arc model of a resource entity as it moves between idle and active states.

The symbols used (see Figure 3.1) for each modeling element are designed to be simple and helpful in developing schematic representations of the construction operation being modeled. Two basic shapes (squares and circles) are used to model active and passive resource states; together with directed arrows (arcs) for resource flow direction, they help to provide a quick visual grasp of the structure of a construction operation.

As mentioned previously, it is convenient to distinguish between the unconstrained (i.e., normal) work task and the constrained (i.e., requiring the initial satisfaction of conditions) work task. While all work tasks are modeled schematically as square nodes, the constrained work task is modeled as a square node with a corner slash. Thus a total of four symbols is required for the modeling of the structure and resource entity flow of construction operations (see Figure 3.1).

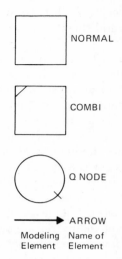

NORMAL   The normal work task modeling element, which is unconstrained in its starting logic and indicates active processing of (or by) resource entities.

COMBI   The constrained work task modeling element, which is logically constrained in its starting logic, otherwise similar to the normal work task modeling element.

Q NODE   The idle state of a resource entity symbolically representing a queueing up or waiting for use passive state of resources.

ARROW   The resource entity directional flow modeling element.

Modeling Name of
Element   Element   Description of modeling element

FIGURE 3.1   Basic modeling elements.

The active working-state models are the NORMAL and COMBI modeling elements. Both have a square node format and model work tasks. Since the work task is the basic component of a construction operation, it should be chosen so that its name or description is sufficient to convey to a crew member or foreman the nature, technology, work content, and resources needed to fulfill the work task.

Simple examples of work task activities are breaking open brick pallets, preparing column formwork, and loading trucks with front end loaders. The definition of a work task thus implies a verbal description of the work task, an indication of the resource entities involved, and an explicit (or implied) awareness of the time commitment of the resource entities to the work task.

A graphical model of a NORMAL work task should have the following ingredients (see Figure 3.2).

1. A unique graphical format [i.e., a square node ($\square$)]. The square node model of a work task can thus be imagined to be an abbreviated form of the bar chart representation of an activity.

2. A descriptive label.

3. A user-defined time delay function that prescribes the resource entity transit time through the work task.

4. An indication of the resource entities that must transit the work task.

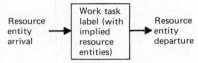

Work task time delay function

FIGURE 3.2   The NORMAL work task modeling element.

The square node models the "NORMAL" work task. A NORMAL work task has no ingredience constraints on its resource entities. For NORMAL work tasks, the arrival of a resource entity at the input side of the work task meets no logical resistance to its commencement of the work task; consequently, there is no queueing up of resource entities before a NORMAL work task. Furthermore, since the NORMAL work task is not ingredience-constrained, several resource entities can traverse it simultaneously. For instance, in modeling a precast concrete element plant, a normal work task model can be used to represent the time spent by a concrete slab in passing through the curing tunnel (see Figure 3.3). This type of model may be used because several slabs can be passing through the work task simultaneously (i.e., in parallel), and the slabs may not be constrained by a processor or ingredience condition.

The more general model of a NORMAL work task shown in Figure 3.4 indicates that the work task can be processed by either resource entity A or resource entity B. That is, the work task can commence (or be realized) by the arrival of a resource entity along any of the arcs incident on its input side;

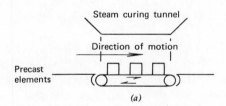

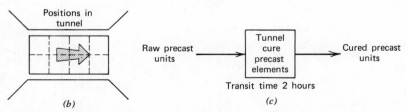

FIGURE 3.3   Work task model of precast element tunnel curing. (*a*) Elevation view of tunnel. (*b*) Plan view of tunnel. (*c*) Work task model.

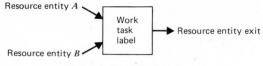

Work task time delay function

FIGURE 3.4   The NORMAL work task modeling element.

thus there is no requirement to wait until ingredience constraints are met. Once the NORMAL work task is commenced, the activating resource is captured for the duration of the user-defined delay. In this way the NORMAL element models the time involvement of resources in work tasks.

A simple example is the loading of a material hoist on a masonry operation. The hoist can be loaded with mortar or bricks, whichever load arrives first (see Figure 3.5). Once the hoist is loaded, a signal is sent and the hoist is made ready for hoisting. Thus two resource entities (i.e., the loaded hoist and the informational signal) can be said to flow from the work task. Work tasks can have any number of output resource entity flows to suit the requirements of the situation or the modeling detail required.

The COMBI modeling element is similar to the NORMAL modeling element with the additional logical requirement of meeting a prescribed set of input resources on its input side before the work task can be initialized. Thus all of the logical requirements needed for the COMBI work task must be available for the work task to proceed. An essential modeling feature of the COMBI element is therefore the identification and definition of the ingredience set of input resources required to start up the work task being modeled by the COMBI element.

A graphical model of a COMBI work task is shown in Figure 3.6. A simple example is the loading of a truck with dirt using a front end loader. This work task requires a truck, a front end loader, and dirt (see Figure 3.7). These resources define the ingredience constraint or the logical conditions required before the work task element "LOAD TRUCK WITH DIRT USING FRONT END LOADER" can begin.

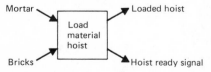

FIGURE 3.5   A multi-input-output use of the "NORMAL" modeling element.

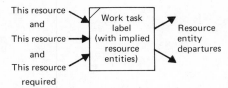

FIGURE 3.6   The COMBI work task modeling element.

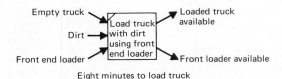

FIGURE 3.7   The use of the "COMBI" modeling element.

The idle or passive state of a resource is modeled by a circle with a slash (Q) to denote a QUEUE node (or Q node) (see Figure 3.8). The Q node is the graphical form; it has an obvious similarity to the concept of units being idle in a waiting line queue. Usually the time a resource entity is in the idle state is unknown and depends on external conditions in a construction operation relating to various work task states and durations and on the efficient design and management of the operation.

The QUEUE node acts as a storage location for resources entering an idle state and releases the resources one at a time to the following COMBI node whenever the COMBI node ingredience logic is satisfied. Figure 3.9 illustrates the idle state of a front end loader. QUEUE nodes always precede COMBI

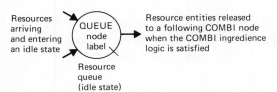

FIGURE 3.8   The QUEUE (Q NODE) resource idle state modeling element.

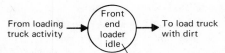

FIGURE 3.9   The use of the QUEUE (Q NODE) modeling element.

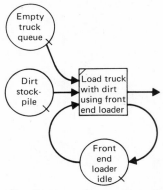

FIGURE 3.10   The COMBI-QUEUE NODE ingredience logic relationship.

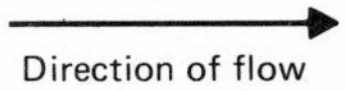

→

Direction of flow

FIGURE 3.11   The ARC Modeling element for resource entity directional flow logic.

nodes, and the set of QUEUE nodes associated with each COMBI node establishes its ingredience logic. Thus the COMBI node of Figure 3.7 is related to three QUEUE nodes, as shown in Figure 3.10.

An arrow or directed ARC (see Figure 3.11) is used to model the direction of resource entity flow between the various active state "square nodes" and the passive state "queue nodes." The ARC modeling element enables the logic and structure of the operation to be developed. The ARC has no time delay properties and simply models entity flow direction. The entity transit time, once initiated, is instantaneous.

The four modeling elements (i.e., the NORMAL, COMBI, and QUEUE nodes and the ARC) are the basic model-building elements. These elements can be combined in several ways to model construction operations, as shown in the next section.

3.3 MODELING THE STRUCTURE OF CONSTRUCTION OPERATIONS

The relative sequence and logic of the work tasks and processes that make up a construction operation constitute the technological structure of the operation. The four modeling elements can be used in a variety of patterns to model construction operation structure. The structure of an operation can be developed in two different ways, either by the continued development of local

situation detail and integration into a final structure or by the initial development of a skeletal frame for the operation structure and the final fleshing in
of detail. The selection of approach depends on personal preference and is
illustrated in later chapters. This section introduces the modeling approach
by first showing that rich local modeling detail can be developed for a basic
labor-intensive situation in construction and by then showing the development
of a skeletal modeling structure for an equipment-heavy construction operation.

Consider the situation where a laborer is shoveling sand from a stockpile
into a sandbox for subsequent pickup by crane for delivery to another location. A simple schematic illustration of the laborers work task (shoveling sand)
is shown in Figure 3.12; a more representative description of this work assignment is shown in Figure 3.13, in which three work tasks have been identified:
the initial delivery of the sandbox; its subsequent filling; and final departure
from the laborers' work face.

A simple initial construct for the laborer-shoveling-sand situation (Figure
3.12) can be developed based on the observation that the active work state of

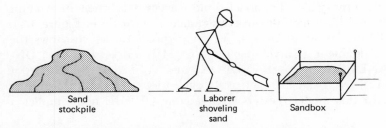

FIGURE 3.12 Schematic model of laborer filling sand box.

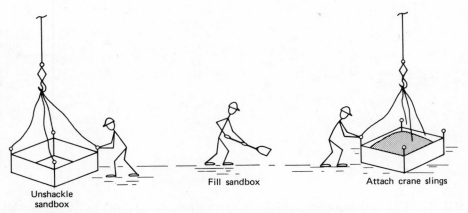

FIGURE 3.13 Schematic model of laborer's work assignment.

the laborer filling the sandbox requires that sand, the empty sandbox, and the laborer be initially available. This concept requires a COMBI node to model the shoveling activity, three preceding QUEUE nodes to establish the resource availability precedence logic, and three ARCS to link (and therefore to define the ingredience logic) the three QUEUE nodes to the COMBI node. This initial construct is shown in Figure 3.14*a*. This model implies that once the three resources are available, the laborer is continuously active in filling the sandbox work task. The model also implies that the three resource entities flow along paths commencing at their initial QUEUE nodes through the COMBI work task, where the SAND entities and the SANDBOX entity are combined into a single FULL SANDBOX entity so that only two paths emerge: the FULL SANDBOX path and the LABORER path. These two paths can then be developed further to suit the construction plan and work pattern that is to follow.

A more realistic situation is to assume that the laborer is captured within a work cycle in which he continually shovels sand and rests from time to time, either through fatigue or because the sandbox is full and he is awaiting another empty sandbox. A model for this situation is shown in Figure 3.14*b*.

If the laborer's activity is to be modeled more accurately, then an attempt must be made to establish his full work assignment, as shown in Figure 3.13. A possible model of this situation is shown in Figure 3.15, which shows the laborer's work assignment as involving three COMBI work tasks (i.e., UN-SHACKLE SANDBOX, FILL SANDBOX, and ATTACH SLINGS) of which one, FILL SANDBOX, is repetitively cycled by the laborer during the filling activity. In this model the laborer available QUEUE node implies a

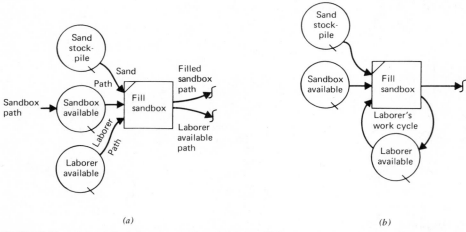

(*a*)   (*b*)

FIGURE 3.14   Models portraying laborer filling sandbox. (*a*) Laborer's work task model. (*b*) Laborer's work cycle model.

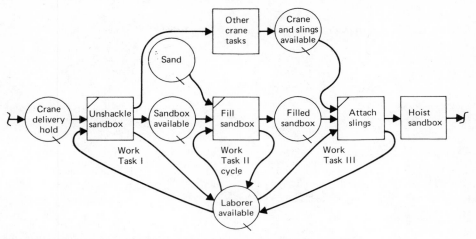

FIGURE 3.15   Laborer's work assignment: filling sandbox.

more complex idle state situation than previously. Figure 3.15 therefore contains explicitly a SANDBOX resource path, a SAND path, a multiloop laborer work assignment, and (implicitly) a CRANE and sling path (even though the NORMAL work task OTHER CRANE TASKS may be deleted from the model).

The SANDBOX path enters with the crane and is involved in each laborer work task and finally is hoisted by the crane. The crane and sling delivers the empty sandbox, is committed to the laborer while he unshackles the sandbox from the slings, is withdrawn to carry out other crane work tasks and then returns to pick up the full sandbox for hoisting. The model requires COMBI elements for each laborer work task since, in each case, several resources are required simultaneously before the work tasks can commence. Each COMBI element has a number of preceding QUEUE nodes that establish the necessary ingredience resource logic. The inclusion of the QUEUE nodes does not necessarily imply, for example, in the case of the laborer that he is lazy, irresponsible, or unwilling to work but, instead, that he may at some time be unable to perform one of his work tasks because another required resource is temporarily unavailable.

The development of "situation" models using the four modeling elements requires a clear understanding of the resources involved and the enumeration of the elemental work tasks. If these work tasks are clearly defined and labeled, meaningful descriptive models result that can be understood by field agents down to the laborer level. The amount of detail used and the level of breakdown of work tasks necessary to describe the construction technology and op-

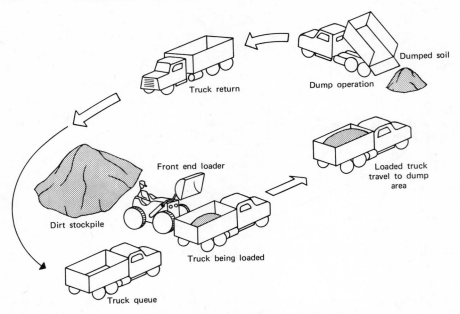

FIGURE 3.16   Schematic outline of earthmoving operation.

eration involved depends on the modeler's viewpoint and the intended model use.

As a second example, consider the development of a model for an earthmoving operation that involves the front end loading of trucks with dirt for transport to a dump area. A schematic outline of the operation is shown in Figure 3.16; it uses a front end loader, some trucks, and dirt.

In order to develop the skeletal framework of the earthmoving operation it is necessary to identify the major resources involved (i.e., trucks, front end loader, and dirt) and establish the various states (i.e., both the active working states and the passive idle states) that the resources traverse in their work assignment paths and cycles. Finally, the integration of the resource paths and cycles establishes the basic structure of the operation.

Each truck, for example, is idle while it queues for loading; it enters active working states when it is being loaded, dumping, traveling loaded to the dump area, and returning empty for another load. A simple model of this work cycle is shown in Figure 3.17a using a single COMBI "LOAD TRUCK" work task that requires dirt and a front end loader for initiation; three NORMAL work task elements "LOADED TRUCK TRAVEL," "TRUCK DUMP ACTIVITY," and "EMPTY TRUCK RETURN"; a single QUEUE element "JOIN TRUCK QUEUE"; and five ARCS indicating the logical relationships between the various truck states.

The front end loader can be initially modeled by a slave unit cycle involving the active state COMBI element "F.E.L. LOADING," the idle QUEUE element "FEL IDLE," and two entity flow directional logic ARCS (see Figure 3.17b).

In Figure 3.17c a soil path model is shown that uses a source SOIL STOCK-PILE and sink destination SOIL DUMP QUEUE nodes together with a

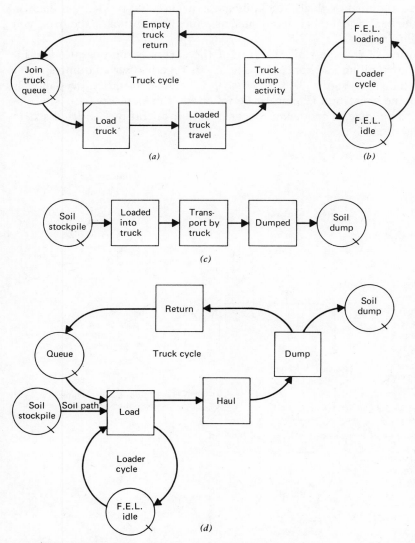

FIGURE 3.17 Development of operational structure. (a) Truck cycle. (b) Loader cycle. (c) Soil path. (d) Earthmoving operation.

COMBI work task LOADED INTO TRUCK and NORMAL work tasks TRANSPORT BY TRUCK and DUMPED to portray the soil involvement in active work states. Finally, four directional ARCS are required to develop the path structure.

The integrated model incorporating the truck and front end loader cycles together with the soil path from stockpile to dump is shown in Figure 3.17*d*. This model can be used as the basis for further development involving dump area spotters and queues, dozer stockpiling operations, and truck maintenance, as well as the basis for further detail such as a fine description of the front end loader loading cycle.

An extension of the skeletal structure of the earthmoving operation to include dozer stockpiling and spreading operations together with a dump spotter foreman is shown in Figure 3.18. Further development of descriptive models depends on the construction plan to be followed; this development is left to the reader. The general modeling of construction operations is discussed in detail in Chapter Five.

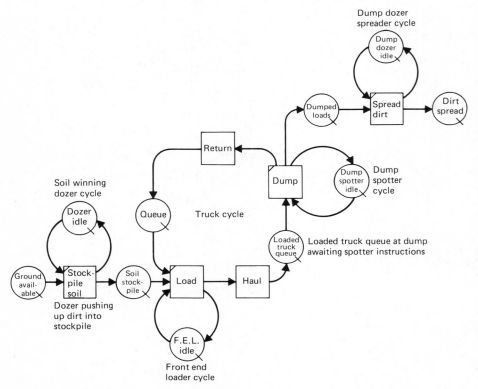

FIGURE 3.18   Model of earthmoving operation.

The foregoing presentation illustrates that the static (or topological) structure of construction operations can be developed and illustrated through the proper use and labeling of the basic modeling elements developed in this chapter. The static structure associated with the modeling of a construction operation is, in fact, a collection of states (active work tasks and idle queue nodes) together with a set of transformation arcs. The topological structure thus defined is the time-invariant or time-independent structural relationship of the network system model developed for the construction operation. If the work task modeling elements are suitably chosen to be intuitively obvious to the field team and management agents involved in the construction operation, the static structure model is capable of portraying the construction technology and construction method of the construction operation.

3.4 DYNAMIC RESOURCE ENTITY FLOWS

The static structure of a construction operation provides the technological blueprint or plan of how the operation is to be performed. The actual resources assigned to the operations and that process through (or are processed by) the various work tasks are required to traverse through the static structure of the operation. The movement of resource entities through the time-invariant static structure is dynamic and introduces the time-dependent properties of performing construction operations.

During the actual performance of the operation, the current status of the operation and location of the various resource entities in the static structure are time-dependent variables. Initially, the resource entities must be made available at the input side of the construction operation to start an activity and, finally, on completion of the operation some resources are freed and become available at the output side of the construction operation for future allocation according to the construction plan. During processing, the actual routes and flow patterns that individual resource entities traverse are defined by the static structure. Consequently, the time-dependent resource movement is governed purely by the work task processing durations and the need for resources to wait before initializing ingredience-constrained work tasks for the necessary ingredience logic to be satisfied. Dynamic resource entity flows are therefore determined by the user-prescribed work task durations and system-generated delays.

Two work task modeling elements, the NORMAL and the COMBI, have user-defined time delay functions that prescribe the time period resource entities are delayed while processing through the work tasks. The QUEUE node has the potential for storing in a waiting, idle, or queue format the resource entities held up by system performance awaiting the satisfaction of COMBI work task ingredience or initializing logic. The ARC modeling element, however, has no time parameters affecting resource entity flow, since it serves purely as a means of indicating resource directional flow logic between the various system NORMAL, COMBI, and QUEUE nodes.

The actual dynamic flow of resource entities through the static structure and their behavior in relation to the modeling elements can be illustrated by considering the resource entities as "blips" traversing the circuitry or network model of the construction operation. Thus, referring to Figure 3.19a, an entity blip "1" arrives at the input side of a NORMAL along ARC B and immediately initiates processing of the work task. The work task duration is modeled by the probability distribution function shown in Figure 3.19b, which has a mean duration of 30 minutes, a duration range between 20 and 40 minutes, and a standard deviation of 5 minutes. Entity blip "1" is captured in the work task for a duration (say of 28 minutes) and initializes the output side of the work task, where it emerges as two simultaneous blips "5" to satisfy the emanating ARC logic for ARC C and ARC D. In general each arriving entity blip "1" along ARC A or ARC B reactivates the NORMAL and requires its own unique duration to process through the NORMAL. Thus entity blip "2" may require 40 minutes transit, blip "3" may require 20 minutes, blip "4" may require 30 minutes, and so forth; consequently, depending on arrival time at the input side of the work task, entity blips may overtake previous arrivals. An example of this situation would occur if the NORMAL element models the transit time of trucks between a dump location and loading point.

However, if arriving entity blips cannot pass one another, then Figure 3.19a must be replaced by Figure 3.20a. The entity blip "1" arrives at the QUEUE node and establishes an entity blip queue at "2." The entity blip "2" looks ahead to see if the square node element that it wishes to process is available; if it is available it proceeds immediately; if another unit is in the square node (i.e., blip "3") the entity blip "2" waits in the QUEUE node queue format until the square node becomes available (i.e., when entity blip "3" leaves the work task). Thus the traditional concept of a queue is established by the QUEUE node logic, and the Figure 3.20a situation implies that the following work task processor has a special flow characteristic that con-

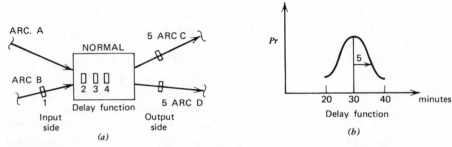

FIGURE 3.19   Resource entity flows through "NORMAL" element.

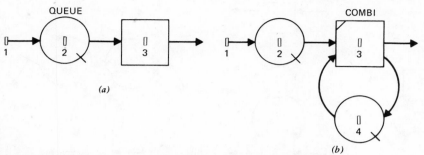

FIGURE 3.20 No passing and single transit work tasks.

strains flow. A more rigorous and informative modeling of the situation is given in Figure 3.20b. Here the square node is replaced by a COMBI node together with a slave unit cycle involving an additional QUEUE node and two ARCS. In this situation entity blip "2" must queue until the slave entity blip "4" is available in its queue position; otherwise the ingredience logic for the COMBI node (i.e., entity blip "2" and one entity blip "4") is not satisfied. Thus Figure 3.20 models the "no passing" and "single entity transit" situation. An example of this situation would be a "POUR ELE-MENT" work task involving only one form, so that a second element cannot be poured until the first and only form is again available. In this case the slave entity could model the form resource unit.

3.5 BASIC RESOURCE ENTITY FLOW PATTERNS

Resource entities move from work task to work task during the performance of a construction operation. The trace or sequence of work tasks traversed by the resource entity together with its idle states indicates the extent of the resource entity's involvement in the construction operation and forms a flow pattern.

A number of basic resource entity flow patterns can be identified (see Figure 3.21). These basic resource entity flow patterns occur frequently in the modeling of construction operations and correspond to unique recognizable sequences in practice. The patterns can be readily modeled using the modeling elements discussed above.

The slave entity pattern

Figure 3.21a models a situation that occurs frequently in practice. The slave entity pattern is produced whenever a resource entity is tied to a single active work state in a cyclic sequence. The resource entity endlessly cycles between the active work state and the idle state. In many cases the model is directly

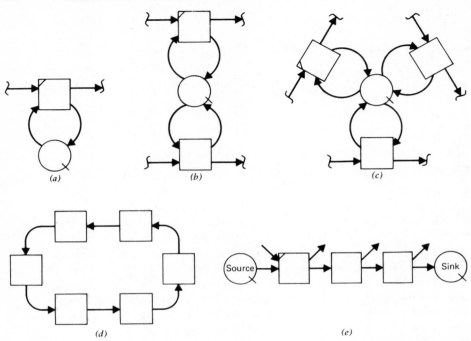

FIGURE 3.21 Basic resource entity flow patterns. (*a*) Slave entity. (*b*) Butterfly. (*c*) Multiloop work assignment. (*d*) Cyclic sequence of work tasks. (*e*) Noncyclic or path sequence of work tasks and states.

valid, while in other cases the resource entity may perform many finer detail work tasks that are all lumped together into the one work task for the purpose of the construction operation analysis. Table 3.1 offers several practical situations that may be modeled by the slave entity pattern.

The butterfly pattern

In many construction operations a resource entity is shared between two (see Figure 3.21*b*) or more (see Figure 3.21*c*) work tasks. In both situations once the resource entity is in the idle state, its subsequent active work task state may depend on any of a number of factors. The availability of the work task and the priority system adopted for the construction operation that gives preference to one work task over another normally control allocation of the resources. Some operational logic may exist for the construction operation that controls the availability of a work task and thus the specific path the resource entity will travel.

Table 3.1 SLAVE ENTITY PATTERN

Resource Type	Operation	Active State 1	Idle State 2
Equipment			
Pump	Dewatering	Pump pumping	Pump stationary
Hoist	Hoisting	Hoist in use	Hoist idle
Front end loader	Loading	Loading trucks	Idle awaiting trucks
Hopper	Containing	Hopper full	Hopper empty
Formwork	Forming	Formwork in use	Formwork not in use
Labor			
Mason laborer	Brickwork	Supplying mason with material	Waiting for orders from mason
Laborer 1	Hand excavation by two-man crew	Using pick to loosen material	Waiting for second member of crew to shovel out loosened material
Laborer 2	Hand excavation by two-man crew	Using shovel to remove loosened material	Waiting for first member of crew to loosen material
Material			
Dirt	Earthmoving	Being pushed to heap	Stored in heap
Space			
Working space	Work face	Working space available	Space cluttered up with items and rubbish
Logic			
Permit	Inspection, order, or management	Permission to proceed	Permission unavailable or withdrawn

Cyclic sequences of work tasks

In many construction operations repetitive processes exist wherein a resource entity continually transits through a cyclic sequence of states. The cyclic sequence of states may be all active work tasks (Figure 3.21*d*) or a mixed sequence of active work tasks and idle states. In Figure 3.21*d* the resource entity processes each work task in turn without delay; in general the resource entity may be unable to process a work task for any of a number of reasons as explained above and thus be forced into an idle Q node situation.

Noncyclic path sequence of tasks

In many construction operations a resource entity is required for a number of work tasks and, when these are completed, the resource entity is no longer needed and exits the operation. Figure 3.21*e* models this situation where the initial and final Q nodes correspond in concept to source (initial availability) and sink (dispatch) nodes of the resource entity in consideration.

Many examples of the above basic patterns exist in practice. Very often the same operation can be modeled in different ways depending on the level of detail being modeled or the purpose of modeling. The following chapters give practical examples of the basic flow patterns introduced above.

PROBLEMS

1. Brick pallets are picked up from the supplier and transported by truck to the job site where they are off-loaded and stockpiled. Draw a model of this process similar to the one developed for the earth hauling operation shown in Figures 3.17 and 3.18.

2. Visit a job site and select a process for investigation. Draw a schematic diagram of the process and/or site layout. Identify the major flow units in the process selected and list the active and waiting states through which each unit passes.

3. On a paving job, the concrete is batched at a tower located two miles from the paving site. Five-at-a-time dry batches are carried in an open bay truck (with appropriate compartments) to a mixer near the paving site. The batches are then dumped individually and sequentially into the skip of the mixer and mixed sequentially, starting with batch 1 and ending with batch 5. As each wet batch exits the mixer, it is dumped into a concrete bucket and lifted by a crane to the placement location where it is dumped, spread, vibrated, and finished by a concrete crew. Identify the active and waiting states through which a batch passes from the time it is dumped, spread, vibrated, and finished by a concrete crew. Identify the

active and waiting states through which a batch passes from the time it is initialized at the tower until it is incorporated into the pavement. Which of the active states would be NORMAL and which would be COMBI elements? Assume that there is only one mixer and one crane.

4. Develop work assignment models (similar to Figures 3.13 and 3.15) for the following labor and craft situations.

   (a) The erection of column formwork by a carpenter and laborer crew.
   (b) The field operation of a drilling machine.
   (c) The placement of concrete in a slab using buggies and vibrators with a concrete crew made up from laborers, cement finishers, and a supporting ironworker and carpenter.

5. Develop work assignment models (similar to Figures 3.13 and 3.15) for the pipe loading crew activities illustrated in the crew balance chart of Figure 2.18.

6. Steel sheet piles 12 feet long are being driven using a double acting compressed air hammer. The steel is positioned initially using a driving template. A mandrel is placed on the pile once it is in position and driving commences. When the pile has been driven 8 feet, the hammer and mandrel are withdrawn and another 12-foot section is welded to the first. After this, driving continues. This process continues until four sections have been welded to the original and driven. The last section is trimmed to a uniform elevation using a cutting torch. The next section of wall is started by positioning the initial sheet pile in the template with interlock to the just completed drive. Assume that the sheets are stacked initially in a stockpile location. Identify the active and waiting state through which each sheet pile must pass from beginning to end of the process. What resource units might constrain the movement of the pile from stockpile to final driven location?

7. Do you feel that the modeling elements introduced in this chapter form a suitable basis for the modeling of construction operations? If not how would you go about improving the capability, relevance, accuracy, and informational content of the proposed modeling methodology?

CHAPTER FOUR

CYCLONE System Definition

The resources available to management on a given project site are broadly defined in the preplan phase of construction management, confirmed in detail by site management as work is initiated, and modified from time to time as work progresses and field situations change. The preplan phase establishes the major construction methods and their requirements of special, general, and supporting equipment and implicitly establishes the framework for site management. Site management confirms the availability and feasibility of major equipment items and establishes in detail an inventory of project resources. This includes minor plant as well as the nature, size, and skill composition of the labor work force. These resources constitute the pool of project resources that management must use over time to complete the project.

The basic problem of construction management at the site level is the proper and timely allocation and application of project resources to site operations throughout the project life. This dynamic commitment of project resources consists of the movement of management entities (i.e., crews and equipment) to various tasks. The changing focus of activity at various site locations and the level and intensity of effort are such that the various resources can be considered to flow from operation to operation, site area to site area, and so forth. The complexity of construction operations generates the need for simultaneous use of labor, equipment, material, supporting cranes and hoists, and decision criteria, so that a variety of interrelated resource flows results. In this way project management can be viewed as the manipulation of networks of interacting resource flows.

At the field site, many resource flows can be identified, especially those associated with the site delivery of materials and subsequent handling and hoisting to the work face area. In these situations limited delivery areas, hoist and crane capacities, and congested access or work areas constrain the rate of material flow. The overloading of these facilities by poorly scheduled deliveries, the competition for hoisting, or the establishment of priorities for material movement, crane use, or work areas present obvious cases where resource flow interactions occur. In many cases congested work site areas drastically affect worker movement within a single crew or between crews using a common access. Finally, tardy approvals and decisions indicate a more subtle form of restricted resource flow.

The application of flow concepts to the allocation and assignment of resources and the nature of relationships between resource flows in the construction area has always been implicitly understood by head office and field personnel, but it has not been extensively documented. The concept of analyzing the flows in networks representing a real-world situation to aid in the evaluation of management decisions is not a new idea. However, a system of notation for modeling and communicating information regarding various processes has not been generally available.

Using the concepts of resource entity flow and active-idle states established in Chapter Three, it is possible to develop a graphical notation that defines these ideas more rigorously and provides the basis for dynamic modeling of construction operations in terms of flow networks. This chapter presents a set of modeling elements that can be used for describing process interaction and implementing analysis of complex construction situations. The modeling system defined by these elements will be referred to throughout the remaining portion of the text as the CYCLONE system. The name CYCLONE system is an acronym for *CYCL*ic *O*perations *NE*twork system.

4.1 THE CYCLONE SYSTEM

The CYCLONE system is based on a set of six modeling elements designated the NORMAL, COMBI, QUEUE, ACCUMULATOR, and FUNCTION nodes and the ARC (see Sections 4.2 to 4.7), four of which were introduced in Chapter Three. This group of elements constitutes a fundamental set of building blocks in terms of which large system configurations can be constructed. By combining the elements together in various configurations, the modeling of work task sequences, processes, and assignments becomes possible. Each element contributes to the development of a network system model of a construction operation that provides insight into flow unit interaction. When this model is properly labeled and drawn, it enables a full representation of the most complex operations and allows a visual understanding of basic process dynamics.

The six modeling elements that form the basis of the CYCLONE system are shown in Table 4.1 together with their symbolic representation and the intrinsic and defined functions associated with each. Each element (with the exception of the FUNCTION node) has an associated function that is intrinsically or automatically defined when it is incorporated into the flow network topology. In addition to the intrinsic functions, three elements (i.e., QUEUE node, FUNCTION node, and ACCUMULATOR) have auxiliary defined functions associated with them. These defined functions are in addition to any basic or intrinsic functions associated with the element.

Intrinsic and defined functions can be compared with the function key on an electronic pocket calculator. If the *cos x* key is pressed, the cosine of the number in the register is automatically calculated. The calculation of the cosine function can be considered intrinsic to the circuitry of the calculator, since no special programmed steps must be entered in order to implement its calculation. By contrast, the defined function is similar to the *program* key, which starts the calculation of a sequence defined by the user in the form of an entered program.

Since the FUNCTION node is a special node incorporated into the process model for the purpose of defining functions, it has no associated intrinsic functions. Therefore, when it is used, a defined function associated with it must be stipulated. If an associated function is not specified, its use in the network model is meaningless. The ARC can also have a user-defined "probability-of-transit" function that influences the directional flow of transiting units.

In this chapter, each of the modeling elements will be discussed in detail. The intrinsic and defined modeling functions associated with each element and the system use of the elements in constructing system models will also be described.

4.2 THE NORMAL MODELING ELEMENT

The NORMAL modeling element is the CYCLONE element with an intrinsic entity flow delay function and with free access input properties. The magnitude and characteristics of the unit flow delay are established by the modeler.

The graphical symbol for the NORMAL modeling element node as previously noted is a square (see Figure 4.1a). The square graphical format emphasizes the relationship of the elemental work task to the rectangular bar chart representation of the much larger project activity, since the activity can be considered as being made up of a sequence of many work task elements.

The NORMAL modeling element is used to logically model unconstrained work tasks. A resource entity arriving at the input side of a NORMAL

Table 4.1 CYCLONE SYSTEM MODELING ELEMENTS

Name	Symbol	Intrinsic Functions	Defined Functions	Remarks
NORMAL Work task	(square)	User-defined time delay	None	
Combination (COMBI) Processor	(square)	1. Ingredience constrained 2. User-defined time delay	None	
QUEUE Node	(circle with mark)	1. System-generated ingredience delay 2. Monitoring system generated ingredience delays 3. Entity initialization 4. Delay statistics	Entity generation Entity initialization Statistics (other) Marking	The GENERATE function associated with a QUEUE node is stipulated by the abbreviation GEN below the symbol
ARCS	(arrow)	Entity flow logic	Transit probability	
ACCUMULATOR (Counter)	(triangle with circle)	Accumulation of system productivity	Experiment Duration Control	
FUNCTION Node	(circle)	None	Consolidation Statistics collection Marking Counting	The defined function associated is indicated by abbreviation CON = CONSOLIDATE STAT = STATISTICS MARK = MARKING KOUNT = COUNTING

Note. Numerical labels are associated with each symbol as specified in the network.

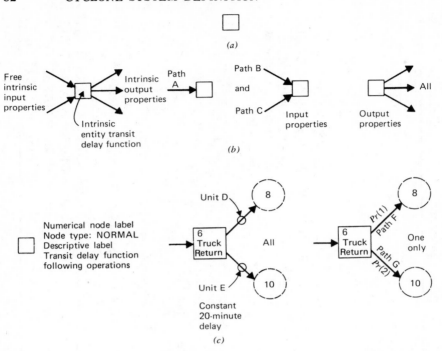

FIGURE 4.1   The NORMAL modeling element. (*a*) NORMAL Graphical Symbol. (*b*) Intrinsic functions: free access input-output; entity transit delays. (*c*) User-defined functions: identification only and characteristics of entity transit delay function. (*d*) Topological syntax for NORMAL modeling element. (*e*) Use of NORMAL.

modeling element is given free access to the element and initiates work task processing for a period of time that is user-defined. In this sense the NORMAL is said to capture the resource entity for active processing. The NORMAL, therefore, models the active working state of a resource. The NORMAL modeling element is said to be logically unconstrained because it has no logical conditions to be met at input. Therefore any and every resource entity arriving at the input side of the NORMAL modeling element initiates (or reinstates) the work task activity, and a specific mix of input resource types is not required. This intrinsic behavior is referred to as the free access input property of the NORMAL modeling element.

Entities arriving along any of the input side arcs (A, B, or C) will initiate the NORMAL and cause the commencement of a unit transit sequence. After a time period defined by the unit transit delay, the output side of the NORMAL is activated, causing a unit to exit along each of the emanating ARCS. (Figure

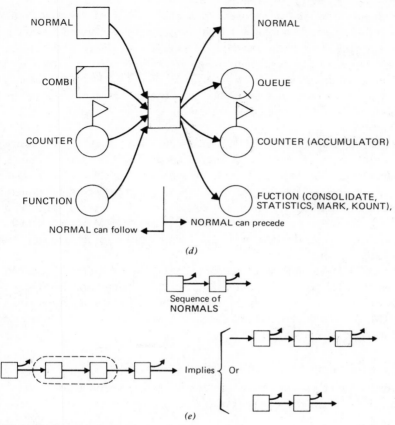

FIGURE 4.1   (*Continued*).

4.1*b* shows this intrinsic output property of the NORMAL.) The intrinsic input-output properties of the NORMAL can be summed up as "ANY" entity initializes the input side,* and "ALL" exiting ARCS are activated unless specifically restricted by the modeler in a probabilistic ARC statement.

If a single entity is required to exit from the output side of the NORMAL, then either a single ARC is provided or a set of probabilistic ARCS is established (see Section 4.5). The number of units released is, therefore, controlled by the number of emanating arcs and the properties associated with the ARCS. Referring to Figure 4.1*c*, in the first case, entities D and E are released to nodes 8 and 10, respectively; in the other case, either entity F is released to node 8 with probability Pr(1) or entity G is released to node 10 with probability Pr(2).

*This is referred to as a logical OR.

The user-defined functions associated with the NORMAL require node identification (labeling) and topological location, together with the specification of the magnitude and characteristics of the transit delay function, as indicated in Figure 4.1c. The processing delay associated with unit transit is specified in the same manner for both NORMAL and COMBI elements. A user-defined delay duration can be specified as either a constant (i.e., deterministic) or as a random variable consistent with a set of probabilistic distributions.

The topological syntax for the NORMAL is shown in Figure 4.1d. Since there is no requirement to wait until ingredience conditions are satisfied, there is no requirement for a QUEUE node preceding a NORMAL work task. Furthermore, since the NORMAL work task is not ingredience-constrained, several entities can traverse it simultaneously. That is, several units can pass in parallel and even overtake other entities in a NORMAL work task. The entities are not constrained to wait until a processor returns to accompany each unit in transit.

In general a NORMAL modeling element is used for each nonconstrained work task element defined in the system. A typical sequence of NORMAL modeling elements is shown in Figure 4.1e, which indicates two output entities for each NORMAL in this case. Three conditions are required for the use of a NORMAL work task element.

1.   The work task modeled has an associated time delay.

2.   The work task start or commencement is not ingredience-constrained but can be initiated immediately following completion of the element preceding it.

3.   It is not possible to consolidate the work task modeled with a preceding element, since a flow unit must exit the preceding element following its completion.

If these three conditions are attained, a situation exists in which the NORMAL can be utilized.

4.3 THE COMBINATION (COMBI) MODELING ELEMENT

The COMBINATION modeling element is referred to as the COMBI node and models a work task that has ingredience constraints on the mix of resource entities that must be available before the work task can commence.

The graphical format symbol for the COMBI modeling element is a square with a corner slash, as shown in Figure 4.2a. The square format indicates that it is an active state and requires a user-defined time delay function.

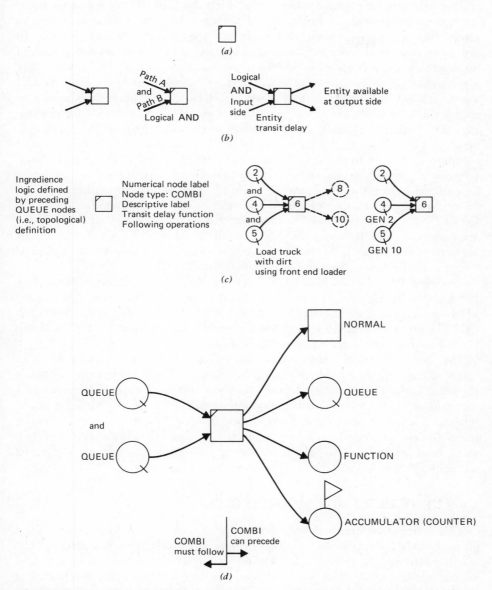

FIGURE 4.2   The COMBI modeling element. (*a*) Graphical symbol for COMBI. (*b*) Intrinsic functions: ingredience-constrained input logic; entity transit delay. (*c*) User-defined functions; node and ingredience logic identification, definition of entity transit delay function, entity meshing logic and generation. (*d*) Topological syntax for COMBI.

Two intrinsic functions are associated with the COMBI modeling element: its ingredience-constraint function and the need for a unit delay function to model the time period that resource entities are captured by the element. In order for a COMBI work task to begin, units must be available along each of its incoming arcs. The arcs incident on the input side of a COMBI work task always link QUEUE nodes to the COMBI work task. Therefore, the COMBI work task is always preceded by a set of QUEUE (idle state) modeling elements that define its associated ingredience set. The requirement for realization (i.e., commencement) of the COMBI work task is that at least one unit be available at each of the preceding QUEUE nodes. This is the requirement commonly referred to in network theory as the logical AND (see Figure 4.2c). In network theory, it is normally stated that all arcs incident on a node must be realized before the node may be realized. This definition implies activity-on-arrow notation with nodes used as event or milestone markers. Since the CYCLONE system is defined in precedence notation and unit or entity flows are defined in the network, the AND requirement stated above becomes the same as requiring at least one entity to be available at each of the QUEUE nodes preceding an ingredience-constrained or COMBI work task.

The ingredience set associated with a given COMBI processor is defined by the user as he establishes the network logic between elements. The logical network structure causes definition of the QUEUE nodes preceding each COMBI processor. These user-defined elements then automatically establish the ingredience requirement (see Figure 4.2d).

The second function intrinsic to a COMBI processor work task is the associated processing delay, which establishes how long a unit requires to transit the work task. This is a user-defined duration and can be specified either as a constant or as a random duration consistent with a set of probabilistic distributions. Various probabilistic distributions may be appropriate, depending on the situation, and can be used to define this processing delay function. When using random delays, a set of values that stipulates the parameters associated with the selected distribution is also specified by the user.

4.4 THE QUEUE NODE MODELING ELEMENT

This element defines a delay or waiting location that is associated with a given COMBI element. It models the idle state of a resource entity. The graphical format of the QUEUE node is a circle with a slash (i.e., a Q), as shown in Figure 4.3a. Entities arriving along any incident ARC are placed in the QUEUE node waiting state.

QUEUE nodes have four intrinsic functions that are related to network flows (see Figure 4.3b); they include system-generated ingredience delays, entity arithmetic, entity initialization, and the monitoring and statistics collection aspects of ingredience delays.

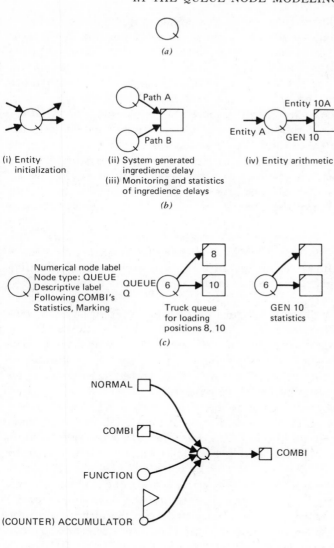

FIGURE 4.3 The QUEUE modeling element. (*a*) Graphical symbol for QUEUE. (*b*) Intrinsic functions: queue list structure and tagging. (*c*) User-defined functions: node identification; entity initialization; entity generation; statistics collection and initialization (markings). (*d*) Topological syntax for QUEUE.

The QUEUE node differs from other elements in that units are released along only one emanating arc at a time. Two cases are possible. A QUEUE node can have a single arc or multiple emanating arcs. The criteria in either case for release of a unit is that the ingredience requirements for the follower COMBI have been realized. In the case of a QUEUE node with more than one emanating arc (i.e., followed by more than one COMBI element), it is possible that the ingredience requirements of more than one follower COMBI are realized simultaneously. In this case, if only a single unit is available in the QUEUE node, it is routed to the COMBI element with the lower numerical label (i.e., number). Priority for units in the QUEUE nodes with multiple follower COMBI elements is always given to the lower-numbered COMBI. This allows the use of the QUEUE node as a switching mechanism and forms the basis for the operation of the Class V control structure to be discussed in Section 8.8.

The function that is of greatest interest relates to the QUEUE node's role as a waiting or holding location. COMBI processors are preceded by a set of QUEUE nodes that defines the associated ingredience requirement. Pending realization of the ingredience requirements, units are delayed at these QUEUE nodes.

Queue nodes allow for development of information regarding system operation by providing a convenient location for maintenance of delay statistics. Such statistics provide data regarding the percent of time the nodes are occupied with waiting entities and the time-integrated average of the number of entities delayed at waiting states. This will be discussed later in Chapter Ten, which considers system performance reporting.

Both resource and processed units are initialized in the CYCLONE system at QUEUE node locations. They can be defined at a QUEUE node preceding any COMBI processor in the process network. The initial position of units flowing in the system may be indicated in a tabular form, describing the entity type, name, and initial QUEUE location, or by using alphabetic letters representing each entity category. These letters are shown on the system network model at the initial entity locations. An associated integer value indicates the number of units in the entity category. Since units can be initiated into the system only at QUEUE nodes, these flow unit markers always appear on a circle format preceding a COMBI work task (see Figure 4.10). The QUEUE node has another important function that can be utilized by the user: the GENERATE function. The GENERATE function generates N units at the associated QUEUE node for each unit arriving along ARCS incident on the QUEUE. It amplifies or multiplies the incoming units by N, an integer number specified by the modeler. The GENERATE function of a QUEUE node thus performs entity multiplication arithmetic on the number of entities in the system (see Section 4.10).

4.5 THE ARC MODELING ELEMENT

The CYCLONE system networking format uses precedence notation (i.e., work task activity on node), since this format gives a better visual presentation of the location and movement of system entity flow units. The directional logic and the state network structure of a system model are defined by ARCS indicating the sequence of work tasks through which flow units pass and by the relationship between the initial and following nodes (see Figure 4.4d).

The ARC modeling element is the arrow, as shown in Figure 4.4a. Its tail end is located at the output side of a CYCLONE node and its head is at the input side of the following CYCLONE node (see Figure 4.4b). The ARC element is the basic modeling element defining network structure, and its intrinsic function is to define flow unit directional logic. No time delay to entity flow is associated with the ARC modeling element, although the ARCS are used by QUEUE and COMBI node elements to define ingredience constraints. The number of ARCS incident on or emanating from any element node is not limited in any logical sense. In CYCLONE, the number of element nodes that can follow a node is limited to 10 in the program implementation (see Appendix C).

Normally, once the entities captured in a CYCLONE node are released, entities flow along all ARCS exiting the node (see Figure 4.4e). This is the same as assuming a 1.0 probability of units being emitted along each of the exit arcs. The path that a unit takes after exiting a work task may be probabilistic. That is, each of the arcs emanating from a work task and tying it to following operational sequences may have a probability associated with it. In such cases, completion of a work task releases a unit along only one of the emanating ARCS. The probability associated with each ARC indicates the percent of the time it will be selected by the exiting flow unit over the other ARCS linking the work task to following CYCLONE nodes. The summation of the probabilities associated with paths exiting an element is equal to 1.0. This is required, since the unit exit probabilities cannot sum to more than or less than 100%.

In cases where unit transit along the exit ARCS of a work task is based on a probability, the probability can be defined as an ARC attribute. The probability function associated with each ARC emanating from a CYCLONE graphical element is shown in Figure 4.4b.

When probabilities are associated with emanating ARCS, a unit is released along only one ARC after completion of the element function. Since this is already the case with QUEUE nodes and since the rule for the path selected is already fixed, PROBABILISTIC ARCS cannot be associated with QUEUE nodes.

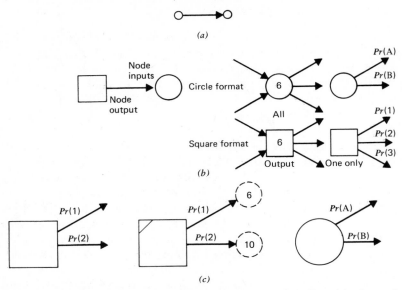

FIGURE 4.4   The ARC modeling element. (*a*) Graphical symbol for ARC. (*b*) Intrinsic function: entity directional flow logic; instantaneous transit, no delay. (*c*) User-defined functions: Probabilistic exits for entity flow from nodes to following nodes. (*d*) Topological syntax for ARC. (*e*) Use of ARC.

4.6 THE ACCUMULATOR MODELING ELEMENT

The ACCUMULATOR modeling element is a CYCLONE system monitoring and control element because of its ability to count the number of units that traverse it and to utilize and compare this entity count with two associated user-defined parameters. In its counting function, the ACCUMULATOR or COUNTER node performs the same function as the FUNCTION node specifier KOUNT (see Table 4.1). However, additional user-defined parameters enable the COUNTER node to produce meaningful information on system performance and to act as a total system control counter to limit further processing after a specified number of system cycles has been completed.

The ACCUMULATOR modeling element is an entity flow counter with both intrinsic and user-defined functions. The intrinsic counting function acts as a simple register that is incremented by one for each entity that traverses through the ACCUMULATOR. Since it offers no time delay resistance to entity flow (and therefore requires no user-defined time delay), it is modeled in circle format with an added flag symbol, as shown in Figure 4.5a. As shown

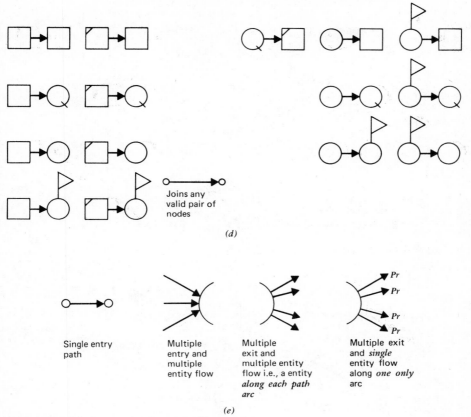

(d)

Single entry
path

Multiple
entry and
multiple
entity flow

Multiple
exit and
multiple entity
flow i.e., a entity
along each path
arc

Multiple exit
and *single*
entity flow
along *one only*
arc

(e)

FIGURE 4.4 (*Continued*).

in Figure 4.5b, the ACCUMULATOR count audits units for the path A total, the path B plus path C total, and the path D total, respectively.

Two user-defined attributes, C and Q, are normally specified. The C parameter is an integer whose value provides a cutoff upper limit for the number of units passing the ACCUMULATOR COUNTER node. Once the cumulative register of transiting units exceeds the C value specified, the CYCLONE model processing is terminated.

The Q parameter of the COUNTER node is defined as a number whose value conveniently scales the current register value to give system productivity values in units (dollars, cubic meters, etc.) that are meaningful to the modeler. Figure 4.5c shows a COUNTER node number 6 that has a preset upper limit

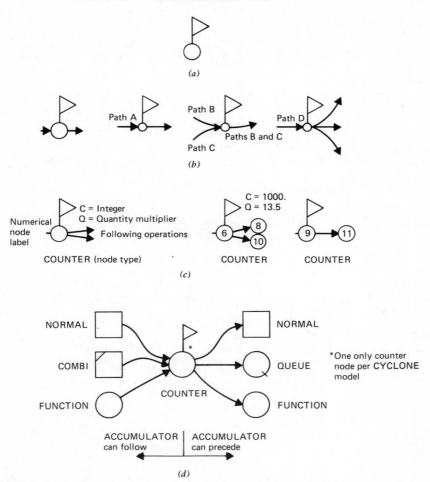

FIGURE 4.5 The ACCUMULATOR modeling element. (*a*) Graphical symbol for ACCUMULATOR. (*b*) Intrinsic function: entity transit flow counter, zero delay. (*c*) User-defined functions: upper limit cycles, quantity multiplier. (*d*) Topological syntax for ACCUMULATOR counter node system.

C of 1000 entities and a scaling quantity Q of 13.5 in the first case and a COUNTER node 9 with no user prescribed C and Q parameters.

Any number of input and output ARCS are permissible for the COUNTER node, as given by the topology requirements of following operations. One COUNTER node must be present in any CYCLONE model. The topological syntax for the COUNTER node is shown in Figure 4.5*d*, which indicates that it cannot be inserted between a QUEUE node and its following COMBI node and that only one COUNTER node is permitted.

4.7 THE FUNCTION MODELING ELEMENTS

Function nodes are CYCLONE system node elements that have no intrinsic or associated functions but that can be inserted into the system model anywhere (except between COMBI and preceding QUEUE nodes) to perform certain special assigned functions.

The graphical symbol for the FUNCTION modeling element is the circle, as shown in Figure 4.6*a*. An entity unit arriving along any incident ARC will activate the function associated with the FUNCTION node. Similarly, once the special function associated with the FUNCTION node is performed, all emanating ARCS are activated (see Figure 4.6*b*). In general, no transiting delay exists except in the case of the CONSOLIDATE FUNCTION node. When a CONSOLIDATE function is defined, exit logic is linked with a user-defined arithmetic division (or scaling down) of transiting entities. The CONSOLIDATE function acts as a flow unit collector or consolidator. One unit is generated along the output ARC(S) of the associated FUNCTION node for each N units arriving along ARCS incident on the FUNCTION node. It operates to divide the number of incoming units by N. The value of N is defined by the modeler. Pending satisfaction of the CONSOLIDATE function, units are detained within the register accounts of the CONSOLIDATE node.

The user-defined function associated with the FUNCTION node relates to node labeling and numerical identification, specification of FUNCTION type and, for the CONSOLIDATE node, the arithmetic division integer value N. In addition to the CONSOLIDATE function, STATISTICS collection, special node MARKING, and counting (KOUNT) functions can be associated with a FUNCTION node so that a multifunction specification can be generated for a FUNCTION node (see Figure 4.6*c*). The topological syntax for the FUNCTION node is shown in Figure 4.6*d*.

In general the CONSOLIDATE node is used to establish the entity meshing arithmetic required when two or more loops or paths are to be joined. Often the CONSOLIDATE node is used in conjunction with the GENERATE function of a preceding QUEUE-COMBI combination, as shown in Figure 4.6*e*. In this case each single parent entity arriving at the QUEUE node is multiplied (by GEN 10) tenfold to activate the slave unit cycle 10 times for each single parent unit, after which the CONSOLIDATE function node restores the identity of the original parent entity (by CON 10) using an entity division process. The use of the GENERATE and CONSOLIDATE functions is presented and discussed in Section 4.10.

The labeling of a node as a STATISTICS or MARK node (as shown in Figure 4.6*e*) in conjunction with a defined statistics set (see Table 10.2 for

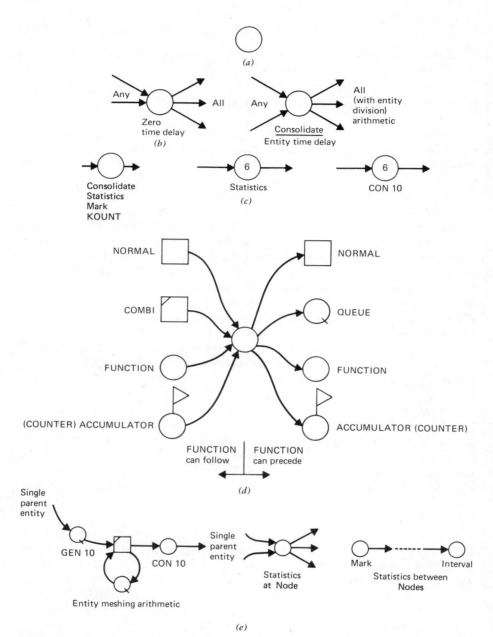

FIGURE 4.6 The FUNCTION modeling element. (a) Graphical symbol for FUNCTION. (b) User-defined functions: entity consolidation arithmetic, statistics collection. (c) User-defined functions: entity consolidation arithmetic, statistics collection and initiation (marking); node identification. (d) Topological syntax for FUNCTION. (e) Use of FUNCTION.

discussion) implements the collection of relevant statistics on units transiting the FUNCTION node. The labeling of a KOUNT FUNCTION node establishes a COUNTER type node, but without the features associated with C and Q parameters.

4.8 NETWORK STRUCTURE

Four modeling elements have intrinsic functions that are used to define the network structure and topological routing of units through the CYCLONE system model. These functions are associated with the COMBI processor work task, NORMAL work task, QUEUE node, and the logical ARC.

The network topology is defined by the way these four modeling elements are used in relation to each other. Although these four elements are sufficient to establish the basic structure of the network from the point of view of flow topology, the inclusion of the FUNCTION and ACCUMULATOR COUNTER nodes obviously affects the appearance of the final CYCLONE model.

The CYCLONE modeling formats are designed to provide maximum freedom in assembly logic. The interdependence among elements is reduced to a minimum, and the rules regarding sequential assembly are consequently simple. Table 4.2 summarizes the topological syntax sections of Figures 4.1*d* to 4.6*d*. The letter in each cell of the tableau indicates the feasibility of an "A" element (indicated to the left of each row) preceding a "B" element (indicated at the top of each column). The only elements that are logically dependent are the QUEUE node and COMBI processor. A COMBI processor can only be preceded by a QUEUE node and, conversely, the only element that can follow a QUEUE node is a COMBI processor. If the COMBI processor

Table 4.2   PRECEDENCE TABLEAU OF ELEMENT A PRECEDING ELEMENT B

A \ B	◁	□	Q	○	♂
◁	N	I	I	I	I
□	N	I	I	I	I
Q	M	N	N	N	N
○	N	I	I	I	I
♂	N	I	I	I	N

M = required or mandatory
I = immaterial
N = nonfeasible

column and QUEUE node row are removed, it can be seen that any precedence relationship among the other elements is acceptable and immaterial with the exception of the ACCUMULATOR element, which cannot precede itself.

The actual appearance of the CYCLONE model will depend on the identification and definition of the network elements (i.e., the NORMAL and COMBI) together with the associated QUEUE nodes, ARCS, and logical relationships. However, the network structure depends largely on the way a work task sequence is modeled. Thus, for example, several units can pass through a NORMAL work task in parallel and are not constrained to wait until a processor returns to accompany each unit in transit.

This is very helpful in modeling situations that are not ingredience-constrained. Consider again the curing tunnel situation shown in Figure 4.7 and originally presented in Figure 3.3. Precast elements are cured using a steam tunnel into which they are loaded and from which they are extracted using a crane. They move from left to right (in the schematic shown) for a predetermined period of time until they are ready for removal. A logical model representing this curing process is shown in Figure 4.7c.

The model illustrates the use of a NORMAL work task to provide the transit function representing the precast elements passing through the tunnel. From a systems viewpoint, this is a delay function without ingredience constraint. From a process viewpoint, there is no constraint that prevents the precast element from starting its transit through the tunnel following the loading work task. However, because of the flow pattern of the crane, it is necessary to separate the transit work task from the loading work tasks. The crane must return to its idle state following loading. Therefore, the transit work task cannot be combined with the loading work task into a single COMBI element. Insertion of the NORMAL work task allows the precast elements to begin transit immediately following loading, while the crane is returned to its idle state for reallocation. Thus, the network topology is directly affected by the degree of accuracy required in the model.

A similar situation occurred after the LOAD or DUMP work task in the earth hauling process (see Figures 3.16 and 3.17). A schematic diagram and a CYCLONE model for a closed loop truck haul situation with a spotter at the dump site is shown in Figures 4.8a and 4.8b. In this case, NORMAL work tasks are used to model the TRAVEL TO DUMP and RETURN TO LOAD work tasks. Since transit can commence immediately following both loading and dumping, there is no need to constrain the start of these work tasks. However, following loading and dumping, a resource must exit to an idle state. Therefore it is not possible to combine the load and transit or the dump and transit into a single COMBI element. Figure 4.8c shows an extension of the haul system problem to include probabilistic ARCS that model breakdown situations that can develop on the truck work tasks "TRAVEL TO DUMP

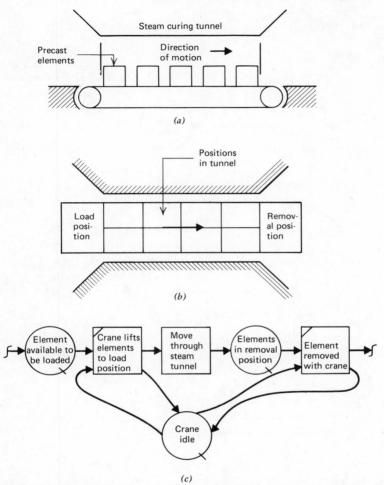

FIGURE 4.7   Precast plant tunnel. (*a*) Elevation view of tunnel. (*b*) Plan view of tunnel. (*c*) Model of tunnel.

SITE," and "RETURN TO LOAD." Clearly, the addition of probabilistic ARCS builds in alternative sequences or courses of action and thereby affects the topological structure of the operation model.

4.9 FLOW UNIT INITIALIZATION AND CONTROL

Two elements have associated unit control functions. The QUEUE node defines the points in the system at which flow units can be initialized. Further-

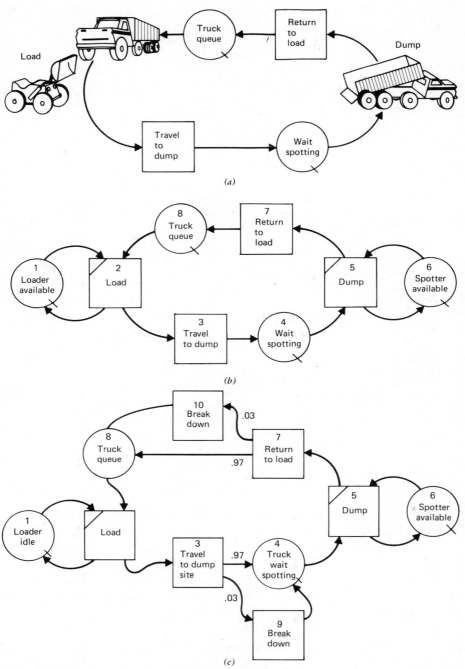

FIGURE 4.8 Haul operation. (a) Schematic diagram. (b) CYCLONE model. (c) Haul system with breakdown probabilities.

more, units can be generated (as required during the process) at QUEUE nodes by defining a special GENERATE function at the appropriate node. Any idle state location can be used to generate system flow units based on the arrival at the QUEUE node of an incoming unit. As previously mentioned, it is also possible to aggregate units at a FUNCTION node by defining a CONSOLI-DATE function. The location of units in the system at the beginning of a production process establishes the initial conditions of the system. If a loading hopper is full of material at the start of a process, the time required for the system to reach a steady state of operation and the level of production is different from that experienced if the loading hopper is empty. In Figure 4.8, if the trucks are all located at the loading queue when the hauling process starts, the system performance during the early phases of operation will be different from that achieved in a system in which two trucks are located at the "LOAD" queue and the other two trucks are located at the "WAIT SPOTTING" queue.

As discussed above, units start at a waiting position (i.e., QUEUE node) preceding some processor (e.g., trucks waiting at a shovel to load). For this reason, resource as well as processed units are initialized in the CYCLONE system at QUEUE node locations. They *must* be defined at a QUEUE node preceding a COMBI work task in the network model. The initial position of units flowing in the system is normally indicated in a tabular form that describes unit type (e.g., truck, crew, welder), number of units of each type, and initial QUEUE node location. For clarity, alphabetic characters may be used to indicate each unit type. The letters are shown on the system network model at the initial locations of each unit category or type. Figure 4.9 illustrates the definition of units in graphical format.

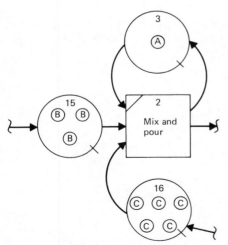

FIGURE 4.9   Entity graphical representation.

Units are initialized in this system as follows.

1. One unit of category A (labor crew) at node 3.

2. Three units of category B (concrete batches) at node 15.

3. Five units of category C (slab forms) at 16.

Figure 4.10 gives another example of flow unit initialization in the model introduced in Figure 4.8a.

The following units are initialized in the haul system model.

1. One loader unit at QUEUE node 1 (LOADER AVAILABLE).

2. Four truck units at QUEUE node 8 (TRUCK QUEUE).

3. One spotter (i.e., grade foreman to "spot" the trucks for dumping) at QUEUE node 6 (SPOTTER AVAILABLE).

Again, it should be emphasized that it is illegal to define a unit as starting in an active state. All units must be initialized so as to be in position to commence some work task.

As defined in the system of Figure 4.10, all trucks start in the system in the TRUCK QUEUE waiting to load. The trucks could have started all at the "WT SPOTTING" QUEUE node. This would imply that all trucks are loaded with material prior to the start of the system and are all waiting at the DUMP location to be spotted. Variations of these two initial conditions are possible. For instance, two trucks could be started in the "TRUCK QUEUE" (element 8) and two at the "WAIT SPOTTING" (element 4). The initial phases of the system's productivity are a function of the initial conditions defined by the positioning of flow units, as shown in Figure 4.11.

Initial system response for the haul system with three sets of initial conditions is illustrated. It is obvious that the productivity of the system with all

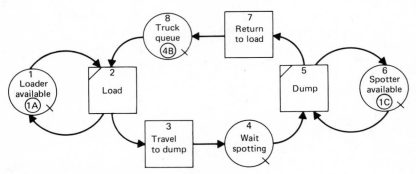

FIGURE 4.10   Haul unit initialization.

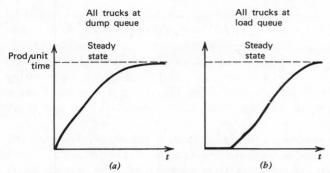

FIGURE 4.11   Haul system response to different initial conditions. (*a*) Four trucks at 4. (*b*) Four trucks at 8.

trucks loaded and waiting at the DUMP location reaches high productivity earlier than the other system. Once the system has been in operation for a certain period of time, it reaches a state in which the variation of productivity with time ($\Delta\text{Prod}/\Delta t$) is relatively small. That is, after a given "transient" period, the system settles down to a steady or nonvariant productivity in which $\Delta\text{Prod}/\Delta t = 0$. When this system has reached this time-invariant state, it is said to be in a "steady-state" condition. The time required to reach steady state is a function of the manner in which the units are initialized in the system. This accounts for the variation in productivity levels in the transient period of the haul system.

Care must be taken to insure that all required units in a system are defined. At least one unit must be generated in each of the flow cycles defined in the system. If a required flow unit is not initialized, the system will at some point call for the required unit, and system execution will be terminated because of failure to meet ingredience requirements. To illustrate, consider again the system of Figure 4.8*a*. If units are initialized, as shown in Figure 4.10, but the spotter unit at 6 is not, all four trucks will process the LOAD operation and move to 3. Following this, they will all travel to the WT SPOTTING idle state at 4, at which point they are unable to proceed because no spotter is available. Then, system execution will stop, since ingredience requirements at 5 cannot be met.

4.10 CYCLONE FUNCTIONS

In modeling some productive systems, it is useful to be able to generate or consolidate flow units at certain points in the system. That is, sometimes units representing master flow units break into component units for processing and then the subunits are reconsolidated. A truck, for example, may arrive at an off-loading point where the pallets it is carrying are processed and then shipped further. The truck prior to its arrival at the off-loading point can be considered a single flow unit or master unit. On arrival, it breaks into 20 units that represent the pallets it is carrying and that are to be processed. After processing, the subunit pallets are reloaded onto 10-pallet-capacity trucks and transported further. In order to model this situation a method is needed for generating 20 units to be processed for each arriving truck. Following processing, there is a requirement to aggregate or consolidate the processed units in groups of 10 for further shipment.

In certain instances, an action is initiated after a certain number of cycles of the system or a system subcomponent have occurred. For instance, after a crane has completed five lift cycles of steel framing, it is to be reassigned. A mechanism for counting the cycles and sending a triggering unit to initiate rerouting of the crane is required. This reduces to the counting and consolidation of five cycle pulse units into a single signal unit. To consolidate units, a defined function is available. It is associated with FUNCTION nodes at appropriate points in the network model (see Section 4.7).

Units can be generated into the system by defining a GENERATE function associated with a selected QUEUE node. Entity generation as well as initiation can only take place at a QUEUE node. Therefore, a GENERATE function can only be associated with a QUEUE node and not with a FUNCTION node. A simplified version of the pallet system is shown in Figure 4.12. In the system just described only one station is available to process the incoming pallets. The model has two processing stations and consists of two QUEUE nodes, a COMBI processor, and a FUNCTION node. Trucks carrying 20 pallet loads arrive at QUEUE node 1.

A defined GENERATE function is associated with QUEUE node 1; it splits or breaks the arriving truck unit into 20 pallet units (e.g., GEN 20). The pallet processors (two each) cycle between 2 and 3. The use of two processor flow units results in parallel processing. Parallel processing is handled by a class VI control structure and will be discussed in Section 8.8. The COMBI processor 3 is ingredience constrained by QUEUE nodes 1 and 2. Following

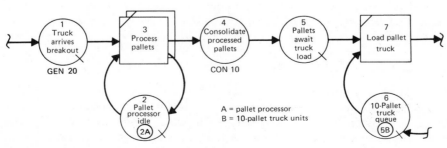

FIGURE 4.12   Pallet processing model.

processing, pallet units pass to FUNCTION node 4, where the function CONSOLIDATE operates to group them into 10 parallel loads. After the consolidation of 10 units into one load unit, this single unit moves to QUEUE node 5 and awaits the arrival of a 10-pallet truck at 6.

Graphically, the functions are defined by writing GEN "N" or CON "N" under the appropriate QUEUE or FUNCTION node. Therefore, the GEN-ERATE function may be thought of as a discrete entity multiplier and the CONSOLIDATE function as a discrete entity divider. In Figure 4.12, the arriving units are multiplied by 20 and then divided by 10. The amount by which incoming units are multiplied or divided is established by the value of N as defined by the modeler. The ability to multiply and divide units at various points in the system leads to added flexibility and efficiency in model-ing, since single units are only expanded as required (e.g., truck to pallets) for processing. Following processing, the expanded units can be reconsolidated to reduce the number of units that must be kept track of at a point in time. This amounts to a means of controlling the population of flow units in the system.

To illustrate this, consider a truck loading problem. Upon arriving at the loader, the truck is empty and has a 10-unit capacity (e.g., cubic yards or meters). If the loader has a two-unit bucket size, truck capacity is equal to five cycles of the loader. Therefore, upon arrival, the truck must generate five load orders to cause the loader to cycle five times. The load orders or commands can also be thought of as five space units representing the empty bay of the truck. Figure 4.13 illustrates schematically (a) the arrival of the truck, (b) its transformation into five load commands, (c, d) the cycling of the loader lead-ing to five loaded spaces, (e) and the transformation of these spaces back into

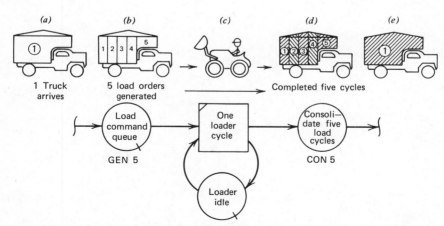

FIGURE 4.13   Unit control using GEN and CON function.

a single truck. The five space or command units are only present in the system at the time of processing. At other times, only a single flow unit (the truck) is present. This greatly simplifies study of the flow of units in the system by keeping unneeded detail to a minimum. In simulation of the unit flows in the system, great economies can be realized using GEN and CON functions.

Consider another situation in which the CONSOLIDATE function is used alone to trigger an action after a certain number of messages have been received. In many instances, a unit must be rerouted after it has completed a predefined number of cycles. Assume that a crane lifts 10 precast elements into place. After these elements have been placed, a deck-forming sequence can begin. In this case, each time an element is lifted and placed, a count is maintained. After 10 lifts are completed, forming begins. In this case, the CONSOLIDATE function acts to count incoming units and initiate an action when N = 10 units (messages) have been received. In the representation of this system component shown in Figure 4.14, the FUNCTION node at 5 generates two triggering units. One travels to 7 to signal release of the DECK-ING sequence and the other travels to 8, initiating the rerouting of the crane to the DECKING sequence through COMBI PROCESSOR ONE. In this case, 8 initiates the "capture" and rerouting of the crane. Capture mechanisms will be discussed in Chapter Nine.

Often two cycles interact, and there is a requirement to make the cycles compatible. In the haul situation in Figure 4.13, it was necessary to make the

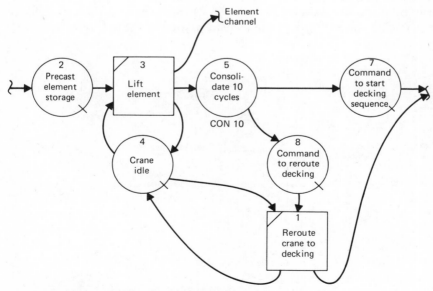

FIGURE 4.14   Triggering sequence.

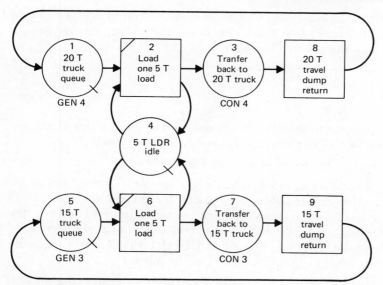

FIGURE 4.15   Cycle compatibility.

truck cycle compatible with the loader cycle. The truck represents five load cycles. Therefore, the truck unit had to be multiplied five times for its inter-action with the loader and then divided again to make it compatible with the logic of its own cycle. Consider a situation in which the loader cycle serves two different size trucks and must be made compatible with two differing cycles. Assume that a 5-ton loader services both 15-ton and 20-ton capacity trucks. Obviously, loading of the 15-ton trucks requires three loader cycles and loading the 20-ton trucks requires four loader cycles. In order to meet the compatibility requirements of each cycle, the GENERATE and CONSOLI-DATE functions can be used in tandem between each truck cycle. This is shown in Figure 4.15.

The 20-ton truck must be subdivided into four load commands, while the 15-ton truck represents three load cycles of the loader. The generation of the load commands is accomplished at QUEUE nodes 1 and 5, respectively, using GENERATE functions. The loaded trucks are redefined as single units at FUNCTION nodes 3 and 7 using the CONSOLIDATE function. The loader cycles between load positions at COMBI work tasks 2 and 6. The routing of the loader is established by giving priority to the lower-numbered COMBI processor. Therefore, if units are available at 1, COMBI work task 2 will be commenced before COMBI work task 6. This leads to a situation in which the loader may be loading a 15-ton truck and stops before completing this job to load a 20-ton truck. This can occur in this model representation since any time a load command is available at QUEUE node 1, priority is given to load operation 2 over 6. In Section 9.5, which describes capture mechanisms, a method of offsetting the priority feature will be discussed.

PROBLEMS

1. What would happen to the flow unit population of the systems illustrated in Figures P4.1a to P4.1c after 10 cycles? After 30 cycles?

2. Simplify or correct networks of Figures P4.2a to P4.2f.

3. A pallet truck picks up 35 brick pallets at the vendor's location and transports them to the job site, where they are off-loaded and stockpiled as 35 individual units. Draw a simple model for this situation utilizing the CONSOLIDATE and GENERATE functions to minimize the number of units flowing in the system. Locate an ACCUMULATOR so as to measure the production rate in pallets stockpiled at the job site.

4. Monitor a simple one-man (or two-man crew) labor-intensive construction operation and develop, in increasing detail and number, lists of work tasks and resources for the operation. Then prepare a series of CYCLONE models for the construction operation. Comment on the need for increased logical constructs as a function of the level of detail used in the models.
   Typical labor-intensive construction operations are the hand excavation of shallow foundations, door hanging, painting, rebar cage preparation, and the like.

5. Develop a CYCLONE model for a construction operation involving a crew and the use of equipment. Then, by direct observation of the crew at discrete times, locate the resources entities (both labor and equipment) in the various work tasks and idle states of your model. Can you use this field data to check your model?

6. There are four columns to be poured, all of which are 24 inches in diameter and 10 to 12 feet in height. Therefore, each column requires approximately 1 to $1\frac{1}{2}$ cubic yards of concrete. Prior to placing concrete, the columns are formed with circular metal forms and scaffolding is assembled in position around each column form. The concrete placement sequence is as follows.

   (a) The crane lowers the concrete bucket (1 cubic yard capacity) into position at the transit mix truck, where a laborer positions the chute so that the bucket may be filled. The truck operator (teamster) regulates the feeding of concrete into the bucket.

   (b) After the bucket is filled, the crane then swings the bucket to a position over one of the columns, where it is positioned into place by two laborers working on the scaffolding.

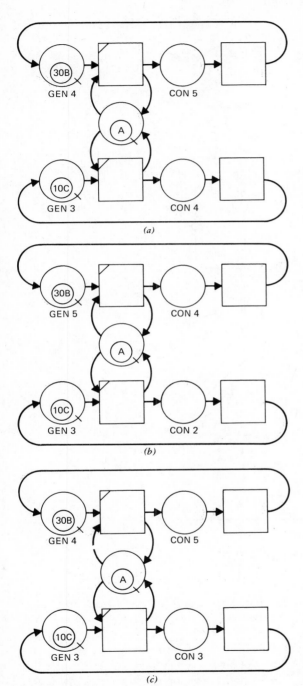

FIGURE P4.1   GEN-CON models.

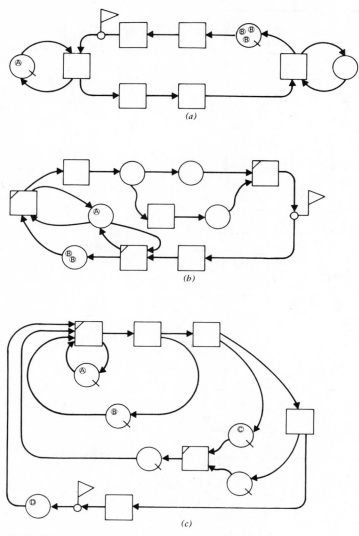

FIGURE P4.2   Networks for simplification.

(c) Once the bucket is positioned, the concrete is released and, after settling in the form, it is vibrated by a third laborer working from the scaffolding.

(d) After the bucket is emptied and a signal is given by one of the laborers on the scaffolding, the crane operator swings back to the transit mix truck where the procedure begins anew, or the bucket is swung to another column if there is concrete remaining in the bucket.

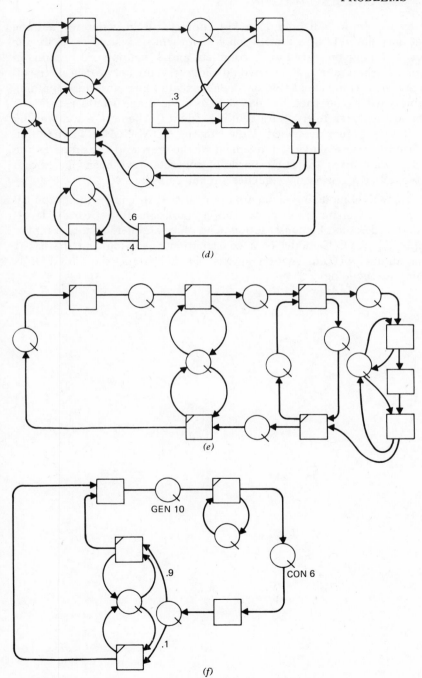

FIGURE P4.2   (*Continued*).

Since each column will hold 1 to 1½ cubic yards of concrete, and the bucket only holds 1 cubic yard, each column will be visited twice by the bucket. If a pour on a column is complete and ½ cubic yard remains in the bucket, the crane operator would move the bucket to the next column, the crew would reposition on the next scaffolding, and the remaining ½ cubic yard would then be placed before the crane would return the bucket to the concrete truck for refilling. Identify the active and waiting states in the system described. How can the CONSOLIDATE function be utilized to implement the movement of the crew and the crane to the second column after 1½ cubic yards of concrete have been placed and vibrated? (Use a triggering sequence similar to Figure 4.14.)

7. The CYCLONE modeling elements are designed to enable the graphical modeling of the structure of construction operations to be readily developed and understood by construction head office and field agents. Prepare a simple CYCLONE model of a construction operation and discuss it with a number of construction agents. Then critique the CYCLONE modeling methodology.

CHAPTER FIVE

Model Formulation

Models are strictly limited representations of real-world situations that are often extremely complex. Consequently, the areas of interest to be modeled and the purpose or intended use of the model should be carefully defined. The specific nature, form, and content of a model depends on the model purpose, the complexity of the real-world situation, the level of abstraction desired, and the skill of the modeler. Finally, models contain data that must be interpreted according to specific rules before useful information and insight become available to the model user.

This chapter introduces the model-building methodology. This will be done in the context of several relatively simple construction operations. Model formulation in this chapter focuses on the portrayal of construction operation structure and logic (i.e., on the development of the static structure). In later chapters more complex construction operations and processes will be discussed, and considerations of the practical level of representation, purpose, and use of models will be introduced.

5.1 THE MODELING PROCEDURE

The procedure for modeling a given construction process involves four basic steps. The steps, as shown in Figure 5.1, are as follows.

1. *Flow Unit Identification.* As a first step, the modeler must identify the system resource flow units that are relevant to system performance and for

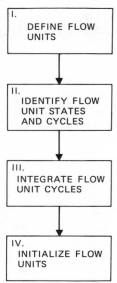

FIGURE 5.1   Steps in model formulation.

which transit time information is available or obtainable from the field. The selection of the flow entities is very important, since it dictates the degree of modeling detail incorporated into the operation model.

2.   *Development of Flow Unit Cycles.* Having identified the flow units that appear relevant to the process being modeled, the next step in model formulation is to identify the full range of possible states that can be associated with each flow unit and to develop the cycle through which each flow unit passes.

3.   *Integration of Flow Unit Cycles.* The flow unit cycles provide the elemental building components of the model. The structure and scope of the model are obtained by the integration and synthesis of the flow unit cycles.

4.   *Flow Unit Initialization.* In order to analyze the model and determine the response of the system model, the various flow units involved must be initialized, both in number and initial location.

Models developed using these basic steps must also be modified to provide for monitoring of system performance. This leads to a fifth stage of system design in which special elements for determining system productivity, flow unit characteristics, and other pertinent information are included in the model structure. These aspects of the system design of construction operations are discussed in later chapters.

5.2 PROCEDURES FOR DEFINING FLOW UNITS

As mentioned above, the first step in model formulation is the identification of the flow units that are relevant to the operation or process to be modeled. The proper selection of flow units requires an intimate knowledge of construction operations and establishes the basis for model validity.

Normally the units selected are physical items such as production units (i.e., front end loader, truck) or resources such as materials, money, and space. Other flow units are informational in nature and indicate that certain conditions have been realized, thus allowing the start of an activity. For instance, a flow unit might indicate that a given inspection has been passed and that the inspected unit can be released for further processing. The availability of a certain amount of money can be represented as a logical unit that allows for the release of a certain amount of work.

The more obvious types of flow units that are of interest in construction are as follows.

1. Machines.
2. Labor.
3. Materials.
4. Space or location.
5. Informational or logical permits (inspection releases, etc.).

All systems at the field operation level involve flow units of these types. A basic step in analyzing any construction operation or process and in developing the appropriate approach and level of detail is to determine which flow units are relevant from the above categories.

The machine unit category is self-explanatory and comprises the construction equipment appropriate to the operation being modeled (e.g., tractor dozers, cranes, and trucks). In almost all material flow process models, some type of equipment unit is required for movement of materials. In some cases, the equipment elements involved in a construction operation are not critical in constraining the flow of the processed material or flow units and, therefore, need not be included in the model. Model detail, for instance, seldom extends to the level of hand tools as a constraining force in construction processes modeled by management. Normally, only larger pieces of equipment are considered in realistic systems models.

Labor resources of all types play a very active role in constraining operations and thus affect the flow of materials and processed units through construction system models. These resources may take the form of an individual laborer or an entire crew. The number of tasks to which labor resources are assigned is also important in defining the associated flow path. The laborer may be assigned only one work task, which leads to the simple "slave" pattern discussed previously. On the other hand, a single operator may be associated with several work tasks in a given operation and constitute a shared resource. Such resources that cycle between various work tasks have a "butterfly" pattern associated with them. A laborer supporting brick masons, for instance, cycles between activities such as:

1. Tend the mortar mixer.
2. Move bricks to scaffold.
3. Carry mortar to scaffold.
4. Elevate scaffold.

Figure 5.2 shows a "butterfly" flow path for this type of situation. Operators of machinery, crews, and support laborers are examples of the labor type of flow units.

Material flow units are, typically, the units that are processed by a construction system and that, after being processed, become integral parts of the final construction product. The definition of material units is closely tied to the type of system being modeled. If the construction operation under consideration is concerned with the placement of concrete, for example, then the material units may be defined as batch loads or even as cubic yards of concrete. If a road paving process is being examined the selected productive units may be extended to the level of sections (50 feet, 100 feet) of the road. In any case, these units are discrete quantities and are important in their relation to the productivity measure of the system. The production of the system is measured as some multiple of the units that pass system productivity measuring elements.

Location or space-type flow units usually constrain the access to certain work processes and thus constrain the movement of other unit types. For example, in a brick plant, a large tunnel is used in which the bricks are cured. The bricks are loaded on transit wagons that move slowly through the tunnel,

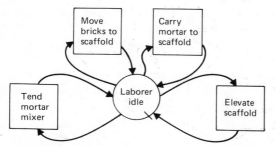

FIGURE 5.2   "Butterfly" pattern model of work task assignment.

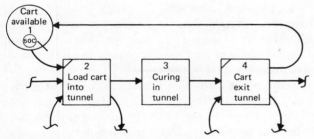

FIGURE 5.3   Brick plant tunnel process.

allowing the bricks to be exposed to curing steam and heat for the prescribed amount of time. The flow rate of the wagons through the tunnel and the size of the tunnel structure govern the number of cart positions or spaces available in the tunnel. A simple model of the brick plant tunnel process is shown in Figure 5.3. In this case, 50 cart positions or space units are associated with the tunnel curing activity and limit the number of brick carts that can be in transit in this activity simultaneously. Therefore, the number of brick carts that can be in activities 2–3–4 is constrained to 50 by the feedback loop 4–1–2. No cart load of bricks can be processed through 2–3–4 until a cart unit is available at QUEUE node 1.

Similarly, informational flow units are often used to link subsystems and as feedback control mechanisms. In many cases, a unit being processed cannot begin transit of a given set of work tasks until a previous unit has completed the sequence. In this situation, an informational unit can be fed back from the last work task in the set to the initial work task, releasing the sequence and allowing the waiting unit to proceed. An example of this is shown in Figure 5.4.

The logical flow unit is the most versatile of all flow units and, in a sense, encompasses the function of all the other categories. Definition of machine, labor, material, and space units establishes the conditions required for a certain task to begin. The conditions amount to the presence or absence of some re-

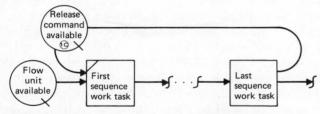

FIGURE 5.4   Feedback using informational unit.

source or release. In some cases, physical items such as money (cash or credit) or documents (purchase orders or invoices) are involved in the realization of the condition. The presence or absence of such units indicates that a condition has or has not been realized and a release or permit may or may not be granted.

As the first step in construction of a model, the manager must analyze the process to be modeled and determine the flow units from each of the categories described that are relevant and important to the modeling of the situation at hand. Once this is accomplished, the next step in the modeling process concerns the determination of the flow unit cycles and repetitive paths transited by the flow units defined. This begins to establish the cyclic structure of the process network model.

5.3 MODELING FLOW UNIT CYCLES

The identification of all the possible states, both active and passive, that a resource flow unit can occupy establishes the basic framework for modeling the flow unit cycle. The active working states of a resource are usually readily associated with a construction process, whereas the resource idle states are rarely considered. In the initial modeling phase, however, care should be taken to ensure that resource idle states are properly identified. This is particularly important if resource idleness is conditional on the availability of other resource flow units.

The flow unit cycles establish the basic set of model components required in a construction operation. The level of detail with which the flow unit cycle is defined establishes the size, complexity, and usefulness of the final system model. Consequently, at this stage in the model formulation procedure, the purpose of the model should be considered. In some cases, for example, a sequence of work tasks can be readily combined into a single work task. If, however, the behavior of an associated critical resource is thereby hidden, it may become necessary to incorporate the finer detail instead of the grosser representation.

In the initial flow unit cycle modeling phase it will be assumed, until otherwise established, that all work tasks associated with a given flow unit are ingredience-constrained; that is, assume that an active state with its associated time delay is a COMBI element. In this way attention is automatically focused (later during flow unit cycle integration) on the checking of ingredience logic associated with COMBI elements. This results in confirmation of the COMBI node or its replacement and simplification by a NORMAL element.

Since all active states are assumed to be COMBI element, each one must be preceded by a QUEUE node (see Table 4.2). This means that for every active state defined in a unit flow cycle, there will be an associated preceding QUEUE node in which the flow unit could be potentially delayed. Following integration of individual cycles into a composite model, a check of each active state is made to determine if any active state is preceded by only one QUEUE node. If this is the case, the modeler reviews the active state defined to establish whether the flow unit is delayed by the requirement for another flow unit before commencing transit of the work task. If another unit is required, this is included by defining its cycle and integrating it into the composite model. If no such constraining unit is required, the work task is redefined as a NORMAL work task, and its preceding QUEUE node is removed.

The flow unit cycles are established by defining the sequence of states met by each resource. For instance, if the work tasks through which a batch of concrete transits are being considered, the sequence of work tasks might be as follows.

1. Place concrete from bucket.
2. Vibrate placed concrete.
3. Screed placed material.
4. Finish concrete.
5. Cure concrete.

Sequentially ordering the work tasks establishes that the concrete units pass through the "vibrate placed concrete" work task before transiting the screed and finish work tasks. All of these work tasks are considered resource-constrained until otherwise established. In some cases, the cyclic movement of a flow unit through its active states will be easily established. For instance, the cyclic movement of a truck on the haul through the load, travel, dump, and return sequence is not too difficult to identify. In other cases, the system is made cyclic by taking units that are exiting the process (e.g., concrete batches or precast panels) and recycling them in order to reenter them into the system. Having units exit without returning is the intuitive feeling regarding their flow. For instance, if the construction operation is that of placing concrete on the upper floors of a building, the operation involves the concrete arriving at the site in a transit truck and being handled through various work tasks until it is placed in a building component. At this point, it essentially exits the system. In modeling construction operations, however, certain benefits accrue by recycling "used" units. This operation of closing the cycle and the advantages derived will become clearer after a few examples have been examined.

5.4 A MASONRY OPERATION EXAMPLE

As an illustration of model formulation, consider the development of a simple masonry operation model. Masonry operations are typical of many building situations that are worked with a crew of mixed crafts. A highly skilled mason whose output paces the job is supported by unskilled laborers who assist the mason by supplying bricks, mixing mortar, supplying mortar, and resetting the scaffold. Although less skilled, the laborer's activities tend to control the rate of productivity of the skilled masons. There is a balancing problem between the crafts that must be considered. If the system is not properly balanced, either the laborer or the masons served by the laborer are idle for significant periods of time. If the system is properly balanced, the idle time of both laborer and masons is minimal. A prime objective of the model of the masonry operation is, therefore, an understanding of the masonry crew structure and the influence of site conditions on crew productivity.

The masonry operation to be modeled is typical of many construction operations that can be described as being based on the interaction of two work task sequences. In its simplest form the masonry operation can be modeled as the interaction of two elemental cycles: a mason cycle and a laborer cycle. The full development of models descriptive of the masonry operation, however, is more complex and will be treated later (see Chapter Six).

In the masonry example, a laborer supports three masons in a bricklaying operation. The masons, in effect, require the service of the laborer in resupplying them with bricks and mortar. The laborer provides this resupply service and, therefore, acts in the capacity of a server. It will be assumed that the scaffold on which the masons are working is too high for the laborer to stack bricks and that the procedure is for the mason to place 20 bricks and then to let down an empty pallet that the laborer replaces with a full one. This process is depicted in Figure 5.5.

The first step in the modeling procedure requires the definition of the flow units involved in the masonry process. Three types of units can be initially identified as candidate categories: (1) machine, (2) labor, and (3) materials. However, the problem as described does not include any machine resources except, perhaps, the scaffold and concrete mixer. In the initial modeling effort assume that the scaffold is already erected and adjusted and that the laborer focus is on maintaining brick supply. Therefore the modeling will concentrate on the labor and materials categories.

In this particular problem, the candidates for flow units will be defined as the masons, the laborer, and the brick pallets. Each individual brick could have been defined as a flow unit, but it is normally more efficient to combine individual units into flow "packages" wherever possible.

To determine whether these candidate units are sufficient for the construction of a process model, it is necessary to determine the sequence of work tasks that the flow units transit. The masons are actively involved in two work tasks. They let down the empty pallet for resupply and place the bricks. They can enter an idle state if the laborer is involved in serving another mason. For our purposes, they cycle between the resupply and bricklaying activities. At this stage, there is no need to break down and define the bricklaying work task sequence further. Therefore, the cycle will consist of the two active states

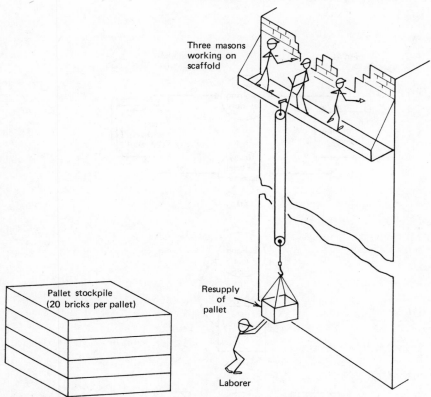

Three masons
working on
scaffold

Pallet stockpile
(20 bricks per pallet)

Resupply
of
pallet

Laborer

FIGURE 5.5   Schematic layout of masonry problem.

(RESUPPLY PALLET and PLACE BRICKS) and two associated QUEUE nodes (MASON AWAITING RESUPPLY and MASON READY FOR BRICK PLACEMENT). This is shown in Figure 5.6a.

Three squares are shown for the bricklaying state, since it is possible for all three masons to be placing brick at the same time. Only one resupply state is shown, however, since only one laborer is available for the resupply activity. The system is constrained by this fact, and this leads to the possibility of a mason entering the idle state while waiting for resupply. This control structure situation will be discussed in detail in Section 8.9, which considers Class VI Structures. Notice also that two QUEUE nodes have been used in accordance with the model development rule of initially assuming that each work task is a COMBI node.

The laborer cycle is quite simple since, for the purposes of this illustration, he can enter only two states. He is either active in the RESUPPLY PALLET

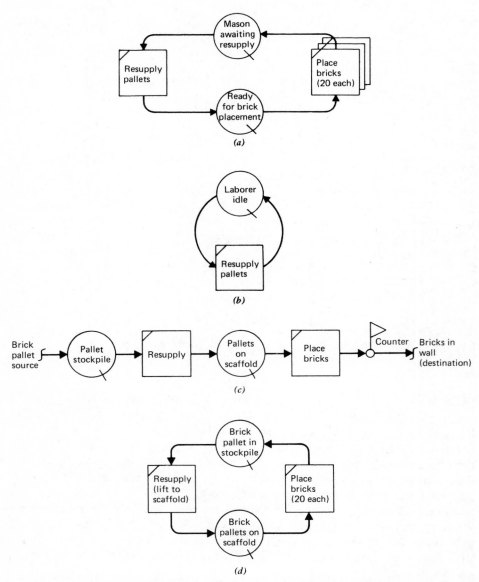

FIGURE 5.6   Individual flow unit cycles for masonry problem. (a) Mason cycle. (b) Laborer cycle. (c) Brick pallet flow path. (d) Brick pallet cycle.

work task or idle LABORER IDLE awaiting demand for resupply on the part of one of the masons. This results in a simple slave cycle, as shown in Figure 5.6*b*.

The brick unit flow through the system also results in two active states and two associated QUEUE nodes. The brick pallets (consisting of 20 bricks) are stacked in a stockpile PALLET STOCKPILE (i.e., idle state) pending their being lifted to the scaffold during the RESUPPLY work task. They are then successively involved in the resupplying and the bricklaying work tasks in that sequence, after which they exit the system. A simple path model that represents this situation is shown in Figure 5.6*c*.

In order to maintain the cyclic aspect of the system's operation, it is possible to close the brick pallet cycle by routing them back to the stockpile, where they can be reinitialized into the system.

As previously mentioned, benefits result from this recycling of processed units. This procedure reduces the number of units that must be defined at any one time, and it is often used in modeling situations where units enter the system from an infinite population and exit. If sufficient units are flowing in the cycle so that the input population does not become constrained, using cyclic instead of open paths for processed units can be very advantageous. This will be referred to in greater detail when discussing hand and computer simulation of the CYCLONE systems. The closed cycle for the processed units (i.e., brick pallets) consisting of two COMBI elements and two associated QUEUE nodes is shown in Figure 5.6*d*.

5.5 INTEGRATION OF FLOW UNIT CYCLES

The next step in model formulation is the integration of flow unit cycles to form the structured model of a construction operation.

In the masonry operation three flow unit cycles and paths have been developed. In order to develop the model, each model component must be integrated through the work tasks that they have in common. Consider initially the integration of the mason and laborer cycles. The RESUPPLY PALLET work task is common to each cycle, and integration results in the model shown in Figure 5.7. This amounts to superimposing the two cycles at the RESUPPLY PALLET work task node. If more COMBI elements had been common to the two cycles, these would have also been superimposed, forming additional points of contact between the two cycles.

At this point, the next flow unit cycle is selected (i.e., the pallet path of Figure 5.6*c* or the pallet cycle of Figure 5.6*d*) and integrated into the composite model that results from the integration of the mason and laborer cycles. The pallet model has two COMBI elements common to the composite model ("RESUPPLY PALLET" and "PLACE BRICKS"). These elements are superimposed on the multicycle model of Figure 5.7, and they give models

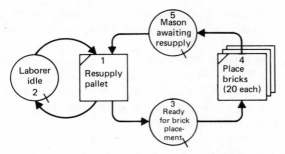

FIGURE 5.7   Integration of laborer and mason cycles.

shown in Figures 5.8a and 5.9a for the pallet path and pallet cycle components, respectively.

This integration leads to parallel QUEUE nodes suspended between the
"RESUPPLY PALLET" and "PLACE BRICK" work tasks. This condition
is simplified by combining the parallel QUEUE nodes, which yields the simplified system models of Figure 5.8b and 5.9b. QUEUE node 2, "LABORER
IDLE," is not included in this simplification, since it was not in parallel between COMBI elements 1 and 4. It is, in fact, required to route the laborer
unit to and from the "RESUPPLY PALLET" work task.

Having now completed integration of all individual flow unit cycles, any
withdrawn cycles are investigated. Cycles withdrawn are those that have no
common COMBI element with any of the other cycles. In some situations,
the modeler may have defined cycles that have no common point of contact
(i.e., COMBI work task) as defined. In such cases, the relevance of the withdrawn cycles must be reconsidered. If they provide worthwhile information
or have an important impact on the system's operation, they must be linked
to the model by redefining work tasks, elaborating work sequences, or constructing connecting cycles. In this simple model no withdrawn cycles exist.

The fact that the COMBI work task (4) in the integrated models in Figures
5.8b and 5.9b is preceded by only one QUEUE node causes consideration of
whether the "PLACE BRICK" work task is constrained by any flow unit
other than those included in the model. In other words, is an important flow-
constraining unit missing? In this case, the level of detail involved in the model
does not warrant consideration of a delay in brick placement following the
"RESUPPLY PALLET" work task. Therefore, the QUEUE node preceding
work task "PLACE BRICK" is deleted and 4 is redefined as a NORMAL work
task. The model with this further simplification is shown in Figures 5.8c and
5.9c.

The requirement for both bricks and a mason to be present for the bricklaying work task "PLACE BRICK" to commence is implicit from the preceding COMBI element (1) where they were combined. Therefore, as modeled,

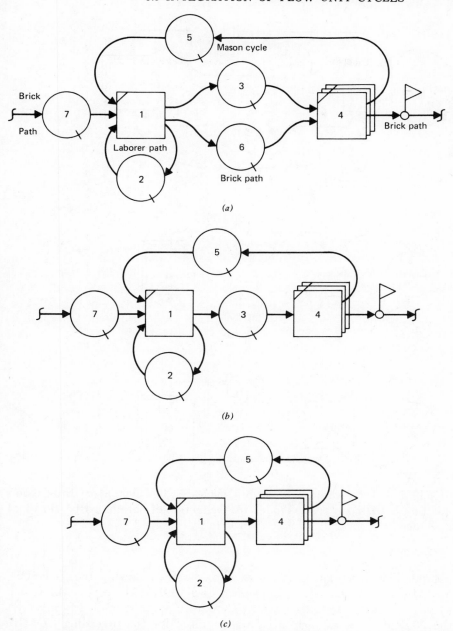

FIGURE 5.8   Integration of flow unit cycles (pallet path). (*a*) Integration of pallet path with composite model. (*b*) Reduction of parallel paths to final process model (pallet path). (*c*) Final process model.

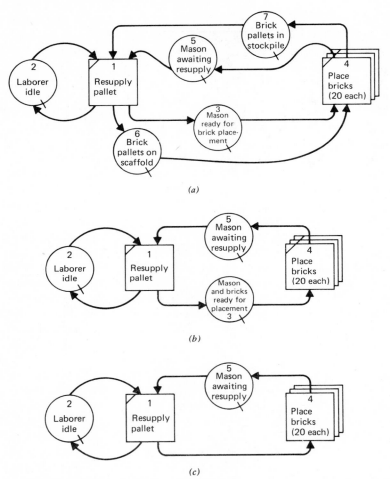

FIGURE 5.9   Integration of flow unit cycles (pallet cycle). (a) Integration of pallet cycle with composite model. (b) Reduction of parallel paths. (c) Final process model.

the bricklaying operation commences immediately following the "RESUP-PLY PALLET" work task and is shown as a NORMAL work task. It has no slash and is not preceded by QUEUE nodes. The units implicitly involved in the NORMAL work task are those required for the preceding COMBI processor minus any units routed out of the COMBI element.

Looking at the model developed thus far from a strictly informational point of view, idle states exist from which information regarding delays and interactions in the system can be determined. From QUEUE node 2, "LABORER IDLE," information regarding the percent of the time the laborer is inactive can be developed. QUEUE node 5, "MASON AWAITING RESUPPLY," information indicates the delays experienced by masons awaiting resupply.

5.6 THE MASONRY OPERATION WITH STORAGE

Consider now a modification of the masonry operation that introduces the use of a storage space, or logical flow unit. In the previous definition of the masonry operation no allowance was made for the stacking of bricks on the scaffold. This led to the requirement for letting down a pallet and waiting for resupply. Now assume that the laborer can stack up to five 20-brick packets on the scaffold. In other words, the scaffold is large enough and accessible enough that bricks can be stored on it.

This site condition modification leads to the concept of a storage or buffer. Using a storage, the productive effort of a resource, such as the laborer, can be stored and transferred, thus reducing the idle time of the resource. The most common example of this storage concept is the hopper in earthmoving operations. The front loader can load the hopper, when no trucks are available for direct loading, thus reducing its idle time and expediting loading of the trucks. Because the hopper is representative of storage of effort on the part of a resource, systems that allow storage of effort in one cycle for transfer to a second cycle are sometimes referred to as "hopper" systems.

The hopper essentially operates as a buffer or storage location between two productive cycles. This helps greatly to minimize the imbalance that might exist between interacting cycles. In the case of the hopper used on the earthmoving operation, the loader remains active, even though no truck is available for loading. The loader continues to load the hopper. The load time from the hopper into the truck is less than the time required for the loader to load the truck directly (i.e., without going through the hopper). This is because the hopper allows loading without the cycle delays inherent in the loader's operation. This concept reduces the idle time of the server unit and increases the productivity of the served units by reducing the delay in service.

The extended masonry operation (defined above) calls for the use of a hopper, since the possibility of stacking bricks on the scaffold provides a storage. Figure 5.10 shows a schematic drawing of the hopper operation in the earthmoving operation and the stack in the masonry problem. In Figure 5.10a the loader operating as a server loads the hopper. The hopper is capable of holding five truck loads. This is shown by the five triangles, which represent individual truck loads. Similarly, the laborer carries bricks to the scaffold and stacks them, as shown in Figure 5.10b.

The question that is posed in modeling a buffered masonry system is: "How is the simple masonry system developed in Sections 5.4 and 5.5 modified to

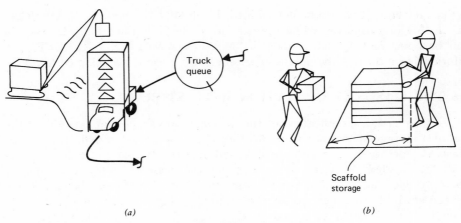

(a)   (b)

FIGURE 5.10 Typical hopper systems.

include the idea of the stacked bricks on the scaffold?" The same three flow units used for modeling the basic masonry problem are still present; they are a mason cycle, a laborer cycle and, explicitly in Figure 5.8c and implicitly in Figure 5.9c, a brick cycle. A new flow unit, however, must be defined to handle the addition of the hopper or brick stack. This is a space flow unit representing the availability or lack of a position in the storage area. In the case of the earthmoving operation, this flow unit represents the positions (i.e., the triangles) in the hopper for one truck load of storage. In the masonry problem, this flow unit defines the position in the stack. Those positions can be in either an occupied state or an empty state. That is, the stack position can be either occupied by a packet of bricks or empty. In order for the flow units to move from one of these states to the other, an action must occur. This means that the flow unit must pass through an active state. If the position in the stack is initially empty, the position becomes occupied when the laborer performs a resupply operation. If the position is occupied, it can pass to the empty state by a mason removing a packet. This cyclic movement of the position unit is shown in Figure 5.11. By initializing five position units at idle state 4, it is possible to include the effect of stacking five brick packets on the scaffold into the original masonry operation. The final model structure would appear as shown in Figure 5.12.

The position flow unit cycle is integrated into the system by breaking apart the "RESUPPLY PALLET" work task in the original model. The brick flow unit is explicit in Figure 5.12a and implicit in the structure of Figure 5.12b, where it enters 1 and passes through 6, 9, and 4. The structure of Figure 5.12b is similar to the link-node systems discussed in Chapter Two. The ultimate selection of model type (i.e., Figure 5.12a or 5.12b) depends on the interest of

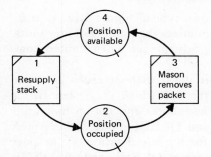

FIGURE 5.11   Position marker cycle.

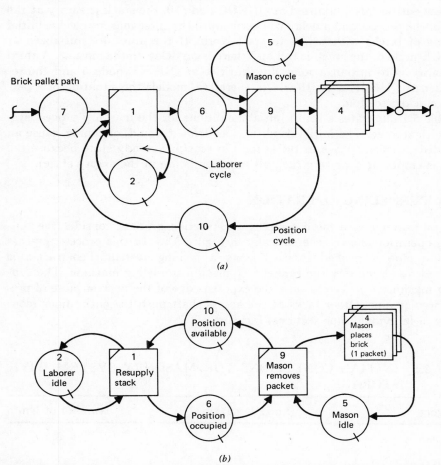

FIGURE 5.12   Total masonry system with storage. (*a*) Pallet path explicit.
(*b*) Pallet cycle implicit.

the modeler. As mentioned previously, the model type of Figure 5.12a is intuitively clearer, whereas that of Figure 5.12b minimizes the number of flow units, is more cyclic, and has computer processing advantages when limited computer storage exists.

As noted in Section 4.4, flow units are initialized without exception in an idle state (i.e., QUEUE node) of the cycle through which they pass. In the example above, this limits initialization of the laborer and mason flow units to QUEUE node 2, "LABORER IDLE," and QUEUE node 5, "MASON IDLE," respectively. Two possibilities exist for the initialization of the space or position units. They can be initialized either at QUEUE node 6, "POSITION OCCUPIED" or QUEUE node 10, "POSITION AVAILABLE." If the flow entities are initialized at QUEUE node 10, the stack is empty at the start of the process and a delay will occur until the three masons receive initial packets of bricks with which to start work. If the units are initialized at QUEUE node 6, the stack is full and there is no delay to the masons. A third possibility is to initialize some flow entities at QUEUE node 6 and the remainder at QUEUE node 10 which, therefore, models the situation of a partially stacked scaffold.

Table 5.1 illustrates a set of initial conditions for the masonry process system with storage. The initial conditions as defined would, of course, cause an extended transient phase in the system in reaching steady state because the stack is empty at time zero (i.e., all space units are in the empty state).

5.7 A TUNNELING OPERATION

As an example of a more complex construction process, consider the tunneling operation shown schematically in Figure 5.13. In this process, precast concrete pipe is pushed forward using a jacking system. Penetration is achieved by excavating the tunnel face using a tunneling machine. The tunneling machine is jacked against the exposed edge of the nearest piece of precast circular pipe. Other jacks at the access shaft push the entire liner (consisting of individual pipe sections) forward.

Table 5.1   INITIAL CONDITIONS FOR MASONRY SYSTEM WITH
            STORAGE

Category	Start Location	Number of Units
Masons	5	3
Laborer	2	1
Space in stack	10	5

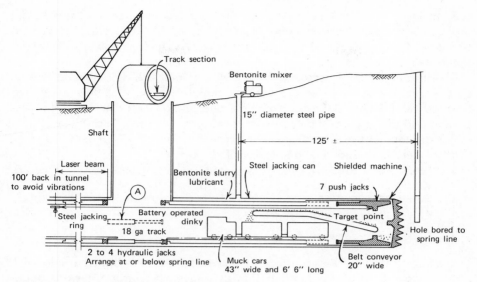

FIGURE 5.13   Tunneling process schematic.

Assume that the tunneling machine operates continuously 24 hours a day with a production rate of 2 feet per hour, and maintenance breaks are only allowed during the period of time that a new section of pipe is being lowered into position. The concrete pipe sections are 6 feet long, and the jack rams also travel 6 feet. The interaction of the crane, the pipe sections, the jack system at the bottom of the access shaft (at A) and the tunneling machine are of interest.

Suppose a model is to be developed of the tunneling operation for estimating the production in linear feet per day. The initial step in the modeling procedure is to identify the relevant system flow units. Certain of the units are obvious.

The following flow units seem to be good candidates.

1.   Pipe sections.
2.   Crane (for lowering pipe sections).
3.   Jack set.
4.   Jacking collar.
5.   Crew for positioning pipe sections.

It also appears that some type of informational unit will be required to link the pipe lowering section of the system with the jacking operation. The message to lower the next section of pipe is triggered by the jack rams having pushed the previous section as far as they can. When this occurs, an informational unit will be required to signal the crane to start the lower pipe section sequence.

The units that are common to both lowering and jacking sequences are the pipe sections. These are the processed or mainline units that flow completely through the system. As such, their idle and active states indicate the positioning and definition of other resources required for the process. The level of detail specified in the definition of the processed unit cycle also establishes the total model detail and the complexity of the model.

The work tasks that are clearly associated with the pipe section flow are:

1. Lower and position pipe.
2. Replace jacking collar and jacks.
3. Jack section forward 6 feet.

It is assumed that sections of precast concrete pipe are stockpiled at a location near the top of the access shaft. Following a message from the jacking operation that movement of the previous section has proceeded as far as possible (i.e., 6 feet), the crane maneuvers, picks up a new section and lowers it into position. A crew working in the tunnel helps to position the lowered section properly.

Following the positioning of the pipe, the jacking collars and the jack set are placed and prepared for further operation. When this is completed, the pipe section is jacked 6 feet forward, and the sequence repeats itself. For both activities, the work crew used for positioning is actively involved.

The cycle for the pipe section flow is shown in Figure 5.14a. The section is shown exiting the system into a queue or idle state. However, in accordance with the convention established previously, it is routed back to the pipe stockpile in order to achieve a closed path.

Having now developed the processed unit cycle, examine the cycles for resources required in moving the processed unit through the cycle. The crane cycle is simple, since the assumption is that its only work task is the lowering of the pipe into the access shaft. This means that it is "slaved" to this work task and passes through only two states, as shown in Figure 5.14b.

If the crane's work tasks included the off-loading and stacking of incoming pipe in the stockpile, the slave pattern would be expanded to a butterfly pattern. This revised listing of work tasks would lead to the cyclic flow shown in Figure 5.14c. The expanded model illustrates better and in more detail how the crane is actually committed. It does, however, imply the inclusion in the model of a truck cycle that interacts with the off-loading work task. For this example, consider that the crane's activity is limited to that of lowering the pipe sections.

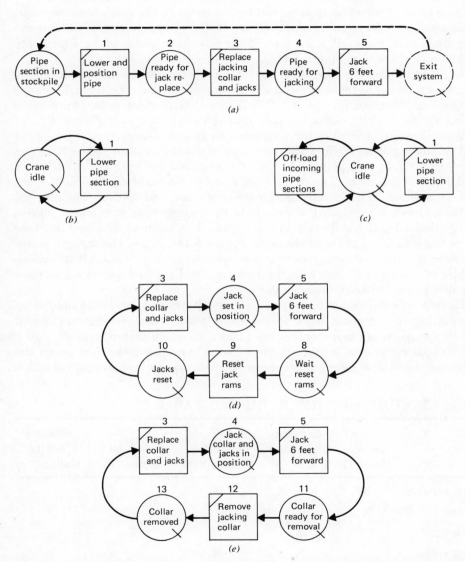

FIGURE 5.14   Model segments for tunneling process. (a) Pipe section cycle. (b) Crane cycle. (c) Dual task pattern for crane. (d) Jack set cycle. (e) Collar cycle.

The jack set has two work tasks that are common to the pipe section cycle: (1) REPLACE COLLAR AND JACKS, and (2) JACK 6 FEET FORWARD. These two work tasks must be included in the jack set cycle. The question is: "Are there any additional work tasks to be considered?" Following the extension of the jack rams to their full length (6 feet), the rams must be reset. This provides a location for the insertion of the next pipe section. The resetting work task, however, is not part of the pipe cycle and must be added only to the jack cycle (see Figure 5.14d).

The flow units that interact with the jack set cycle can be developed by examining the work tasks and the waiting states that precede them. The pipe sections are required for the REPLACE and JACK work tasks (work tasks 3 and 5). The jacking collar is replaced at the same time as the jacks, so that it is also required for all of the operations in this sequence.

The jacking collar cycle work tasks can be developed from the three cycles developed up to this point. The collar is required for both the REPLACE and JACK work tasks. The collar must also be removed from the jacked section in addition to being replaced on the newly inserted section. The removal of the collar is not included in any of the cycles developed previously. Therefore, it must be unique to the "collar" cycle. The collar cycle based on these three work tasks appears as shown in Figure 5.14e. Again the waiting states indicate the array of resources that are required for the COLLAR to negotiate its cycle. The pipe sections, the jack sets, and the jacking crew constrain the flow of the collar through its cycle.

Having now developed four cycles, a tentative model can be assembled by integrating the common work tasks. These work tasks represent points at which the cycles are tied together (i.e., points at which they interface).

The four cycles lead to the specification of five work tasks that define the columns of a composite ingredience matrix. This matrix is shown in Table 5.2.

Table 5.2 COMPOSITE INGREDIENCE TABLE

	Lower Pipe (1)	Replace Jacks and Collar (3)	Jack 6 Feet Forward (5)	Reset Jack Rams (9)	Remove Jacking Collar (12)
Pipe section	X	X	X		
Crane	X				
Jack set		X	X	X	
Jack collar		X	X		X
Crew 1	X	X			X
Crew 2			X	X	

The flow units selected define the rows of the matrix. Of those units defined, only the crew flow cycles have not yet been investigated. Two crews have been defined to simplify the structure of the initial model for illustrative purposes. However, following initial structuring, an alternate model with only one crew performing all operations can be developed.

By examining the columns of the composite ingredience table, the paths that flow through each work task can be determined and, consequently, the waiting states that precede the work task can also be determined. The idle states (i.e., QUEUE nodes) that precede a given work task are uniquely associated with it and, as discussed in Chapter Three, they constitute the ingredience set required to allow commencement of the work task. Thus three elements, (1) pipe sections, (2) crane, and (3) crew 1 are required for the commencement of work task 1 "LOWER PIPE." In graphical notation, this requirement results in a model segment, as shown in Figure 5.15.

The composite ingredience table enables the ingredience set associated with a work task to be determined by reading down the columns. By reading across each row, the work tasks through which a unit must flow can be determined. The network segment for each column in the composite ingredience table and the total composite (integrated model) are shown in Figure 5.16 and 5.17. Since the flow cycles for the crews have not yet been defined, they are left to be completed.

Three of the flow paths exiting work task 3 pass on to work task 5. In order to maintain the preceding idle state logic developed in Figure 5.16c, three separate idle states (one for each of the flow units) are shown. The three idle states can be consolidated into one QUEUE node. That is, the three flow units are considered to be combined for the transit between work tasks 3 and 5. This leads to a less cluttered graphical model, as shown in Figure 5.18. Since

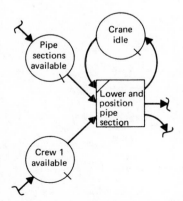

FIGURE 5.15   Network segment for work task 1.

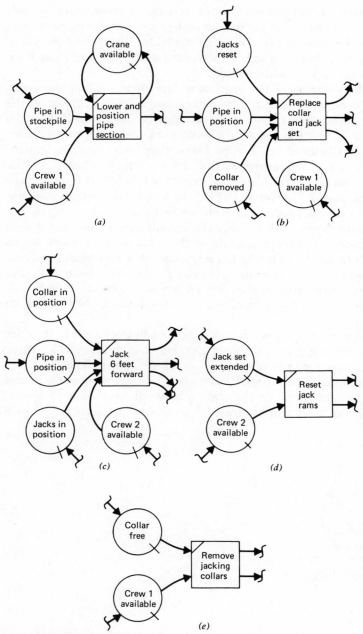

FIGURE 5.16   Network segments. (*a*) Lower pipe. (*b*) Replace jacks and collar. (*c*) Jack 6 feet forward. (*d*) Reset jack rams. (*e*) Remove jacking collar.

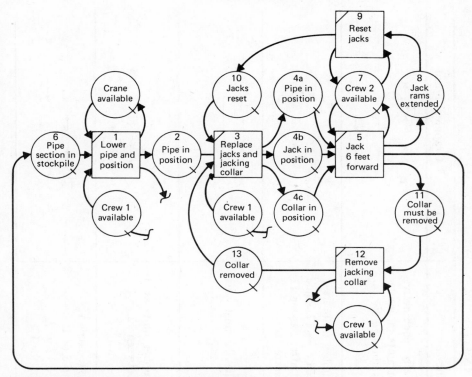

FIGURE 5.17   Integrated model for tunneling process.

the pipe, jacks, and collar have been combined at 3 and flow together to work task 5, "JACK," the ingredience set can be represented by a single unit instead of by three separate units.

5.8 CREW UNIT FLOW PATTERNS

At this point a decision must be made concerning the flow of the crew units. Examination of the crew 1 row of the composite ingredience table (Table 5.2) indicates that this crew unit is required for the commencement of three work tasks. Two subsets of these three work tasks occur sequentially. Work task 3 follows 1 directly without another intervening active state. Work task 3 also follows 12 directly (i.e., without any intervening active states). Five flow patterns for CREW ONE are possible. These are shown in Table 5.3.

Obviously, pattern one would not be appropriate to the tunneling operation; as long as there are pipe sections in the stockpile, the crew one unit will be continuously rerouted back to COMBI element 1, since priority is given to

Table 5.3 POSSIBLE CREW 1 FLOW PATTERNS

	Flow Pattern	Implied Constraints	Graphical Form
1	1, 3, 12	Unit returns to be rerouted after each active state. No topological constraints on movement. Only constraints established by labeling.	
2	1, 3 → 12	Caught in loop 3 → 12 Implies units at 1 must wait if unit is between 3 → 12, since it can only be rerouted to 1 after passing 12 and returning to the idle state (16).	
3	1 → 3, 12	Caught in loop 1 → 3 Implies 12 must wait if unit is between 1 and 3.	
4	1 → 12, 3	Caught in loop 12 → 1 Implies units at 3 must wait if crew unit is between 12 → 1.	
5	1 → 3 → 12	Constrained by sequential flow pattern. Between 3 → 12 can't go to 3 or 1. Between 12 → 1 can't go to 12 or 3. Between 1 → 3 can't go to 1 or 12.	

136

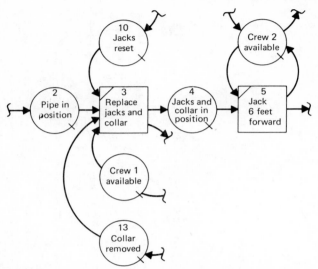

FIGURE 5.18 Consolidated unit waiting state.

lowest-numbered work tasks. This would lead to a continuous lowering of pipe
sections into the access shaft with no forward tunneling. Having piled all the
pipe sections in the access shaft, the crew could quit and go home. A situation
would result similar to the action of the wizard's broom in the well-known
fairy tale.

Flow pattern one is desirable from the modeling standpoint, since it requires
the addition of only one modeling element, the idle state 16. However, in
order to remedy the problem described above, a control unit must be added
to keep the crew unit from continuously cycling back to work task 1 (i.e.,
from continuously lowering pipe sections). This can be achieved by defining a
message or informational flow unit that is sent from the RESET JACK RAMS
work task (9) to the LOWER PIPE AND POSITION work task (1). This
leads to the following ingredience requirements for work task (1).

Lower and position Pipe section	Pipe	Crane	Jack	Collar	Crew	Message
	Yes	Yes	No	No	Yes	Yes

The revised model structure to include the newly defined informational flow
unit is shown in Figure 5.19.

The definition of the message unit leads to the addition of a new idle state
(17) to indicate the informational ingredience requirement at work task 1.
Therefore two elements must be added to control the flow of the pipe sections

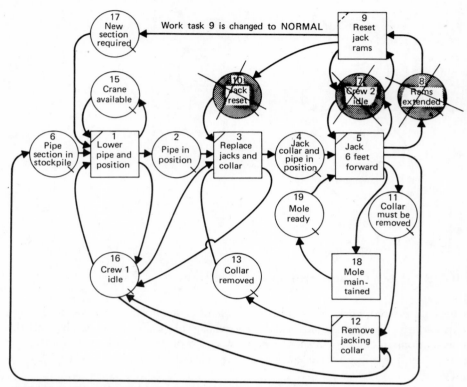

FIGURE 5.19 Total system with informational unit.

through the system. The addition of the message flow cycle leads to an economy, however, since its path from 9 to 17 to 1 renders the idle state at 10 superfluous. The "RESET JACK" QUEUE node completing the jack cycle is nested inside the newly defined message cycle. Therefore idle state 10 constraining 3 is not needed. Element 3 is constrained through work task 1 by idle state 17. Looking at it from a slightly different point of view, it is obvious that work task 1 requires a pipe section and a message unit. If only one message unit is defined (as would be the case in this instance), this constrains the number of pipe sections that can be flowing between work tasks 1 and 5 at any given time to one. As previously modeled, the commencement of work task 3 requires a pipe section, the jack unit, which indicates that rams have been reset (10), and a crew. Since, however, the jack unit becomes immediately available following 9, as does the message unit, its presence is implicit if a message unit is available at 17.

The net effect of these additions and deletions is that the model has been increased in size by one element. Idle state 10 can now be deleted. Therefore the net addition is one element. A further economy can be realized if it is assumed that the requirement for crew two is not constraining and that it is always available at work tasks 5 and 9, as required. Because of the sequential nature of work tasks 5 and 9, and because the same unit processed at 5 is then processed at 9, this constraint is superfluous as long as only one pipe section is processed at a time. This is insured by the message cycle, as discussed above. The deletion of these idle states associated with crew two and the jack set cycles is indicated in Figure 5.19 by shading over these idle states 7, 8, and 10. Work task 9 can now be modeled as a NORMAL work task.

One item must be considered before the model is complete. In the original statement of the problem maintenance on the tunneling machine will be performed during the time a new section of pipe is being lowered into position. This means that a unit representing the tunneling machine cycle must be added to the model. The tunneling machine (or "mole," as it is commonly called) cycles between the jacking work task (5) and the maintenance work task (18). The new composite ingredience table for the final model is as given in Table 5.4.

Referring to Figure 5.19, it can be seen that the number of idle states preceding an active state do not always agree with the number of units shown in the ingredience table. For instance, work task 5 (jack 6 feet forward) requires four flow unit types according to the ingredience table. It is preceded by only two idle states. This occurs since two of the required units (i.e., the message and pipe sections) were combined at work task 1 and have remained combined

Table 5.4 FINAL COMPOSITE INGREDIENCE TABLE

	Lower and Position Pipe	Replace Jacks and Collar	Jack 6 Feet Forward	Reset Jack Rams	Remove Jacking Collar	Mole Main- tenance
	1	3	5	9	12	18
Pipe Section	X	X	X			
Crane	X					
Jacking collar		X	X		X	
Crew	X	X			X	
Message	X	X	X	X		
Mole			X			X

through active state 3 and idle states 2 and 4. A third ingredient (the jacking collar) was also added to the original combination at work task 3. Therefore the composite unit waiting at idle state 4 is, in fact, a combination of three flow units: the pipe section, the lower pipe message, and the jacking collar. Following the jacking work task (5), the composite unit breaks apart into its components. Four paths diverge from COMBI element 5. The message unit moves to activity 9 (RESET JACK RAMS), the pipe section is recycled to the stockpile (idle state 6) as discussed above, the jacking collar passes to idle state 11 in its cycle, and the mole enters the maintenance activity.

Since work tasks 9 and 18 require only single units, no QUEUE nodes preceding them are required. The single arrow leading to them satisfies the ingredience requirement. This means that when the preceding work task is completed, the appropriate flow unit can move directly to the following work task. Therefore, work tasks 9 and 18 are modeled as NORMAL elements and require no preceding QUEUE nodes.

The solution presented for the tunneling problem uses the message informational unit to control the lowering of new pipe section into the excavation. This causes the definition of an additional flow unit. It is possible to control the flow of pipe sections into the tunnel by utilizing flow pattern 5 from Table 5.3. This utilizes the CREW flow unit as the control unit, thus reducing the total number of units defined. Design of this system as well as the explanation of flow patterns 2 to 4 for the crew unit are left as exercises for the reader. A formal presentation of the model-building procedure to include flow graphs is presented in Appendix B.

The preceding examples illustrate the modeling methodology for the development of meaningful static structure models of construction operations. The further development of models to incorporate field practice know-how, logic, and procedures will be considered in later chapters.

PROBLEMS

1. Extend the mason model with storage (Figure 5.12b) so that the laborer's activities include those shown in Figure 5.2. In particular, design a segment that implements the "ELEVATE SCAFFOLD" sequence. Assume that the scaffold is a prefabricated steel pipe structure that is built up in sections and that the scaffold must be elevated three times during the wall construction activity.

2. Extend the tunneling model of Figure 5.19 to include one or more of the following operational sequences.

    (a) Hauling of concrete pipes from the supplier's yard to the job site stockpile.

(b) Constraint introduced by operation of the betonite mixer.

(c) Delay sequence introduced to realign tunneling machine using laser and target.

(d) Constraint introduced by operation of muck train in removing material.

3. In asphalt paving operations, a paving train consisting of a spreader, a "breakdown" roller, and a finish roller proceed linearly along the area to be paved, as shown in Figure P5.3a. Trucks haul hot mix asphalt from the plant to the job site and dump the material into the spreader skip. The following resources are to be considered in modeling an asphalt paving operation.

(a) Spreader.

(b) Breakdown roller.

(c) Finish roller.

(d) Asphalt plant.

After 15 spread cycles, the spreader must reposition to make a new pass (i.e., parallel to the just completed pass). After 5 spread cycles, the section spread is released to the breakdown roller for compaction of the spread hot mix asphalt. The initial cycles have been identified as shown in Figure P5.3b to P5.3f.

Integrate the above unit cycles into a composite model and, using CONSOLIDATE function nodes, design the triggers required to activate the reposition work task for the spreader and the release of a section for breakdown rolling.

4. Consider the labor-intensive operation of the manual delivery of concrete on a construction site using wheelbarrows, as shown in Figure 13.1. Using the procedures outlined in Section 5.2, identify the system resource unit flows, cycles, and resource unit initializations. Then develop an ingredience table formulation of the CYCLONE model similar to Table 5.2. Can you develop the model to incorporate further space and informational flow units?

5. Using an approach similar to that developed above in Problem 3, design a model for an operation in which a slip-form paver is being used to place concrete deck on a high-speed highway. The operation to be modeled begins with the materials required for the concrete (aggregate, cement, etc.) stockpiled at the site. A front end loader brings materials from stockpiles to a bin system. From the bins, the materials are carried by a conveyor system to a second set of bins capable of weighing materials. Here materials are weighed out according to the requirements of each per 1

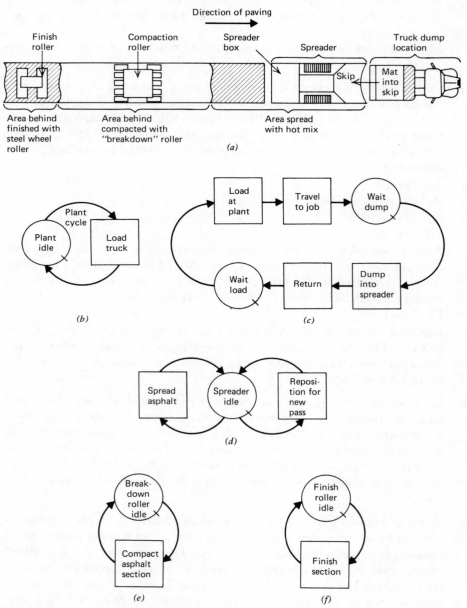

FIGURE P5.3 (a) Asphalt paving train. (b) Plant cycle. (c) Truck cycle. (d) Spreader cycle. (e) "Breakdown" roller cycle. (f) Finish roller cycle.

cubic yard of concrete. The ingredients in 1 cubic yard of concrete are as follows:

541	pounds of cement
1060	pounds of sand
2145	pounds of stone
32.8	gallons of water
4%	air entraining agent
21	ounces of water admixture

The materials are combined in a mixer batching out 8 cubic yards per truck. The truck moves under the mixer and is then filled with concrete. The truck makes a hauling trip to the location of the pour. Once there the trucks queue up, waiting to dump and fill the concrete spreader. The truck moves alongside the spreader and, in two movements, dumps its load. As the truck returns to the concrete plant, the paving process begins. The spreader spreads a swath of concrete with reinforcing in place from 12 to 24 feet wide. The spreader moves on slowly, continuously pouring out the concrete. Following this spreader is the actual "slip-form" finisher. This machine moves along finishing off the concrete surface and giving a smooth-edged form to the side of the slab. As this machine moves on, another machine, the tube finisher, comes along with a steel tube to finish off the concrete. A texturizing machine follows the tube finisher and gives a texture of $\frac{1}{8}$ inch grooves, as required by the Department of Transportation.

6. An estimate of production on the following project must be made.

   (a) Type of Construction. Emplacement of sewer pipe from Chattahootchee River sewerage treatment plant to Hemphill site.

   (b) Types of Activities

   (1) Pavement Breaker. Precedes ahead of excavation equipment, breaks up pavement for excavation.

   (2) Excavation. Clamshell bucket excavates and clears ditch for pipe.

   (3) Haul. Debris cleared from ditch, broken pavement, and material is hauled from site and dumped.

   (4) Manual Labor. One crew performs final grading of the bed, aids in placing pipe, and does moisture proofing. The second crew does hand backfill and compaction and then forms and pours concrete paving.

   (5) Backfill. Aside from a small amount done by laborers, the clamshell bucket does the majority of the backfilling operation.

   (6) Compaction and Preparation for Paving. Done by small roller and manual labor.

   (7) Pavement with concrete deck.

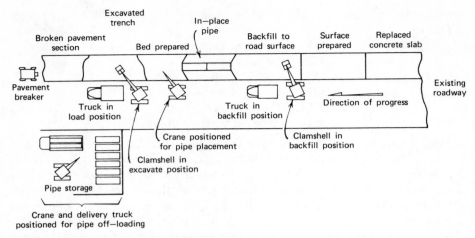

FIGURE P5.6   Sewer line job layout.

The layout of the construction site is given in Figure P5.6. Assume production is to be measured in 50-foot sections of completed roadway (i.e., following installation of pipe and paving). Develop a CYCLONE model of the process considering the following flow units.

(a) Pipe sections.

(b) Cranes—one for excavation and one for pipe placement.

(c) Trucks.

(d) Pavement breakers.

(e) Work crews.

First develop the individual flow unit cycles for each unit type and then integrate the individual cycles into a composite model. Place the AC-CUMULATOR element in the model so as to measure 50-foot section productivity.

7. Identify a number of relevant (but significantly different) model purposes that construction agents may adopt in relation to a specific construction operation, and then develop a variety of CYCLONE models to meet or satisfy these specifications. Comment on the various work task breakdowns and specifications and on the need for different numbers and types of resource flow units.

CHAPTER SIX

Work Task Transit Times

At the construction operation level, labor, equipment, and material resources are applied to the technology of the operations according to the skill and motivation of the men involved. In this way, units of work are achieved over time in a sequence and to a level of quality specified by the construction plan. During the execution of the construction operation, resources are committed to and captured by the process and are then not available for other tasks.

The time commitment of resources to construction operations is of paramount importance to project field staff. Field management is interested in the efficient use of resources to achieve work production within schedule at an acceptable quality. Consequently, useful models must focus on attributes relevant to management problems; these models must provide insight into ways that management goals can be achieved.

In CYCLONE models, labor, equipment, and material resources are modeled explicitly as flow units. The technological characteristics of a construction operation are captured in the structure of the process model, its breakdown into work task sequences, and the flow patterns of units involved in the process. By associating work task transit times with the various work tasks in the CYCLONE model, the model is capable of representing the time period during which resource units are involved with work task sequences. In this way predictive models can be developed for the determination of operational output and productivity over time. It is also possible to determine resource idle times and the influence on productivity of different resource allocations so that management can plan and control the construction operation. Finally, it is possible to determine the work load assigned to specific labor and equipment resources so that meaningful assessments can be made of work quality levels that may be achieved.

This chapter considers how work task durations are determined and used in the CYCLONE system. It investigates the types of time durations that are of interest in modeling construction processes and discusses the manner in which they are defined.

6.1 WORK TASK DURATIONS

Two of the CYCLONE system elements have user-defined delays or transit times associated with them. The COMBI and NORMAL work tasks provide points within the system being modeled at which time durations are defined. These two elements implement the transit time delays input to the system. In this capacity, they function as input elements, allowing the input of system time parameters. None of the other four elements provide for the input of time parameters specified by the modeler. Since the actual system performance is a function of defined element times and system logic, the definition of the COMBI and NORMAL work tasks and their associated time duration is extremely important in capturing the essential features of the real-world system.

Work task transit times determine the time durations that resource flow units are captured by, or are involved in, a work task. These times are determined by estimation or measurement.

Once the operational technology has been decided for a work task, the basic factors that influence work task durations are:

1. The magnitude of the work content involved in the work task.

2. The extent to which equipment is used in the work task. Thus the size and efficiency of equipment, equipment characteristics, and functions performed have a dominant effect in equipment heavy operations but less effect in labor-intensive operations.

3. The extent to which labor is used in the work task. Thus the skill level, crew mix, and crew size are significant factors in labor-intensive work tasks. In these cases the intensity of physical effort, team spirit, and motivation directly influence work task productivity and duration.

4. The physical environment of the work site, working conditions, shift hours, weather, and so forth.

5. The level and efficiency of management of the foreman and at the work site.

The specific influence of these factors on the productivity of a construction operation and thus on the duration of a work task will vary from site to site and is difficult to determine. Some factors are random in their occurrence and impact magnitude, while others have a readily evaluated effect. Depending on the relative magnitude and mix of these factors, work task duration estimates may be deterministic, probabilistic around a known mean, or almost completely random.

Work task duration estimates focus on the determination of a specific duration reflecting the influence of relevant and definable job factors. The remaining unknown or "difficult" factors are bundled together for handling by field management expertise as contingency factors. Thus common practice is to determine specific work task durations with an implicit range of time within which the "normal" duration is tolerated. In this way durations can be considered as either deterministic or probabilistic, depending on the purpose for which the data and the model are used.

For example, in earthmoving operations, the duration of a load, haul, dump, and return cycle for a scraper is affected by haul distance, grades, rolling resistance, engine horsepower and efficiency, loading time and operation, altitude, and weather conditions affecting ground surface conditions. Methods exist for determining average (deterministic) cycle times, or probabilistic measures of cycle times (see Sections 6.2 and 6.3).

Depending on the level of detail of the work task definition, the entire earthmoving cycle or a component of it may be considered as a unique work task that requires duration times. In this way deterministic or probabilistic segments may be separated or combined as required for modeling accuracy.

The duration of a labor intensive work task depends on the energy content required by working conditions and its influence on labor fatigue, the depletion rate demand on human energy reservoirs, the influence of rest periods on energy replenishment, the skill level and planning required for the basic components of the work task, the influence of motivation and attitude to work, team spirit, and the foreman-crew relationship. Again, the duration of a labor-intensive work task can be determined deterministically through average work rates and productivities or, more realistically, modeled as probabilistic (see Sections 6.2 and 6.3).

In practice work task durations are determined by one or more of the following methods.

1. *Past Experience.* The agents involved in the work process know the time required by frequent past experience on identical or similar tasks of the same magnitude.

2. *Estimates.* Planning and estimating agents obtain data from previous work on similar work tasks. This data enables them to establish productive rates as a function of the size and mix of resources allocated to the work task. Consequently, work task durations can be determined once resource allocation and work content of a work task are established.

3. *By Fiat.* In some cases, for large and highly repetitive operations, either past experience is unavailable or unusual features of the planned operation reduce the reliability of past data and it becomes worthwhile to establish experience and data by an initial trial or mock-up run on the operation.

4. *Use of Predictive Models.* Often the basic components of an operation are known and the productivity of resource units working these components is also known. Given situations where the relative magnitude and mix of the basic operation components change, predictive models of productivity and duration estimates are very useful and practical.

The selection of method for the determination of work task duration depends on the nature of the operation and the professional skill of the estimators.

6.2 DETERMINISTIC WORK TASK DURATIONS

A deterministic duration assigns a specific fixed value to the duration of the work task. Any flow unit resource entering the work task is captured for the exact value time period defined before it is released to subsequent system work tasks.

Work task durations may be deterministic because:

1. The work task may have a fixed duration or a resource entity is to be captured for a specific time. Simple examples are the mixing time of N revolutions for a concrete mixer; a curing time of one week before stripping; and an 8-hour shift duration.

2. The work task duration may be subject to small variations about a specific mean value so that from any useful time scale the work task duration is constant. Simple examples are hoist time on a building site, scraper cycle times on sites under ideal conditions, and the time to fill a truck with gravel.

3. The purpose of the modeling is such that any probabilistic variation can be ignored. In some cases, as mentioned previously, a broadly defined work task that is subject to large random variation can be broken down into a system set of smaller work tasks, which may localize variations so that some may now be considered as almost deterministic.

Most construction operations performed within a controlled environment, especially those that are equipment-oriented, fall into category 2 above. Most labor-intensive operations fall into category 3 because of individual human factor considerations. Therefore, even in situations where randomness or variability are present, it is common practice to neglect the variation and use a single specific value of time duration when the impact of the variability is considered to be small or insignificant. In such cases, a constant or deterministic value of time duration is selected that adequately represents the time duration of a work task despite any small variation present. Therefore, it is common practice in calculating the productivity of relatively simple construction processes to assume *deterministic* values for the process work tasks, even though the work tasks may have highly random time durations associated with them.

As mentioned in the previous section, work task durations may be determined in a number of ways. Consider the development of predictive models for equipment-heavy and labor-intensive operation. Simple methodologies have been developed for earthmoving operations based on rated equipment characteristics, equivalent grades, and haulage distances. Similarly, for simple labor work tasks, human factor analysis of microactivities enables models to be developed for the durations of many labor-intensive work tasks.

Consider a simple earthmoving haul operation of the type presented in Chapter Three. A simple CYCLONE model for the process is shown in Figure 6.1 where the haul unit is a 30-cubic yard scraper and it is loaded in the cut area with the aid of a 385-horsepower pusher dozer. By reducing the model to only two cycles and using deterministic values for the work task durations, it is possible to solve the model simply.

To determine the deterministic durations for the scraper travel times to and from the fill location, it is necessary to consult the performance handbooks published by most manufacturers; these handbooks show the development of this information from charts or specifically developed nomographs. Assume that in this case the 30-cubic yard tractor scraper is carrying rated capacity and operating on a 3000-foot level haul where the rolling resistance (R.R.)

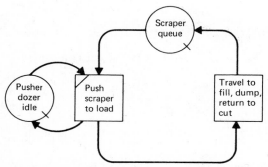

FIGURE 6.1 Scraper-pusher dual cycle model.

developed by the road surface is 40 pounds per ton. Using a standard formula, this converts to:

$$\text{Effective Grade} = \frac{\text{R.R.}}{20 \text{ lb/ton/\% grade}} = \frac{40 \text{ lb/ton}}{20 \text{ lb/ton/\% grade}} = 2\% \text{ grade}$$

By consulting the nomographs given in Figure 6.2, the following travel times can be established.

1. Time loaded to fill: 1.4 minutes.

2. Time empty to return: 1.2 minutes.

Assume further that the dump time for the scraper is 0.5 minutes and the push time using a 385-horsepower track-type pusher tractor is 1.23 minutes, developed as follows.*

$$
\begin{array}{ll}
\text{Load time} & = 0.70 \\
\text{Boost time} & = 0.15 \\
\text{Transfer time} & = 0.10 \\
\text{Return time} & = \underline{0.28} \\
\text{Total} & = 1.23 \text{ minutes}
\end{array}
$$

Using these deterministic times for the two types of flow units in this system (i.e., the pusher and the scrapers) the scraper and pusher cycle times can be developed, as shown in Figure 6.3, as follows.

$$\text{Pusher cycle} = 1.23 \text{ minutes}$$
$$\text{Scraper cycle} = 0.95 + 1.2 + 1.4 + 0.5 = 4.05 \text{ minutes}$$

These figures can be used to develop the maximum hourly production for the pusher unit and for each scraper unit as follows.

Maximum System Productivity (assuming a 60-minute working hour)

1. Per Scraper

$$\text{Prod (scraper)} = \frac{60^{\text{min/hr}}}{4.05_{\text{min}}} \times 30 \text{ cubic yards (loose)}$$

$$= 444.4 \text{ cubic yards loose/hr}$$

2. Based on Single Pusher

$$\text{Prod (pusher)} = \frac{60}{1.23} \times 30 \text{ cubic yards (loose)}$$

$$= 1463.4 \text{ cubic yards loose/hr}$$

---

*See, for instance, page 66 of *Fundamentals of Earthmoving* published by the Caterpillar *Tractor* Company.

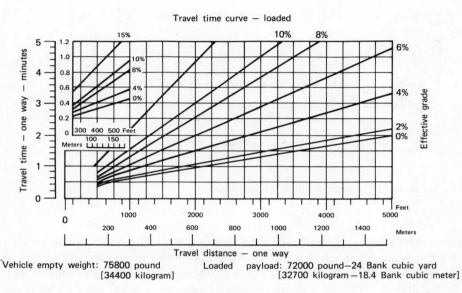

Travel time curve — loaded

Vehicle empty weight: 75800 pound   Loaded   payload: 72000 pound—24 Bank cubic yard
          [34400 kilogram]   [32700 kilogram —18.4 Bank cubic meter]

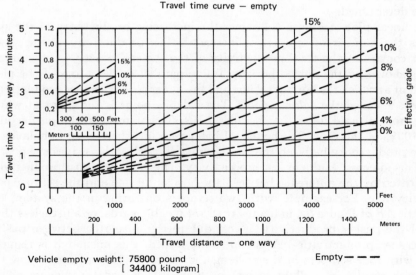

Travel time curve — empty

Vehicle empty weight: 75800 pound   Empty — — —
          [34400 kilogram]

FIGURE 6.2   Travel time nomographs. (Caterpillar Tractor Co.)

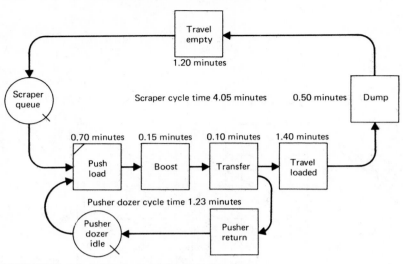

FIGURE 6.3   Scraper-pusher cycle times.

Using these productivities based on a 60-minute working hour, it can be seen that the pusher is much more productive than a single scraper and would be idle most of the time if matched to only one scraper. By using a graphical plot, the number of scrapers that are needed to keep the pusher busy at all times can be determined.

The linear plot of Figure 6.4 shows the increasing productivity of the system as the number of scrapers is increased. The productivity of the single pusher constrains the total productivity of the system to 1463.4 cubic yards. This is shown by the dotted horizontal line parallel to the x-axis of the plot. The point at which the horizontal line and the linear plot of scraper productivity intersect is called the *balance point*. The balance point is the point at which the number of haul units (i.e., scrapers) is sufficient to keep the pusher unit busy 100% of the time. To the left of balance point, there is an imbalance in system productivity between the two interacting cycles; this leaves the pusher idle. This idleness results in lost productivity. The amount of lost productivity is indicated by the difference between the horizontal line and the scraper productivity line. For example, with two scrapers operating in the system, the ordinate AB of Figure 6.4 indicates that 574.6 cubic yards, or a little less than half of the pusher productivity, is lost due to the mismatch between pusher and scraper productivities. As scrapers are added, this mismatch is reduced until, with four scrapers in the system, the pusher is fully utilized. Now the mismatch results in a slight loss of productivity caused by idleness of the scrapers. This results because, in certain instances, a scraper will have to

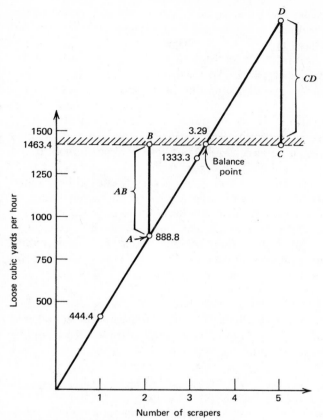

FIGURE 6.4   Productivity plot.

wait to be loaded until the pusher is free from loading a preceding unit. If five scraper units operate in the system, the ordinate CD indicates that the loss in the productive capacity of the scraper because of delay in being push loaded is:

$$\text{Productive Loss} = 5\,(444.4) - 1463.4 = 758.6 \text{ cubic yards}$$

This results because the greater number of scrapers causes delays in the scraper queue of Figures 6.1 and 6.3 for longer periods of time. The imbalance or mismatch between units in dual-cycle systems resulting from deterministic times associated with unit activities is called *interference*. It is due only to the time imbalance between the interacting cycles. It does not consider idleness or loss of productivity because of random variation in the system activity durations. This can be extended to multiple-cycle systems with deterministic durations such as those modeled in Chapter Five. In such systems it is possible to develop the idleness and productivity loss resulting caused by the complex *interference* of interacting flow units.

In many cases, only a deterministic analysis of system productivity is undertaken because it is sufficiently accurate for the purpose of the analyst. However, in some systems, the loss of production because of the random variation of cycle durations is significant enough to justify consideration.

A methodology for the estimation of the duration and productivity of the labor-intensive operations called engineered estimates has been developed based on methods-time measurements (MTM). The MTM system is based on the concept that methods determination preestablishes the time determination. Once a method is determined, a time can be very accurately compiled from the times needed for the elemental operation based on human factors.*

Each elemental operation is subdivided into human factor measurements such as reach, move, turn, apply pressure, grasp, position, release, disengage, eye travel and focus, and body, leg, or foot motions. Each of these is further divided into many micromovements. For each of these there is an elemental time measured in TMU's (time-measured units) where 1 TMU = 0.00001 hour = 0.0006 minutes = 0.036 seconds.†

Some governmental service organizations have been using MTM time standards extensively since the late 1950s. The data manuals for engineered performance standards‡ of the U.S. Navy, for example, are available to the public and cover a wide range of construction operations.

---

*See, for instance, Harold B. Maynard, G. J. Stegemerten, and John L. Schwab, *Methods-Time Measurement*, McGraw-Hill Book Company, New York, 1948.

†John L. Schwab, "Methods-Time Measurement," in Harold B. Maynard (ed.), *Industrial Engineering Handbook*. 2d ed., McGraw-Hill Book Company, New York, 1963, pp. 5-13 to 5-38, and Henry W. Parker and Clarkson H. Oglesby, *Methods Improvement for Construction Managers*, McGraw-Hill Book Company, New York, 1972, Chapter 2 and Appendix A.

‡The U.S. Navy Facilities Engineering Command has issued a series of Engineered Performance Standards (EPS) publications with the following NAVFAC publication numbers and titles: P-700.0, Engineers' Manual, 1963; P-701.0, General Handbook, 1964; P-701.1, General Formulas, 1962; P-702.0, Carpentry Handbook, 1962; P-702.1, Carpentry Formulas, 1962; P-703.0, Electrical and Electronic Handbook, 1963; P-703.1, Electrical and Electronic Basic Supporting Data, 1957; P-704.0, Heating, Cooling and Ventilating Handbook, 1963; P-704.1 Heating, Cooling and Ventilating Formulas, 1963; P-707.0, Machine Shop, Machine Repair Handbook, 1960; P-707.1, Machine Shop, Machine Repair Formulas, 1959; P-708.0, Masonry Handbook, 1963; P-708.1, Masonry Formulas, 1963; P-709.0, Moving and. Rigging Handbook, 1962; P-709.1, Moving and Rigging Formulas, 1962; P-710.0 Paint Handbook, 1963; P-710.1, Paint Formulas, 1963; P-711.0 Pipefitting and Plumbing Handbook, 1962; P-711.1, Pipefitting and Plumbing Formulas, 1962; P-712.0, Roads, Grounds, Pest Control Handbook, 1963; P-712.1, Roads, Grounds, Pest Control Formulas, 1963; P-713.0, Sheetmetal, Structural Iron and Welding Handbook, 1960; P-713.1, Sheetmetal, Structural Iron and Welding Formulas, 1960; P-714.0, Trackage Handbook, 1963; P-714.1, Trackage Formulas, 1963; P-715.0, Wharfbuilding, 1963; P-715.1, Wharfbuilding Formulas, 1963.

Table 6.1   MASON'S LABORER WORK TASKS (PARTIAL EXTRACT
FROM PWD-3)*

Task	Element Description	Analysis Chart Reference	Leveled Time
F	Pick up six bricks from pile and set on scaffold	MTM 12	0.0088
G	Fill 3-gallon bucket with mortar (using shovel)	MTM 12	0.0094
H	Dump mortar from bucket into mortar pan	MTM 13	0.0027
W	Obtain water in pail (3 gallons)	PWD-14-L	0.0493
Y	Pour water from pail (3 gallons)	PWD-4-F	0.0026
B1	Carry pail or mortar to mortar board and return	PWD-2-A1	0.0047
J1	Additional walking required each 24 bricks	MTM 15	0.0012
X	Other work tasks including instruction, planning, scaffold adjustment	—	—
U	Personal activity, including coffee and comfort breaks	—	—
D1	Remix mortar	MTM 15	0.0084

*See Formula PWD-3, "Lay Common Bricks," Masonry Formulas, Engineered Performance Standards: Public Works Maintenance, U.S. Naval Facilities Engineering Command, NAVFAC, P-708.1, 1963, pp. 9-16.

To illustrate the engineered standards approach to activity duration determination and the CYCLONE modeling of a labor-intensive activity, consider again the masons' laborers work assignment in the typical masonry operation introduced in Chapter Five. An extensive analysis of this situation is given in Formula PWD-3, "Lay Common Brick." *

Table 6.1 gives a partial list of elemental activities for the masons' laborers work assignment of maintaining brick and mortar flow to masons working on a scaffold. It assumes a fairly typical working site layout and environment and includes two additional catchall work tasks; work task X (other work tasks) and work task U (personal activity).

*Formula PWD-3, "Lay Common Brick," Masonry Formulas, Engineered Performance Standards: Public Works Maintenance, U.S. Naval Facilities Engineering Command, NAVFAC P-708.1, 1963, pp. 9-16.

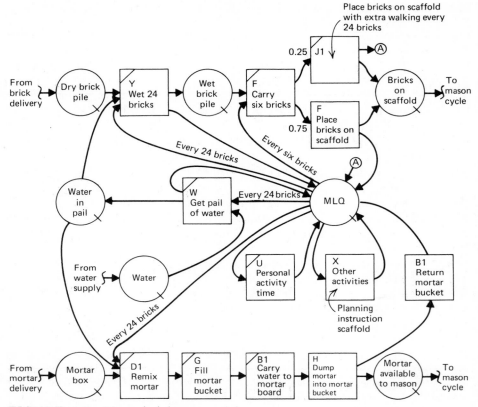

FIGURE 6.5   Mason's laborer model.

A CYCLONE model of the mason's laborer based on this list is given in Figure 6.5; it shows cycle frequencies based on the handling of groups of 24 bricks.

In the engineered standards approach, each of the elemental work tasks of Table 6.1 is analyzed in terms of MTM microstandards based on human factor measurements to develop "leveled" time estimates. These time estimates shown in Table 6.1 relate to an assumed job layout and refer to basic work efforts independent of the total requirements of the job, social environment, and human fatigue. The time estimates have to be interpreted and supplemented for any specific situation and time estimate.

An example of the breakdown analysis for the elemental activities is shown in Table 6.2 for the activity G, "FILL 3 GALLON BUCKET WITH MORTAR USING A SHOVEL." These breakdown and time estimates are supported by further human factor analysis (see reference formula PWD-3) in

Table 6.2   MICROANALYSIS OF MASON'S LABORER WORK AC-
TIVITY: FILL 3-GALLON BUCKET WITH MORTAR, USE
SHOVEL*

Description of Micro Work Task	Time Measurement Units (TMU)
Turn to 3-gallon bucket	18.6
Bend to pick up bucket	29.0
Reach to handle	6.4
Grasp handle	2.0
Stand up	31.9
Turn to mortar box	37.2
Set pail down	29.0
Release pail	2.0
Step to get shovel	18.2
Reach to shovel	11.5
Grasp shovel	2.0
Stand up with shovel	31.9
Move to mortar	19.4
Bend to mortar	154.0
Move shovel	44.5
Push shovel into mortar	53.0
Arise	159.5
Move shovel to bucket	100.5
Turn shovel	27.5
Hit shovel on side of bucket	43.5
Move shovel to mortar	73.0
Move shovel aside	14.6
Toss aside	10.6
Release shovel	2.0
Turn to bucket	18.6
TOTAL TMU's	940.4
Leveled time 0.0094 hr	

*See Formula PWD-3.

terms of basic microstandards. In the example shown, activity G has a leveled
time of 940.4 TMU (or 0.0094 hours).

Using the leveled time estimates of Table 6.1, an analysis of the masons'
laborers work assignment shown in Figure 6.5 (excluding work tasks X and U)
gives a time estimate of 0.1135 hours (6.81 minutes) for the basic support
flow cycle time for 24 bricks and relevant mortar flow. This basic job cycle

time must be complemented by estimates for the masons' laborers' involvement in other work tasks before any final time estimate or productivity is established.

Table 6.3 gives a typical breakdown of work tasks performed by the mason's laborer in a total work assignment situation. Here the basic job component B.J. refers to the maintenance of brick and mortar supply described and analyzed above. Auxiliary work refers to the planning and skill level of the trade involved and, in this case, could include additional work sequences relating to scaffold erection and adjustment. Delays are divided into those imposed by the operational system (i.e., mismatches in supply of materials and demands from the masons) and those that are personally dictated (rest periods, comfort breaks, etc.).

Table 6.3 also gives typical percentage ranges of time spent in the various categories that pertain in practice. Obviously the percentage time breakdown of any specific job will be unique and will depend on site conditions, crew mix, attitude, and performance.

The specific assumed breakdown shown in the estimate of Table 6.3 gives the basic job activity as 67% of the day work period. On this basis the laborers' productivity of bricks and mortar supplied to the scaffold can be calculated as:

$$\frac{(\text{B.J. } 0.67 \times 8 \text{ hr})}{(\text{average cycle time per 24 bricks} = 0.1135 \text{ hr})} \times 24 \text{ bricks} = 1133 \text{ bricks}$$

This throughput productivity is supported by current masonry crew structures and productivities. Table 6.4 gives typical crew composition and productivities for several common masonry activities. Assuming that two laborers ensure the input supply to the model of Figure 6.5, then each mason's laborer supports two masons who thus lay between 1050 and 1150 bricks a day. Therefore, the engineered standards estimate coupled with realistic field work task breakdowns gives realistic productivity and work task duration estimates.

Table 6.3   DAILY BREAKDOWN FOR MASON'S LABORER

	Typical Ranges (%)	Estimate (%)
Basic job (maintaining brick and mortar flow).	40–80	67
Auxiliary work (planning, scaffold work).	40–5	15
Delays (idle time because of crew and lack of materials).	20–5	8
Personal activity delays (comfort breaks, etc.).	2–10	10

Table 6.4   TYPICAL MASONRY CREWS AND PRODUCTIVITIES

Eight masons	Six laborers	Standard face brick	4200 bricks per 8-hour day
Eight masons	Six laborers	Common brick backup	4600 bricks per 8-hour day
Eight masons	Two laborers	Mortar	16 cycles per 8-hour day
(Basic crew ratio: 2 masons to 1 laborer)			

The information given in Figure 6.5 and Tables 6.1 to 6.4 must be carefully interpreted. The idle time delay estimate of Table 6.3, for example, need not apply in any specific case, since the laborer may schedule his rate of work to anticipate crew demand; therefore, by working slower, he can eliminate idle time for the same effective production. Alternatively, he may prefer to work faster and rest and recuperate more frequently to suit his individual temperament and environmental conditions.

The importance of the engineered estimate approach lies in its ability to develop, preplan, and design models for construction operations. In this respect the methodology is useful for deciding deterministic work task durations for most labor-intensive operations.

6.3 RANDOM WORK TASK DURATIONS

The influence of mismatches in equipment fleets and crew mixes on system productivity was discussed in the last section in terms of deterministic work task durations and cycle times. In systems where the randomness of work task durations and hence of cycle times is considered, system productivity is reduced further. The influence of random durations on the movement of resource flow entities is to ensure that the various entities eventually get bunched together and thus, by arriving at and swamping COMBI-type work tasks, delay the productivity of cycles and operations by increasing the time that resource units spend in idle states pending release to productive work tasks.

Consider the scraper-pusher problem and assume that the effect of random variation in cycle activity duration is to be included in the analysis.

In simple cases such as the two-cycle system model of Figure 6.1, mathematical techniques based on *queueing theory* can be used to develop solutions for situations where the random arrival of scrapers to the dozer can be postulated. In order to make the system amenable to mathematical solution, however, it is necessary to make certain assumptions about the characteristics of the system that are not typical of field construction operations (see Appendix A). The solution technique that is more general in its application is *discrete unit simulation.* The concepts employed in discrete unit simulation will be introduced in the next chapter.

Figure 6.6 indicates, in a heuristic manner, the influence of random durations on the scraper fleet production. The curved line of Figure 6.6 slightly below the linear plot of production based on deterministic work task times shows the reduction in production caused by the addition of random variation of cycle activity times. This randomness leads to bunching of the haul units on their cycle. With deterministic work task times, the haul units are assumed to be equidistant in time from one another within their cycle.

In deterministic calculations, all three of the haul units shown in Figure 6.7a are assumed to be exactly 1.35 minutes apart. In this system, there are three units, and the hauler cycle time is taken as a deterministic value of 4.05 minutes. In systems that include the effect of random variation of cycle times, "bunching" eventually occurs between the units on the haul cycle. That is, the units do not stay equidistant from one another but are continuously varying the distances between one another. Therefore, as shown in Figure 6.7b, a situation often occurs in which the units on the haul are unequally spaced apart in time from one another. This bunching effect leads to increased idleness and reduced productivity. It is intuitively clear that the three units that are "bunched" as shown in Figure 6.7b will be delayed for a longer period at the scraper queue, since the first unit will arrive to load only 1.05 minutes instead of 1.35 minutes in advance of the second unit. The bunching causes units to "get into each other's way." The reduction in productivity caused by bunching is shown as the shaded area in Figure 6.6 and is in addition to the reduction in productivity caused by mismatched equipment capacities.

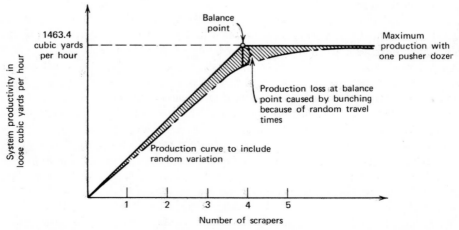

FIGURE 6.6   Productivity curve to include effect of random cycle times.

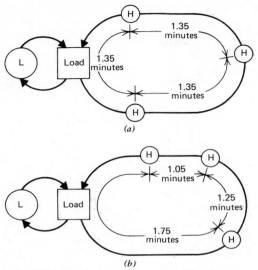

FIGURE 6.7   Comparison of haul unit cycles.

This bunching effect is most detrimental to the production of dual-cycle systems such as the scraper-pusher process at the balance point. Several studies have been conducted to determine the magnitude of the productivity reduction at the balance point because of bunching. Simulation studies conducted by Morgan and Peterson of the research department of the Caterpillar Tractor Company indicate that the impact of random time variation is the standard deviation of the cycle time distribution divided by the average cycle time. Figure 6.8 illustrates this relationship graphically.

As shown in the figure, the loss in deterministic productivity at the balance point is approximately 10% due to the bunching; this results in a system with a cycle coefficient of variation equal to 0.10. The probability distribution used in this analysis was lognormal. Other distributions would yield slightly differing results. The loss in productivity in equipment-heavy operations such as earthmoving is well documented and recognized in the field, mainly because of the capital-intensive nature of the operation and the use of scrapers in both single unit operations and fleet operations. To some extent, field policies have emerged to counteract this effect by occasionally breaking the queue discipline of the scrapers so that they self-load when bunching effects become severe. The resulting increased load and boost time for the scraper adds little to the system productivity, but it does break down the bunching of the scrapers.

Many cases exist in construction of the loss of productivity because of the interacting of randomly perturbed cycles. For example, in the masonry oper-

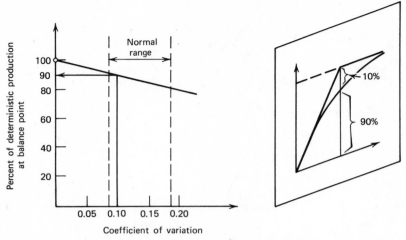

FIGURE 6.8   Plot of cycle time coefficient of variation.

ation, initial conditions relating to the status of scaffold stacks of bricks affect mason productivity until a transient "workup" phase has elapsed. A common solution to this situation is to have the masons' laborers workday begin earlier than that of the masons or to have some laborers work later than the masons to restock bricks at the end of a day.

Often the material handling and supply routes on construction sites provide situations where serious interaction develops between apparently totally independent activities. Competition for material hoists and transport space on congested sites reduces productivity, introduces random perturbations in activity durations, and sometimes reaches crisis magnitude. Very little information exists for estimators on the magnitude, influence, and cost of these interactions on construction sites. As will be shown later, CYCLONE models can be developed that focus on idle, queue, or delay aspects of such system problems.

6.4 RANDOM DURATION DISTRIBUTIONS

The impact of random work task durations on multicycle systems is not easily analyzed. It is a function of the numerous configurations that dictate cycle interactions and the types of distributions used to approximate the random variation of activity durations. Probability distributions are useful in describing observed variation of work task times in the field, since they are easily defined in a mathematical sense by relatively few parameters.

Probabilistic functions are used to define the populations from which random time durations can be taken for simulation of a system. A probability density function associates a probability with each of the values of X along the X axis of the function plot. In the case of random time duration, the X variable is the random time delay to be potentially selected. The probability of selection of a particular time duration, X, is given by the area under the curve (i.e., between the curve and the X axis) associated with a specific value.

In the function shown in Figure 6.9, the probability density values are shown along the Y axis. The probability that a value will fall at or between 1.0 and 2.0 (i.e., $1.0 \leq x \leq 2.0$) is the shaded area. Areas can be approximated by multiplying small segments of the interval between 1.0 and 2.0 by the probability density values that form the height of an inscribed rectangle (see Figure 6.10).

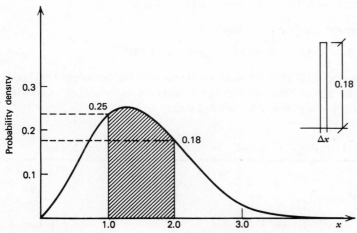

FIGURE 6.9   Typical probability distribution.

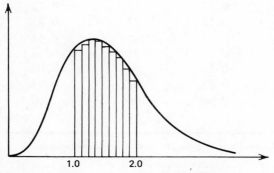

FIGURE 6.10   Inscribed rectangles.

To get a gross approximation of the area under the curve, simply approximate it as a trapezoid and calculate the area:

$$A = \tfrac{1}{2}(h_1 + h_2) = \tfrac{1}{2}(1.0)\,(0.25 + 0.18) = \tfrac{1}{2}(0.43) = 0.215$$

This calculation indicates that the probability that a duration between 1.0 and 2.0 will occur is approximately 21.5%, or a little better than 1 in 5. The probability that a duration of exactly 2.0 occurs is the very small area:

$$A = (0.18)\,(\Delta x)$$

where Δx is a very small base area defined by the point occupied by 2.0 on the X axis. This base is, of course, infinitesimally small, and the best that can be done is to take a small base and multiply it by the density value as follows.

$$A = 0.18 \times (0.1) = 0.018 = 1.8\%$$

or, using an even smaller base dimension,

$$A = 0.18 \times (0.01) = 0.0018 = 0.18\%$$

which indicates the probability that the duration will be between 1.95 and 2.05 is approximately 1.8 in 100, or that the chances are around 0.18 in 100 that the duration will be between 1.995 and 2.005.

Table 6.5   FIELD OBSERVATIONS OF BRICKLAYING TIMES

(1)	(2) Number of Obser-vations	(3) Relative* Frequency	(4) Prob-ability† Density	(5) Cumulative Number of Observations	(6) Cumulative Distribution
Interval					
0–2.99	0	0	0	0	0
3.0–3.99	12	0.12	0.12	12	0.12
4.0–4.99	15	0.15	0.15	27	0.27
5.0–5.99	18	0.18	0.18	45	0.45
6.0–6.99	21	0.21	0.21	66	0.66
7.0–7.99	16	0.16	0.16	82	0.82
8.0–8.99	12	0.12	0.12	94	0.94
9.0–9.99	6	0.06	0.06	100	1.00
Total	100	1.00			

*Relative frequency = Number of observations in class internal ÷ total operations.

$$\dagger\text{Probability density} = \frac{\text{relative frequency}}{\text{class interval}} = \frac{\text{relative frequency}}{1.0}$$

The rationale for handling the selection of durations in this manner derives from the practice in statistics of organizing observed data in a histogramatic format. That is, observed data is organized in class intervals. The number of observations in a given interval dictates the height of the histogram segment that has the class interval as base. For example, suppose the times required for a mason to lay a packet of 10 bricks as observed 100 times in the field are summarized in Table 6.5. The observed data have already been arranged into class intervals in the table.

The data plotted as a histogram are presented in Figure 6.11. The Y axis of the plot indicates the number of observations associated with each class interval. The number of times an observation occurred between 3.0 and 3.99 is 12. Therefore the height of the element of the histogram with the base 3.00–3.99 has a height of 12 observations. Another way of viewing this is to say that 12/100 is 12% of the 100 observations that fall in the 3.00–3.99 interval, and the area of this histogram element is 12% of the total area in the histogram. Based on the sample of data collected, it is possible to say that the probability that future observations will fall in this interval is 12%. This is not true in an exact sense, since the sample taken comes from a larger population of values that can be thought of statistically as defining a smooth and continuous function—the probability function of the population of observations consisting say of measurement of the time for placing every 10 brick packet placed in the United States in 1975. Of course, it is impossible to have all of these times, since they are not recorded.

An approximation of the probability function defining all of these times (i.e., the total population) using our sample of 100 observations can be obtained by drawing a smooth curve through the midpoints of histogram ele-

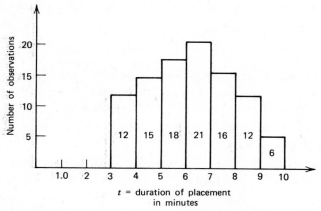

FIGURE 6.11   Histogram of bricklaying times.

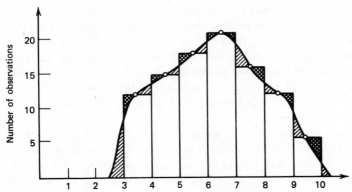

FIGURE 6.12   Smooth curve approximation of population probability function.

ments, as shown in Figure 6.12. In drawing the smooth curve through the element midpoints, assume that the percent area under the curve between 3.0–3.99 is still 12% of the total area under the curve. The smooth curve results in loss of the double cross-hatched areas and addition of the singly cross-hatched areas. The assumption here is essentially that if the class intervals are small enough the loss will be balanced by the gain and the area under the curve between 3.00–3.99 will still be approximately 12% of the total area under the curve.

   The sum of the relative frequencies of the observations in each class interval is 1.0. Therefore the area under the curve representing the probability function of all observations should be 1.0. The value of this area must be set to 1.0; therefore:

$$\text{Area} = \int_{x_m}^{x_n} f(x)\, dx = 1.0 \qquad (6\text{-}1)$$

where

x_m = the lower boundary on the X axis of the probability curve
x_n = the upper boundary on the X axis of the probability curve
$f(x)\, dx$ = area of the element under the curve with base dx and height $f(x)$

This is shown schematically in Figure 6.13.

   The actual area between 3.0 and 3.99 that represents the probability that an observation will fall in this interval is given as:

$$\text{Area}\,(x_{3.0-3.99}) = \int_{3.0}^{3.99} f(x)\, dx = P_{3.0-3.99} \qquad (6\text{-}2)$$

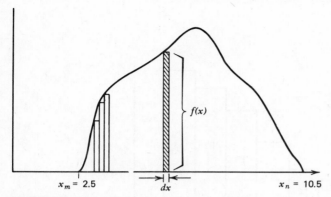

FIGURE 6.13   Schematic of elements that sum to 1.0.

Based on this definition of probability, the value of the probability density function is:

$$\text{P.D.}_{\cdot i} = \frac{P_i}{x_{i+1} - x_i} \qquad (6\text{-}3)$$

where

x_i = the left X-axis boundary of interval i

x_{i+1} = the right X-axis boundary of interval i

$P_i = \displaystyle\int_{x_i}^{x_{i+1}} f(x)\, dx$ = the probability of an observation in interval i

This is the class interval of 3.00–3.99, which reduces to

$$\text{P.D.}_{\cdot 3.0-3.99} = \frac{\displaystyle\int_{3.0}^{3.99} f(x)\, dx}{1.0} \qquad (6\text{-}4)$$

Using the rather large class interval of 1.0 that has been assumed, it is doubtful that $P.D._{\cdot 3.0-3.99}$, or the other values for the smooth curve of Figure 6.13 will equal the histogram probability density values given in Table 6.5. However, if the approach is accepted that the probability density values of the actual population of observations will vary somewhat from the sample, and that the smooth curve is a good approximation for the population, the probability densities given by Equation 6-3 are acceptable.

In the generation of random variates for simulation purposes, the distribution that is of greatest importance is the cumulative probability distribution. In the histogram of Figure 6.11, if a plot of the sum of the observations in each element moving from left to right is made, a cumulative probability representation results as shown on Figure 6.14. The cumulative number of observations is given in column (5) of Table 6.5.

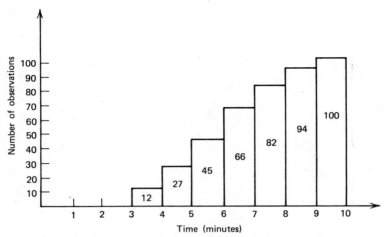

FIGURE 6.14   Cumulative histogram plot.

As in the case of the frequency distribution histogram, a continuous function is substituted for this histogramatic representation. The cumulative probability function is defined as:

$$C_r = \sum_{i=1}^{r} p_i \qquad (r = 1, 2, \ldots n) \tag{6-5}$$

where r = the interval with upper boundary x, to which the probability is to be accumulated $(r \le n)$

Figure 6.15 shows a smooth curve plot of the probability density figure shown in Table 6.5, column (4). The plot is achieved by associating the probability density values with the midpoints of the class intervals with which they are associated. The cumulative probability associated with interval three (5.0–5.99) is then

$$C_3 = \sum_{i=1}^{3} p_i = (0.12 + 0.15 + 0.18) = 0.45$$

The ordinates of this cumulative probability function indicate the probability that an observation (i.e., in this instance, a 10-brick placement time) will be less than or equal to X. Therefore, the ordinate associated with a given X value also represents the area under the probability density curve left of X (since this defines the probability that an observation will occur less than or equal to X). The continuous curve approximating the cumulative probability function for the population of 10-brick placement times is constructed by plotting the cumulative distribution values given in column (6) of Table 6.5

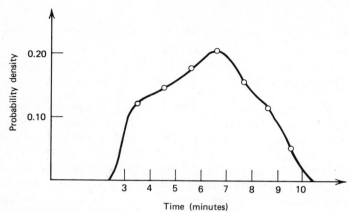

FIGURE 6.15   Smooth curve plot.

at the X value corresponding to the right-hand (upper boundary) point of the class interval with which they are associated. This is shown in Figure 6.16.

This plot is useful, since it relates all of the probability values from 0 to 1.0 along the Y axis to specific values of X (i.e., observation durations). The plot can be used as a nomograph to relate probability values with unique duration values. For instance, if the X axis is entered with the values 0.68, the plot "maps" to a value on the X axis of 7.1 (as shown by the dotted line). Samples can be selected from the distribution plotted in Figure 6.16 by generating values between 0 and 1.0 and entering the cumulative probability distribution on the Y axis. The numbers between 0 and 1.0 used to insure that a truly

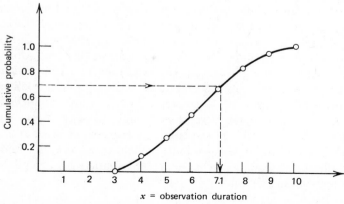

FIGURE 6.16   Plot of cumulative probability density function.

random selection of times is achieved are called pseudo-random numbers. The process of using these numbers to generate random observation times is called Monte Carlo simulation.

6.5 MONTE CARLO SIMULATION

The random numbers generated by the Monte Carlo technique act like a "roll of the dice" in providing a selection method for generation of random system work task durations. This "roll of the dice" concept normally associated with a casino environment leads to the use of the term Monte Carlo. A major interest in the technique has to do with its application in examining and exercising probabilistic or stochastic process models such as those found in construction. In these processes involving random variables, the Monte Carlo process is used to sample the random distributions to generate random time durations or delays. The technique can, however, also be used to evaluate strictly deterministic problems such as the evaluation of complex integrals, which cannot be solved efficiently by analytic mathematical methods.* The validity of response exhibited by systems simulated using Monte Carlo methods depends on the actual randomness of the variates generated. This, in turn, is a function of the randomness of the 0–1.0 range numbers used to enter the cumulative distribution, as shown in Figure 6.16. Random numbers are usually taken from random number lists when performing hand simulations. Such a list of random numbers is shown in Table 6.6. The numbers contained in such tables vary in the number of digits they contain, depending on the source of the table. In this case, the numbers consist of six digits and are scaled to the range 0–1.0 dividing by 1000000. The sequence of random numbers required to produce a "stream" of random variates is taken from the table by reading across the rows, down the columns, or using any logically nonrepetitive series of numbers.

If a simulation is to be performed requiring a large number of random numbers (e.g., 1000), it is tedious and time-consuming to hand simulate and utilize random number tables. In such cases, computer simulation using Monte Carlo methods is utilized for speed and efficiency. It is possible to utilize programs that have large lists of random numbers (e.g., Table 6.6) stored, from which numbers can be called as required. From a computer processing standpoint, however, this is an inefficient approach. It is more common to use a random number generator that generates the latest random number from the previous number. The sequence is derived by "boot-strapping" the new number from the previous one. The first number in the sequence is defined by the user and is called the random number "seed." The sequence of numbers generated is referred to as a random number stream.

---

*See, for instance, Naylor et al., *The Rejection Method*, Chapter 4, 1966.

One of the more common methods used for generating random numbers in this fashion is called the "congruence" method. In this method, the remainder of a calculation involving the previous random number becomes the new random number. Three values are defined for such a generator. Alpha, α, is used to multiply the latest random number, $R(i)$. Beta, β, is added to the product and this sum is divided by K. The relationship between $R(i + 1)$ and $R(i)$ is then:

$$R(i + 1) = \text{the remainder of } \frac{(\alpha R(i) + \beta)}{(K)} = (\alpha R(i) + \beta)(\text{modulo } K) \quad (6\text{-}6)$$

Modulo K referred to the $\alpha R(i) + \beta$ expression indicates the specification of the remainder deriving from a division by K. R_0 is the random number seed, which is the first number in the sequence. The value of K is selected to scale the quantity $\alpha R(i) + \beta$ to the range 0–1.0. Therefore, it is of the form 10^n where n is the number of digits in $\alpha R(i) + \beta$. The size of $\alpha R(i) + \beta$ is constrained to the largest number size (in digits for a decimal system) that can be defined in the computer being used. Therefore, K is a function of the computer, since the number of digits in the largest number is fixed by the computer type used.

The constraints, α and β, are selected so as to make the numbers generated as random and nonrepetitive as possible. Even the best selections of α and β, however, lead to some repetition over large sequences of numbers. Because of this, such sequences cannot be considered totally random. Therefore, such computer generated numbers are referred to as pseudo-random numbers.

6.6 RANDOM VARIATES FROM CONTINUOUS FUNCTIONS

As mentioned previously, random variates can be generated by entering a cumulative probability function with a uniformly distributed number between 0 and 1.0 and mapping this to the continuous variable axis (i.e., the X axis). This implies that the random variable is continuous in the range of interest and that, therefore, the cumulative probability function is also continuous. Now examine the process of generating random variates for some common types of continuous probability distributions. Use of such standard distributions has the advantage that they can be defined uniquely in terms of relatively few parameters. This does away with the requirement to specify the cumulative function interval by interval, as was done in the masonry problem.

Some of the more commonly used distributions are given in Table 6.7. In general these distributions can be specified using the parameters (1) mean value, (2) standard deviation, and (3) upper and lower limits. A discussion of these parameters and their development from observed data can be found in an introductory statistics text.*

*See, for instance, Feller, 1957 or Ang and Tang, 1975.

Table 6.6 RANDOM NUMBERS

258164	244733	824904	959712	284925	062825
547250	466759	943814	751744	707634	376550
279794	797398	656465	505360	241001	256756
676883	778968	934335	028735	444391	538814
056700	668517	599657	172246	663342	229231
339846	006566	593875	032328	975552	373848
036783	039384	559225	193777	846672	240567
220480	236066	351556	161368	074279	441791
321406	414815	106967	967134	445197	647755
926274	486088	641104	796227	668169	882135
551342	913235	842276	771953	004479	286810
304312	473198	047928	626475	026876	718933
823825	835986	287273	754598	161107	308715
937351	010233	721707	522461	965570	850209
617730	061361	325338	131225	786849	095472
702187	367781	949838	786484	715749	572211
208356	204205	692568	713559	289632	429389
248744	223866	150708	276511	735843	573432
490798	341698	903251	657207	410058	436704
941463	047882	413364	938779	457579	617269
642372	286994	477391	626291	742379	699424
849870	720032	861112	753498	449229	191795
093443	315302	160820	515872	692334	149489
560052	889689	963853	091735	149304	895946
356517	332082	776563	549817	894838	369583
136699	990251	654104	295173	362940	215001
819290	934772	920183	769050	175190	288566
910170	602271	514838	609073	049977	729456
454833	609543	085541	650304	299551	371782
725920	653122	512693	897409	795288	228180
350587	914302	072686	378353	766325	367552
101159	479593	435653	267561	592743	202833
606294	874310	610972	603571	552441	215643
633650	239915	661686	617332	310901	292418
797598	437881	965626	699801	863313	752542
780166	624326	787185	194055	174009	510141
675692	741722	717763	163035	042897	057390
049565	445296	301705	977129	257123	343977
297081	668767	808201	856124	541013	061544
780488	008061	843715	130923	242413	368876

(*Continued*)

Table 6.6   (*Continued*)

677624	048345	056556	784673	452850	210769
061141	289772	338980	702709	714037	263205
366460	736682	031592	211482	279375	577461
196284	415086	189369	267476	674370	460850
176396	487709	134951	603059	041642	761984
057200	922951	808817	614263	249601	566725
342842	531427	847407	681412	495933	396506
054743	184960	078683	083843	972241	376355
328117	108529	471591	502517	826831	255591
966490	650465	826352	011698	955365	531832
792363	898373	952494	070140	725692	187387
748793	384127	708486	420392	349224	123071
487669	302166	246104	519512	092989	737614
922713	810962	474975	113551	557332	420673
530000	860263	846638	680563	340216	521195
176412	155727	074070	078756	039000	123638
057296	933327	443946	472031	233763	741017
343418	593614	660673	828993	401014	441071
058191	557661	959553	968321	403375	643445
348781	342185	750789	803337	417523	856303
090331	050801	499633	814559	502313	131999
541400	304492	994413	881820	010477	791123

Fortunately, many of the frequency distributions observed in the field can be described with reasonable accuracy by continuous distribution functions which, in turn, are easily defined in terms of their parameters. Obviously, the first thing to be done is to select a continuous distribution that might possibly be used as the distribution representative of the field data observed. The next item of interest is to solve for the distribution parameters that best fit the data observed. What is normally done, however, is to solve for the parameters of the data actually observed in the field. These values (i.e., mean and standard deviation) are then assumed to be the correct parameters of the candidate distribution. That is, they are assumed to be the parameters of a candidate theoretical distribution such as those in Table 6.7.

Comparisons are then made between the plot of the actual data observed and the plot of the theoretical candidate distribution. This is illustrated schematically in Figure 6.17. Throughout the remainder of this chapter the continuous plot of the observed data is shown as a dotted line. The data in this case are assumed to be normally distributed. This theoretical distribution (normal), with the parameters developed from the observed data, is plotted

Table 6.7 COMMONLY USED PROBABILITY DISTRIBUTIONS

Distribution	Formula Defining Probability Density Function	Description	Schematic
Normal	$f(x) = \dfrac{1}{\sigma_x \sqrt{2\pi}} \exp -\dfrac{1}{2}\left(\dfrac{x - \mu_x}{\sigma_x}\right)^2$	The normal distribution is continuous and symmetric about its mean and is defined by two parameters, the mean, μ, and standard deviation, σ.	
Lognormal	$f(y) = \dfrac{1}{\sigma_y \sqrt{2\pi}} \exp\left[\left(-\dfrac{1}{2}\right)\left(\dfrac{y - \mu_y}{\sigma_y}\right)^2\right]$ for $-\infty < y < -\infty$ and $y = \ln x$	The lognormal distribution is continuous and asymmetric and has a mode that is skewed to the mean value. The distribution is characterized by its modal value, its mean, and its standard deviation.	
Exponential	$f(x) = \alpha e^{-\alpha x}$ $\alpha > 0$ and $x \geq 0$	The exponential distribution is continuous in the range $0 < x < +\infty$. It is used to represent the intervals between distinctly random events in "memoryless" processes.	
Gamma (Erlang)	$f(x) = \dfrac{\alpha^k x^{(k-1)} e^{-\alpha x}}{(k-1)!}$ $\left.\begin{array}{l} \alpha > 0 \\ k > 0 \\ x > 0 \end{array}\right\}$ all nonnegative	"If a process consists of "k" successive events and if the total elapsed time of this process can be regarded as the sum of k independent exponential variates each with parameter α, the probability distribution of this sum will be a gamma distribution with parameters α and k."[*] It is a continuous distribution and may be fitted to many positively skewed distributions of statistical data by varying the values of α and k.	
Poisson	$f(x) = e^{-\lambda}\left(\dfrac{\lambda^x}{x!}\right)$ $x = 0, 1, 2, \dots$ $\lambda > 0$ or $P(n) = \dfrac{(\lambda t)^n e^{-\lambda t}}{n!}$ λ = mean of exponentially distributed interarrival times	The Poisson is a discrete distribution used to describe the probability of x arrivals in a given time interval, assuming the intervals between arriving units are exponentially distributed with mean $= \lambda$.	

*Direct quote Naylor et al; p. 87.

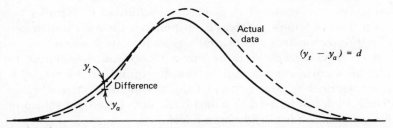

FIGURE 6.17   Comparison of theoretical with actual distribution.

as a solid line. The difference in the observed frequency, y_a, from the estimated frequency as given by the theoretical distribution, y_t, is developed for each observed data point, x_i. Whether or not the assumption that the theoretical distribution is a good or reasonable representation of the observed data is determined by examining d, where

$$d = (y_t - y_a)$$

The distribution of d is evaluated using a chi-square distribution to determine whether the proposed assumption is correct. If the assumption is acceptable, the cumulative density function of the theoretical distribution can be used to generate random variates. If the assumption is not valid, the actual plot of the observed data must be used to develop the random variates. In hand simulation of a system, the efficiencies involved in using the theoretical distribution to generate random variates are not so significant. If the assumption holds, the variates could be generated by substituting the uniformly distributed 0–1.0 random numbers into the expression for the cumulative distribution of the theoretical distribution. For instance, if the distribution of observed data proves to be normally distributed, the random variate required can be generated by taking two uniformly distributed random numbers (range 0–1.0) and solving the following expressions to generate two normally distributed deviates:

$$x_1 = (-2 \log_e r_1)^{1/2} \cos 2\pi r_2$$

$$x_2 = (-2 \log_e r_1)^{1/2} \sin 2\pi r_2 \qquad (6\text{-}8)$$

where

x_i = normally distributed random variate
r_i = uniformly distributed number in the range 0–1.0.

Generally, however, in hand simulation of a system, it is much easier to use the plot of the cumulative frequency as shown in Figure 6.16.

When conducting a computer simulation, however, the use of the theoretical distribution where justified to generate the random variates (by reference to

a formula) is much more efficient. This is because definition of the plot requires storage of a table of values, whereas definition of a theoretical distribution requires only the parameters and expressions such as (6-8) above.

The procedure of checking field data against a theoretical distribution to
see if one might be used is summarized in the flow graph in Figure 6.18.
Consider, for example, the field data given in Table 6.5 and determine whether
a theoretical distribution can be used to model it. A plot of the data shown in
Figures 6.11, 6.12, and 6.15 has already been made.

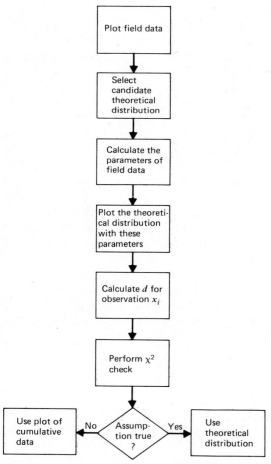

FIGURE 6.18 Procedure for checking field distribution versus theoretical
distribution.

Now a candidate theoretical distribution must be selected. First try the NORMAL distribution to see if it is reasonable to assume the data is normally distributed. In order to plot the distribution (Figure 6.19), the mean and standard deviation of the data must be calculated. The formulas for the mean and standard deviations are as follows.

$$\bar{x} = \frac{\sum\limits_{i=1}^{n} x_i}{n} \qquad s = \sqrt{\frac{\sum\limits_{i=1}^{n} (x_i - \bar{x})^2}{n - 1}} \qquad (6\text{-}9)$$

These formulas apply to ungrouped data. In this case, the data is grouped and it is to be used instead of returning to the original list of observations. In order to calculate the arithmetic mean from grouped data, the following formula should be used.

$$\bar{x} = \frac{\sum\limits_{i=1}^{n} x_i f_i}{\sum\limits_{i=1}^{n} f_i}$$

where f_i is the cell frequency for cell i, and x_i is the mean value for the cell. This yields:

$$\bar{x} = \frac{\begin{aligned}[3.495(12) + 4.495(15) + 5.495(18) + 6.495(21) \\ + 7.495(16) + 8.495(12) + 9.495(6)]\end{aligned}}{100}$$

$$= \frac{623.5}{100} = 6.235$$

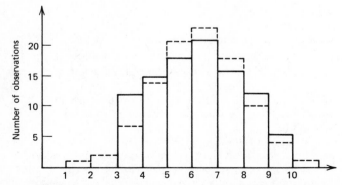

FIGURE 6.19   Observed versus theoretically distributed data. (Theoretical distribution shown with dotted line.)

The standard deviation for grouped data is given by the formula

$$s = \sqrt{\frac{\sum\limits_{i=1}^{n} (x_i - \bar{x})^2 f_i}{\left(\sum\limits_{i=1}^{n} f_i\right) - 1}}$$

Substitution of the values into this expression yields:

$$s = \left\{ \frac{\begin{bmatrix} (3.495 - 6.235)^2(12) + (4.495 - 6.235)^2(15) \\ + (5.495 - 6.235)^2(18) + (6.495 - 6.235)^2(21) \\ + (7.495 - 6.235)^2(16) + (8.495 - 6.235)^2(12) \\ + (9.495 - 6.235)^2(6) \end{bmatrix}}{99} \right\}^{1/2}$$

$$= \sqrt{\frac{297.402}{99}} = \sqrt{2.97} = 1.725$$

In order to apply the chi-square criterion, the normal curve frequencies to be expected for a theoretical normal distribution with mean 6.235 and standard deviation of 1.725 must be calculated. This can be done by using the unit standard normal distribution to calculate the areas within the class intervals on the theoretical normal distribution with these parameters. This is summarized in Table 6.8.

Taking the data from Table 6.8, there results:

$$x^2 = \frac{(12 - 9.7)^2}{9.7} + \frac{(15 - 13.9)^2}{13.9} + \frac{(18 - 20.8)^2}{20.8} + \frac{(21 - 22.6)^2}{22.6}$$

$$+ \frac{(16 - 17.6)^2}{17.6} + \frac{(12 - 9.9)^2}{9.9} + \frac{(6 - 5.5)^2}{5.5}$$

$$= 1.75*$$

The number of degrees of freedom of the data is $(7 - 3) = 4$.† By consulting a table of critical values of x^2 at the level of significance of 0.05, it can be seen that the critical value is 9.488. Since the calculated value of x^2 is less

*Values for the first two $(0 - 2.99, 3.0 - 3.99)$ and the last two $(9.0 - 9.9, 10.0 - \infty)$ cells were combined in the calculation, since the chi-square criterion cannot be used if any expected cell frequency is less than five.

†Note that the number of degrees of freedom is calculated by considering the total number of cells of observed data (seven cells, considering combination of the two upper times and lowermost cells) minus the restriction imposed by the observed data. In this case, three restrictions result because of the utilization of three quantities from the observed data that were used in our calculation of the expected (i.e., theoretical) frequencies: (1) the mean, (2) the standard deviation, and (3) the total frequency. Therefore, the degrees of freedom are given as:

$$(7 - 3) = 4.$$

Table 6.8 THEORETICAL NORMAL CURVE FREQUENCIES

The χ^2 (Chi-Square) Statistic Is Defined by the Expression

$$\chi^2 = \sum \frac{(n_i - e_i)^2}{e_i}$$

where n_i = the observed value and e_i = the estimated value.

(1) Class Interval	(2) Boundaries	(3) $\left\{\dfrac{\text{Column}(2)}{-6.235}\right\}$	(4) $z = \left\{\dfrac{\text{Column}(3)}{1.725}\right\}$	(5) Area	(6) Difference	(7) Normal Curve Frequency	(8) Observed Frequency
0–2.99	2.995	−3.24	−1.88	−0.4699	0.030	3.0	0
3.0–3.99	3.995	−2.24	−1.298	−0.4032	0.067	6.7	12
4.0–4.99	4.995	−1.24	−0.719	−0.2642	0.139	13.9	15
5.0–5.99	5.995	−0.24	−0.139	−0.0557	0.2085	20.8	18
6.0–6.99	6.995	0.76	0.441	+0.1700	0.2257	22.6	21
7.0–7.99	7.995	1.76	1.020	+0.3461	0.1761	17.6	16
8.0–8.99	8.995	2.76	1.600	+0.4452	0.099	9.9	12
9.0–9.99	9.995	3.76	2.180	+0.4854	0.040	4.0	6
10.0–∞					0.015	1.5	0

than the critical value, it can be said that the observed data constitutes a "good fit" of the expected normal distribution. Therefore justification exists for considering the observed data as being normally distributed.

Since the assumption of normality has been supported, equation set (6-8) can be used to generate random variates that are normally distributed and defined by the standard deviation and mean values calculated for the observed data. In this case, the normal distribution was selected as a candidate for verification. The distribution of the observed empirical data may lead to the selection of one of the other distributions. If one of these distributions had been selected and verified as being representative of the observed data, techniques (similar to Equation 6-8) are available for the generation of random variates. A detailed discussion of these techniques is beyond the scope of this book. A comprehensive discussion of methods for generating random variates for commonly encountered probability distributions is presented in Chapter Four, Naylor et al., 1966.

6.7 RANDOM VARIATES FROM EMPIRICAL DATA

If the assumption of normality had not been supported by the observed empirical data, it would be necessary to reject the use of a NORMAL distribution for generation of the random transit delay times. In this case, the observed data and the cumulative plot of Figure 6.16 must be used as the basis for generating the random transit times.

In hand simulation, the procedure is basically that described in Figure 6.16. The steps are as follows.

1. Collect field observations (empirical data).

2. Arrange data into appropriate class intervals.

3. Develop a table containing the information of Table 6.5 to include the cumulative probability densities.

4. Make a cumulative probability density plot similar to Figure 6.16 for the data set under consideration.

5. For each random observation desired:

(a) Get a random number from a random number list such as that shown in Table 6.6.

(b) Scale the random number to the range 0–1.0.

(c) Enter the Y axis of the cumulative plot and map the random number to the X axis, as shown in Figure 6.16.

(d) Record the random variate that is the X value generated by the mapping above.

This procedure is shown in flowchart format in Figure 6.20.

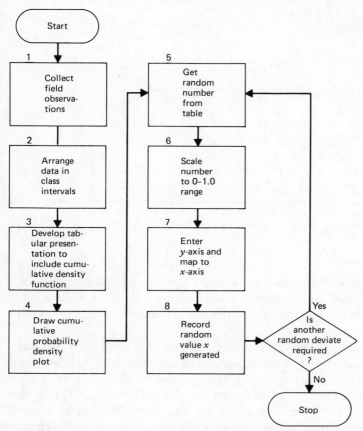

FIGURE 6.20   Procedure for random variates from empirical data.

An alternative method of developing the cumulative density plot can be used as follows.

1. Plot a cumulative histogram plot similar to Figure 6.14. The rightmost cell of the plot will have a height equal to the total number of observations recorded.

2. Scale the values on the Y axis (i.e., the number of observations axis) to the range 0–1.0 by dividing the Y-axis values by the total number of observations. This will result in the height of the rightmost cell becoming 1.0.

3. Draw a smooth curve through the ordinates representing the upper boundary of each class interval.

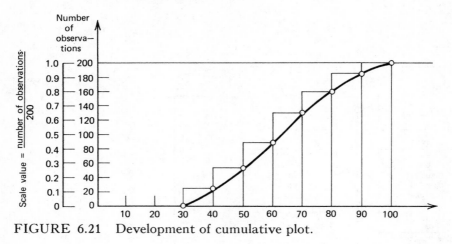

FIGURE 6.21   Development of cumulative plot.

4.   The resulting plot can now be used as the cumulative plot, which is entered with the 0–1.0 range random number (operation 7 of Figure 6.20). Figure 6.21 indicates the development of the cumulative plot using this method.

The generation of random variates from empirical data during computer simulation is essentially an automated version of the procedure just described. Input defining the class intervals and the number of observations in each class is made to the program. This is stored and used to generate a cumulative plot similar to that of Figures 6.16 and 6.21. The plot is defined by a set of linear equations that describe the segments connecting the upper boundary ordinates for each cell. Each cell is handled as if it were a discrete random variable, d_i. A cell probability, $P(X = d_i) = p_i$, is associated with each cell. For instance, in the case of the bricklaying data, the p_i values associated with each cell would be as follows:

d_i	p_i
0–2.99	0
3.0–3.99	0.12
4.0–4.99	0.15
5.0–5.99	0.18
6.0–6.99	0.21
7.0–7.99	0.16
8.0–8.99	0.12
9.0–9.99	0.06

The computer pseudo-random number generator generates a random value, r, and the appropriate cell, d_i, is selected based on the expression:

$$p_1 + \ldots + p_{i-1} < r \leq p_i + \ldots p_m$$

The value m is the number of cells considered.

If the number of cells in the histogram is large, the number of stored segment equations can become very large, and the procedure becomes inefficient from a processing standpoint. That is, considerable computer storage as well as search time during execution will be required. For these reasons, it is most efficient, when possible, to use a standard functional routine.

PROBLEMS

1. Given the data shown in Table A.1 (see Appendix A), calculate the balance point of the system production for systems of 1 to 6 trucks. Is the system working above or below the balance point?

2. The loader and arrival times for a system consisting of four trucks and two loaders are distributed as follows:

Loader Time in Minutes		Back Cycle in Minutes	
Interval	Frequency	Interval	Frequency
0–0.5	2	4.0–4.5	4
0.51–1.0	5	4.51–5.0	10
1.01–1.5	10	5.01–5.5	28
1.51–2.0	20	5.51–6.0	32
2.01–2.5	25	6.51–7.0	20
2.51–3.0	25	7.01–7.5	4
3.01–3.5	10	7.51–8.0	2
3.51–4.0	2		100
4.01–4.5	1		
	100		

These tables represent field data on haul and load times. Calculate the production of this system deterministically and using the appropriate queueing nomograph in Appendix D. Develop a plot of the system production both deterministically and to include bunching for systems from 1 to 10 trucks.

3. High-rise building construction sites are often served by material hoists. Material arriving at the ground level of the hoist and the push-button call for hoist service can be considered as resource flow units entering idle queue states awaiting hoist service work tasks. Material hoist work tasks comprise loading, haulage to the various floors, and unloading. Develop CYCLONE models for the material hoist operation assuming:

   (a) A single common haulage work task for all served floors with a probabilistic duration to represent the various hoist trip times to the floors.
   (b) Individual deterministic hoist travel times for each floor. For this case incorporate into the model a probabilistic switch to allow for the modeling of different floor destinations of the material.

   Hence observe a specific material hoist situation, record hoist operation times, and thus determine the relevant hoist haulage work task durations.

4. Suppose in the mason's laborer model of Figure 6.5 and Table 6.1 the work force is congested and the brick and mortar storage area are located 30 feet away. Assuming that the mason's laborer is responsible for the additional haulage work tasks, update the CYCLONE model and determine the impact on the daily production using the data of Table 6.3. Hence determine from Table 6.4 the new mason-laborer crew mix to ensure no loss in daily production. Base your estimate on your own time estimates for walking empty-handed and loaded work tasks.

5. If the system in Problem 2 were working at the balance point, what production loss would result from bunching using the Morgan and Peterson approach?

6. Given the data shown in Table A.2 (see Appendix A) perform a chi-square test to determine if the arrival times can be assumed to be exponentially distributed ($K = 1$) at the 5% significance level.

7. By field observation and timing, determine the duration of a work task in histogram form for several different time interval segments. Comment briefly on their relation to commonly available head office estimate data.

CHAPTER SEVEN

Hand Simulation

The concept of flow units and their use in the development of network system models for construction operations have been introduced in previous chapters. The identification of work task sequences through flow unit cycles provides the building blocks for the network system model. The network model gives a blueprint (i.e., portrays logic and structure) of a construction operation but gives no indication of the response of the model to the application of specific flow unit resources.

The behavior of a CYCLONE network system model depends on the movement of the resource flow units through the structure and logic of the model. This chapter discusses the mechanics of the flow unit modeling and analysis of CYCLONE models using hand simulation techniques. Later chapters treat the use of computer processing techniques.

7.1 DISCRETE SYSTEM SIMULATION

Movement of units in the real-world system provides the basis on which the logic of the CYCLONE system network is developed. The object of developing the model of a production system is to examine the interaction between flow units, determine the idleness of productive resources, locate bottlenecks, and estimate production of the system as constituted. In order to achieve this objective the movement of the units through the system must be effected in a manner that simulates the movement of the real-world production resources. This allows the study of the process in an environment that approximates a laboratory.

The use of a model to represent a real-world system is an abstraction and as such loses some of the fidelity and much of the detail of the actual situation. Some very important aspects of the real-world system may be lost in this abstraction. However, care in developing models to an appropriate level of detail should offset this difficulty to a great extent.

The advantages to be achieved focus on the simplicity of a paper and pencil modeling that is inexpensive to construct and may provide insights into system operation that would be extremely expensive to gain by observation of the actual system.   These insights often may be gained only after costly mistakes and expensive production runs have been made with the real-world system.

In both cases, the system to be studied must be exercised and observed in order to determine system response and imbalance between resources. Such imbalances result in process bottlenecks and inefficiencies; they should be avoided, or minimized, by a proper selection and balance of resources.

In exercising the model of a construction process that has randomly defined work task duration, Monte Carlo simulation can be used to move the flow units through their cycles and advance them from state to state. As the name implies, the movement of units is achieved by rolling imaginary dice to determine when units are moved and the amounts by which they are delayed. The dice used in this case are the pseudorandom numbers discussed in Chapter Six. These are used to select variates from probability distributions that represent the randomness of system work task durations. If the delays associated with system work tasks are deterministic (i.e., not defined by a probability distribution), the dice are not needed. In such cases, a simple simulation can be performed in which delays at each system work task are predefined.

Whether the durations of work tasks in the construction process to be studied are randomly defined or deterministically defined, the movements of units for purposes of simulation take place at discrete points in time. That is, the simulation work task (e.g., lifting a precast panel with a crane) is defined in terms of its starting event and its end event. During the time between a work task's beginning and end, the units that transit it are captured by the work task and a fix on their location is possible. Therefore, the system network model is only concerned with their movement at fixed points in time (i.e., when they become available for other work tasks). The same is true for those units delayed at QUEUE nodes, where the discrete time at which a flow unit arrives in the QUEUE node and when it leaves are of interest. These points are again discrete events along a line representing the passage of time. Because of the discrete nature of flow unit movements in the systems model, the procedure is referred to as discrete system simulation. Procedures for manually moving units in discrete jumps on a graphical model are called hand simulation. A discrete computer simulation is implied if a computer is used to move flow units through the system in a discrete manner.

7.2 DISCRETE SIMULATION METHODS

One concept basic to all discrete system simulation is that of a simulation clock (SIM CLOCK), which keeps track of simulation time (SIM TIME). The method by which this clock is advanced is of primary importance, since it establishes how and at what points in SIM TIME the system is reviewed to determine whether flow unit movement should take place. Two methods can be used to advance the SIM CLOCK from discrete point to discrete point in time.

In one method, the clock is advanced in even or equal time steps. In this method, the clock advances from discrete event to discrete event in uniform steps (e.g., 1 minute, 2 minutes, 1 time unit). After each step, the system is reviewed to see if any unit movement was scheduled to take place. This reduces to determining whether any work tasks were scheduled to terminate during the interval of the time step. That is, a schedule of events is maintained and interrogated across the interval just elapsed to determine whether any units have terminated the work task in which they were delayed. If so, they are released from the work task and allowed to flow further through the system model, generating appropriate delay durations at the following elements to which they pass. This results in new scheduled events that are recorded for future review. This method of advancing the clock is called the "uniform time step" method.

It is an acceptable method when the scheduled movement of flow units is fairly "uniform." However, in cases where there are several quick movements of units followed by long periods of inactivity, this method can be very inefficient. If, for example, a situation occurs in which several units move within 5 minutes and then no further activity occurs for 2 hours, this method would require 24 reviews of the schedule using time intervals of 5 minutes (i.e., $(2 \times 60$ minutes$/5 = 24)$, during which no system activity would occur. That is, under this method and time interval, the schedule would be consulted 24 times, despite the fact that nothing is scheduled.

A method that is better adapted to the general case is one that advances the SIM CLOCK based on the scanning of end event times of the work tasks as scheduled during the simulation. In order to implement this method, a simple record-keeping system is used to keep track of scheduled work task end event times. This system is like an appointment book designed to indicate when work tasks are to terminate. The required system consists of an EVENT list and a CHRONOLOGICAL list. When event times are generated, they are listed on the EVENT list. Entries from the event list are then recorded in order on the CHRONOLOGICAL list at the time they will actually occur. The EVENT list contains the events that are scheduled, and the CHRONO-LOGICAL list contains the events that have occurred. The last event on the CHRONOLOGICAL list is the event now occurring and therefore represents the time now (TNOW).

The record-keeping aspect of hand simulation keeps track of the event from the time it is generated to the time it has occurred. By examining the EVENT list of Figure 7.1, it can be seen that the event pool consists of four event times that have been generated but that have not yet occurred. Events that have occurred are crossed off the event list and moved to the chronological list. The SIM CLOCK has not advanced to any of these scheduled events, so they constitute a pool of events that will occur in the future as simulated time progresses. Four events have been transferred to the chronological list. Three have already occurred, and one (7.2) represents the simulated time now (TNOW). The time steps to date have been: 0-4.14, 4.14-5.8, 5.8-6.25, and 6.25-7.2.

Each of the event times generated has been recorded in the sequence generated in the event list of Figure 7.1 and is not listed immediately in the chronological listing. Some values generated later in the sequence are chrono-logically earlier than those generated earlier in the sequence. For instance, the third value in the event list is earlier (5.8) than the second value (6.25). If the 6.25 value had been recorded immediately in the chronological listing, it would have been necessary to erase it and insert the 5.8 value.

The transfer of event times from the event list to the chronological listing is the mechanism by which the simulation clock is advanced. The last entry on the chronological listing always indicates the TNOW (time now) value.

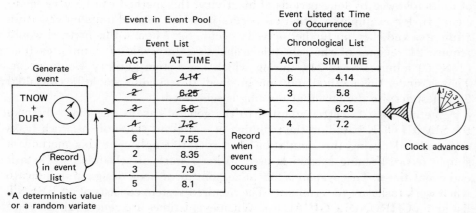

FIGURE 7.1   Discrete event processing.

The event time selected to be moved from the event list to the chronological list is always the earliest event not yet transferred. Once an event is transferred to the chronological listing, it is crossed off of the event list. At the moment it is transferred, the previous simulated time is changed to the SIM TIME of the transferred event. In this way it is not possible to get the entries on the chronological list out of time order from earliest to latest. Since the TNOW value is that of the last event on the chronological listing, any newly generated event times must be later chronologically, since TNOW + DUR $\geq$ TNOW. It will be seen later that there are some instances in which the DUR value is zero and the event time generated is immediately transferred through the event list to the chronological list.

The flow of the simulated event left to right in Figure 7.1 begins at the time the event is generated. This takes place at the time a work task can commence. Since durations must be generated for all elements other than QUEUE nodes and ARCS, a generation of an event time takes place each time a COMBI, NORMAL, FUNCTION node, or ACCUMULATOR can commence. The COMBI can commence when its ingredience conditions are satisfied. The NORMAL, FUNCTION node, and ACCUMULATOR elements can commence when units transit from the element or elements preceding them.

One major phase of the simulation of discrete unit systems consists of the identification of work tasks that can commence, the generation of durations for work tasks that can commence, the calculation of event times corresponding to the termination of these work tasks, and the recording of these termination events in the event pool. This phase is referred to as the EVENT GENERATION PHASE.

When it is determined that a work task can commence, the first task is to move the flow units that are to transit the work task to the graphical element representing it. After this, the delay time is generated, the work task end or terminal event time is calculated, and the generated event time is recorded in the EVENT list (i.e., the event is scheduled). At any given TNOW, all work tasks that can commence are started, therefore they generate END event times, as shown in Figure 7.2.

Once all work tasks that can be started at TNOW are commenced, the second major phase of the simulation procedure is started (i.e., by advancing the SIM CLOCK). This phase is referred to (see Figure 7.3) as the ADVANCE PHASE. This phase essentially reduces to the transfer of the next earliest event from the EVENT list to the CHRONOLOGICAL list. This makes the next earliest event the last entry in the CHRONOLOGICAL list and moves the SIM CLOCK from its previous setting to a TNOW value, which is the SIM time of the transferred event. In other words, the TNOW pointer of Figure 7.1 always drops to the last entry in the CHRONOLOGICAL list and the SIM CLOCK advances to the event time of the newly inserted entry.

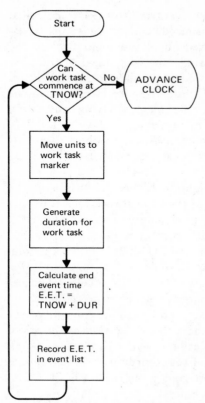

FIGURE 7.2   Event generation phase.

When the SIM CLOCK is advanced, all work tasks that can be terminated are ended and the unit(s) held in transit are released. After all units have been released to the elements following the work tasks in which they have been delayed, the event generation phase recommences. Thus, the continual cycling between the EVENT GENERATION and the CLOCK ADVANCE phases results in the movement of flow units through the simulated system. Since the SIM CLOCK is always advanced to the time of the next earliest schedule event, this procedure for simulation is called the next event method.

7.3 THE NEXT EVENT SIMULATION ALGORITHM

The list in Table 7.1 summarizes the steps implied in the flow diagrams shown in Figures 7.2 and 7.3. A composite flow diagram for this procedure is shown in Figure 7.4. A segment has been included in the flow diagram to differentiate between procedures used, depending on whether the delay is

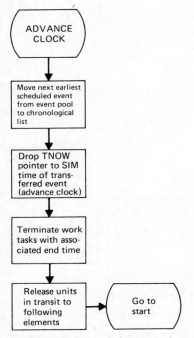

FIGURE 7.3   Advance phase.

deterministic or random. In the case where the work task commenced has an associated random delay, a random variate is generated from the cumulative probability distribution, as described in Chapter Six.

The algorithm described at this point does not contain any provision for developing statistics on performance of the system or the idleness incurred by various units delayed at QUEUE nodes. The algorithm presented provides only for the movement of units through the system defined. Acquisition of statistics will be discussed later in this chapter.

7.4 THE MASONRY MODEL

Consider again the mason laborer problem defined in Chapter Five and perform a hand simulation of the system in which the laborer stockpiles the brick pallets on the scaffold. In this system assume that the scaffold is large enough to allow for the stacking of three 10-brick packets. The CYCLONE model for this system is shown again in Figure 7.5. One laborer and three masons are defined in the system to be hand-simulated. The simulation will be carried out until each of the three masons has completed three brick-placing sequences. No bricks are stacked on the scaffold at the beginning of the shift.

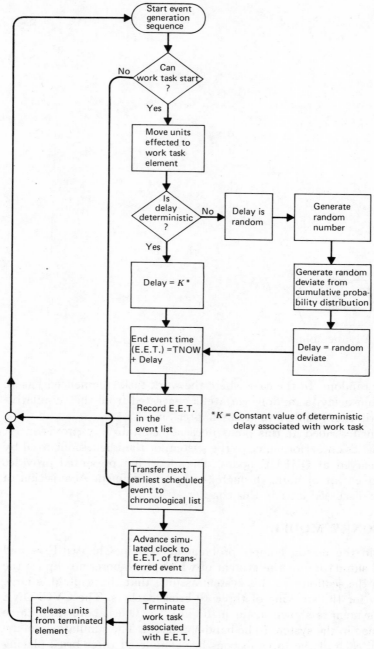

FIGURE 7.4 Simulation flow diagram.

192

Table 7.1 NEXT EVENT ALGORITHM—A—GENERATION PHASE

GENERATE PHASE

1. Can any work task start (i.e., are ingredience conditions met)?
2. If YES, continue; if NO, go to step 8.
3. MOVE units that can begin transit of a work task into work task element.
4. GENERATE duration of transit time (i.e., transit or processing delay).
5. CALCULATE scheduled termination of work task (END EVENT TIME = TNOW + DUR).
6. RECORD E.E.T. in EVENT list.
7. RETURN to step 1

B—ADVANCE PHASE.

ADVANCE PHASE

8. TRANSFER next earliest scheduled event on the EVENT list to the CHRONOLOGICAL list. (Cross out entry on EVENT list.)
9. ADVANCE Sim Clock to time of transferred event.
10. TERMINATE work tasks with associated END EVENT TIME.
11. RELEASE unit(s) to following elements.
12. Return to step 1.

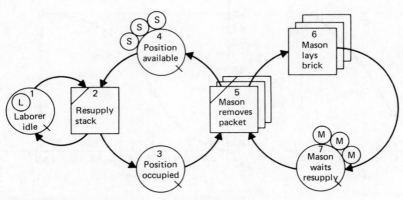

FIGURE 7.5 Mason resupply system.

Table 7.2   TRANSIT TIMES FOR MASON PROBLEM

(a) Placement (Bricklaying) Time for 10 Brick Packets			(b) Carrier Cycle Time (Transfer of Packets to Scaffold)		
Interval	Frequency	Cumulative Plots	Interval	Frequency	Cumulative Plots
3.0–3.99	12	12	0.0–0.49	2	2
4.0–4.99	15	27	0.5–0.99	8	10
5.0–5.99	18	45	1.0–1.49	16	26
6.0–6.99	21	66	1.5–1.99	20	46
7.0–7.99	16	82	2.0–2.49	22	68
8.0–8.99	12	94	2.5–2.99	16	84
9.0–9.99	6	100	3.0–3.49	9	93
			3.5–3.99	4	97
			4.0–4.99	3	100

Parts a and b of Table 7.2 gave the observed time distributions for bricklaying and transfer of brick packets from the ground stockpile to the stack on the scaffold. Markers representing the laborer, the three stacks positions, and the three masons are shown at QUEUE nodes 1, 4, and 7, respectively. Figures 7.6a and 7.6b show the cumulative plots of the transit delay information given in Table 7.2. Assume that the time required by the mason to remove the brick packet (activity 5) has a deterministic value of 1.0 minutes. Table 7.3 gives the EVENT list and the CHRONOLOGICAL list.

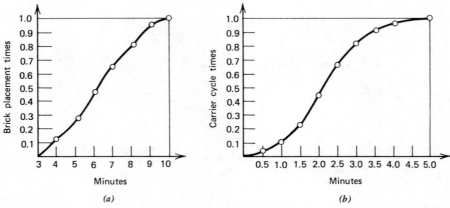

FIGURE 7.6   Cumulative distribution plots.

Table 7.3   EVENT AND CHRONOLOGICAL LISTS (MASON
             PROBLEM)

Transfer	Event List				Chronological List	
	ACT	TNOW	DUR	E.E.T.	ACT	SIM TIME
X	2	0	3	3	2	3.0
X	2	3.0	4.9	7.9	5	4.0
X	5	3.0	1.0	4.0	2	7.9
	6	4.0	4.5	8.5		

In order to understand the simulation process, proceed step by step through the first few GENERATE and ADVANCE cycles. First, after examining which work tasks can commence, notice that the only ingredient requirement that is satisfied is that for COMBI work task 2. Therefore, a stack unit and the laborer unit are moved to work task 2. The time generated for this delay is 3.0 minutes, and the E.E.T. is 3 minutes.* No other work tasks can be started, so move now to the ADVANCE phase.

The E.E.T. for work task 2 is transferred to the CHRONOLOGICAL list and the clock is advanced to a SIM TIME of 3.0. To indicate this transfer, a check is placed in the TRANSFER column of the EVENT list. At this point, the units delayed in 2 are released, and they pass to the following elements. The laborer returns to the idle state at 1 and the stack marker moves forward to the "POSITION OCCUPIED" queue node. Having released these units, simulation control now returns to the GENERATE phase.

The system is again examined to see which work tasks can commence. At this time (SIM CLOCK 3.0), two work tasks can be started. The ingredient requirement for 2 is met because the laborer and a stack position are both available. The "MASON REMOVE PACKET" work task can also begin, since a unit is available at 3 (the unit just released from 2) and a mason is available at 7. Proceeding in accordance with the flowchart of Figure 7.4, the laborer and the stack marker are now moved to 2, generate a random delay time for work task 2, and enter it into the EVENT list. The generated delay is 4.9 and that, added to TNOW, gives an E.E.T. for COMBI 2 of 7.9 minutes. The stack marker (representing a packet available on the scaffold) and one of the mason units at 7 are moved into the "MASON REMOVES PACKET" activity at 5. Again a delay is generated and recorded in the EVENT list. The delay in this case is deterministic, since a constant time of 1.0 minutes

*Since each time a simulation is conducted a new set of random numbers will normally be used, this represents only one of an infinite set of time sequences that could be obtained.

was assumed for the mason to pick up the brick packet. Adding this deterministic delay to the TNOW value, an E.E.T. for COMBI 5 of 4.0 is obtained. Having started all work tasks that can commence, return to the ADVANCE phase.

By checking the EVENT list, it can be seen that two events are scheduled to occur. The earlier of these two events is the termination of activity 5 at 4.0 minutes. Therefore, transfer this event to the CHRONOLOGICAL list and place a check for it in the TRANSFER column of the EVENT list. The SIM TIME value of TNOW becomes 4.0, since the clock is advanced to 4.0. Now check to see which work tasks can terminate. The only work task affected by the new TNOW is 5, which is terminated and therefore releases the mason and stack marker flow units. These units pass on to elements 6 and 4, respectively. The movement of the stack marker to 4 indicates that the stack position is now empty. At this point there are no packets stacked on the scaffold. This concludes the ADVANCE phase and requires a return to the GENERATE phase.

Again examining the work tasks that can commence, notice that only one work task can be started. Work task 6 is a NORMAL work task and can start as soon as a unit from 5 arrives. At TNOW (SIM TIME = 4.0), no other work tasks can start. Therefore, a random delay time is generated for NORMAL work task 6. The random delay generated is 4.5. This added to the TNOW value yields an E.E.T. for work task 6 of 8.5. This is recorded in the EVENT list.

Returning to the ADVANCE phase, it can be seen that of the two scheduled events that have not been transferred, the earlier of the two is E.E.T. = 7.9 associated with work task 2. Transferring this event to the CHRONOLOGICAL list, place a check in the TRANSFER column of the EVENT list, and advance the clock to a SIM TIME value of 7.9. That is, TNOW is 7.9. Now terminate work task 2 and allow its flow units to pass forward to the appropriate elements. The stack marker passes to QUEUE node 3, indicating that a packet is now available on the scaffold, and the laborer returns to his idle position at 1 pending redeployment.

If a snapshot of the system and the flow unit positions was made at this point, it would appear as shown in Figure 7.7. No units are being either resupplied (ACT 2) or picked up (ACT 5). The laborer is idle at QUEUE node 1. Two stack positions are empty at QUEUE node 4. That is, a packet is available at QUEUE node 3, while one mason is placing brick in NORMAL work task 6. This is summarized in Table 7.4.

The discrete time jumps (clock advances) that have taken place to move the system to this configuration are summarized in the TIME LINE diagram shown below the model in Figure 7.7. Three discrete jumps have occurred. In jump one the clock is advanced from 0 to 3.0. Then, the clock moved to 4.0 and finally, it is advanced to the present TNOW value of 7.9. This can also be determined by consulting the SIM TIME column of the CHRONOLOGI-

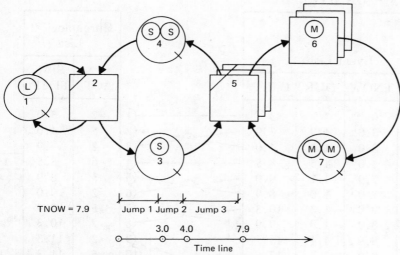

FIGURE 7.7   System status at TNOW = 7.9.

cal list. This column always shows the sequence of discrete jumps used to move from $t = 0$ to TNOW. The TNOW value is the last entry in the CHRONOLOGICAL list. At this point the EVENT list indicates that only one event is scheduled to terminate. This verifies the fact that only one unit is in transit. The only work task containing a transiting flow unit is work task 6, "MASON LAYS BRICK."

Continuing the simulation through three complete mason cycles results in the EVENT and CHRONOLOGICAL lists, as shown in Table 7.5. The system status as of TNOW = 29 is shown in Figure 7.8. By examining the EVENT list, it can be seen that three events are scheduled. Two bricklaying work tasks (6) are in progress. COMBI two will end at 32.0. Furthermore, it can be

Table 7.4   FLOW UNIT POSITIONS AT TNOW = 7.9

Element	Flow units at TNOW = 7.9
QUEUE node 1	One laborer unit
COMBI 2	None
QUEUE node 3	One stack unit
QUEUE node 4	Two stack units
COMBI 5	None
NORMAL 6	One mason unit
QUEUE node 7	Two mason units

Table 7.5 EVENT CHRONOLOGICAL LISTS FOR THREE MASON CYCLES

Transfer	Event List					Chronological List	
	ACT	TNOW	DUR	E.E.T.		ACT	SIM TIME
√	2	0	3	3	①→	2	3.0
√	2	3.0	4.9	7.9	②	5	4.0
√	5	3.0	1.0	4.0	③	2	7.9
√	6	4.0	4.5	8.5	④	6	8.5	1
√	2	7.9	1.1	9.0	⑤	5	8.9
√	5	7.9	1.0	8.9	⑥	2	9.0
√	2	9.0	1.3	10.3	⑦	5	9.9
√	5	8.9	1.0	9.9	⑧	2	10.3
√	6	8.9	4.1	13	⑨	2	11.3
√	6	9.9	7.7	17.6	⑩	5	11.3
√	2	10.3	1.0	11.3	⑪	2	11.8
√	5	10.3	1.0	11.3	⑫	2	12.3
√	2	11.3	0.5	11.8	⑬	6	13.0	2
√	6	11.3	7.0	18.3	⑭	5	14.0
√	2	11.8	0.5	12.3	⑮	2	14.4
√	5	13.0	1.0	14.0	⑯	6	17.6	3
√	2	14.0	0.4	14.4	⑰	6	18.3	4
√	6	14.0	8.0	22	⑱	5	18.6
√	5	17.6	1.0	18.6	⑲	5	19.3
√	5	18.3	1.0	19.3	⑳	2	21.4
√	2	18.6	2.8	21.4	㉑	6	22.0	5
√	6	18.6	3.5	22.1	㉒	6	22.1	6
√	6	19.3	3.9	23.2	㉓	5	23.0
√	2	21.4	2.6	24.0	㉔	5	23.1
√	5	22.0	1.0	23.0	㉕	6	23.2	7
√	5	22.1	1.0	23.1	㉖	2	24.0
√	6	23.0	6.0	29.0	㉗	5	25.0
√	6	23.1	4.7	27.8	㉘	2	25.5
√	5	24.0	1.0	25.0	㉙	2	27.5
√	2	24.0	1.5	25.5	㉚	6	27.8	8
	6	25.0	7.5	32.5	㉛	5	28.8
√	2	25.5	2.0	27.5	㉜	6	29.0	9
	2	27.5	4.5	32.0			
√	5	27.8	1.0	28.8			
	6	28.8	6.0	34.8			

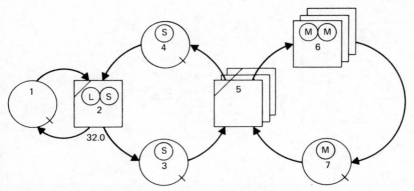

FIGURE 7.8 System status at TNOW = 29.0.

established that another event will be scheduled during the GENERATE phase, since a mason is idle and a packet is available at QUEUE node 3. Therefore, before the clock is advanced from 29.0, a "PICK UP" work task (5) E.E.T. can be generated and added to the EVENT list.

It is also possible to make a rough estimate of the system productivity to this point. Since three mason cycles have been completed, the number of bricks placed to this point is estimated as:

$$3 \text{ masons} \times 3 \text{ cycles} \times 10 \text{ bricks per cycle} = 90 \text{ bricks}$$

This has consumed 29.0 minutes and the projected hourly production rate of the system is:

$$\frac{60 \text{ minutes per hour}}{29 \text{ minutes}} \times 90 \text{ bricks} = 186.2 \text{ bricks per hour}$$

This estimate of productivity is influenced by the fact that the system is still in a transient stage* and has not reached a steady-state level of operation.

An estimate of the length of time that units were waiting in each of the QUEUE nodes can also be determined. By investigating the EVENT list, it can be seen that during the periods 12.3–14.0 and 14.4–18.6, nothing was scheduled at 2. Therefore, the laborer was idle. The fact that work task 2 was busy except for these periods is established by investigating all TNOW and E.E.T. values associated with work task 2. These are excerpted and shown in Figure 7.9. There is a continuous pattern of activity at COMBI work task 2 except for the periods noted. This means that during the first 29.0 minutes of the system simulation, the laborer was idle 5.9 minutes, or 20.34% of the time. Similarly, the mason idle time can be established by examining the schedule

*This was discussed in Section 4.9.

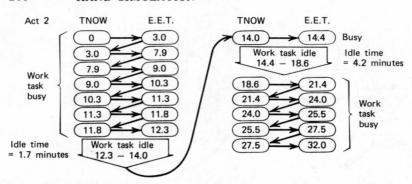

FIGURE 7.9   Schedule pattern for COMBI work task 2.

patterns of activities 5 and 6. By examining Figure 7.10, it can be seen that the only time at which all three masons are idle is during the first 3.0 minutes.

7.5 HAND SIMULATION STATISTICS

Use of the schedule pattern in determining the idle time for QUEUE nodes is cumbersome. It is also desirable to be able to measure other statistics indicative of the system's overall response and performance. In order to do this it is necessary to expand the hand simulation algorithm to encompass the concept of statistics collection.

The simulation clock has been discussed in the context of its functions as controller of system time. It is now useful to define additional clocks that are used for measuring system time statistics. These clocks are associated with each flow unit and travel through the system with their assigned unit. They are essentially stopwatches that can be switched on and off to measure flow unit transit time between selected points in the system. The flow unit clocks are designed to be started as their associated flow unit passes one point in the system and then stopped at a later point. The measured time is then recorded and maintained for statistical purposes. Following this, the clock is reset to zero. The modeler defines the points at which the unit clocks are to be switched on and off. These clocks can be referred to as STAT clocks. As many STAT clocks can be associated with each flow unit as desired. That is, each individual flow unit can be thought of as carrying around as many STAT clocks as are required. However, for discussions at this time, assume that each unit has only one associated STAT clock.

The STAT clock concept can be demonstrated by using it to determine the idleness of the laborer in the mason problem discussed above. In order to determine how much time the laborer spends in idle state 1, place a switch to turn on the STAT clock associated with the laborer at the time that unit enters QUEUE node 1 going to COMBI work task 2. Each time it is switched

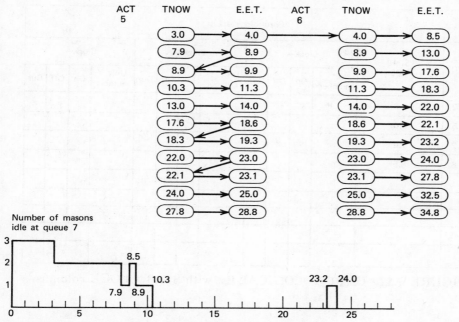

FIGURE 7.10   Schedule pattern of work tasks 5 and 6.

off, the time will be recorded and the clock reset to zero. The CLOCK ON and CLOCK OFF switches are located as shown in Figure 7.11.

In order to implement the action of these switches it is necessary to modify the structure of the Chronological list to include columns for recording STAT Clock statistics. The new format of the CHRONOLOGICAL list is shown in Figure 7.12. Three "CLOCK" columns have been added to handle the pos-

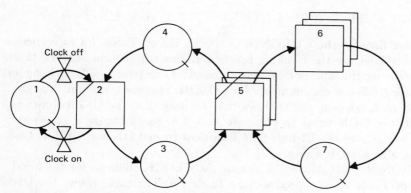

FIGURE 7.11   STAT CLOCK switches.

A C T	SIM TIME	Clock—				Clock—				Clock—			
		Unit number	On	Off	Dur	Unit number	On	Off	Dur	Unit number	On	Off	Dur
2	3.0	L1	√	√	0								
5.	4.0												
2	7.9	L1	√	√	0								
6	8.5												

Chronological List

FIGURE 7.12   CHRONOLOGICAL list with STAT CLOCK columns.

sibility that up to three clocks might be simultaneously turned ON or OFF (i.e., at the same SIM TIME). In each CLOCK column, subcolumns to record the flow unit number with which the clock activated is associated, ON and OFF, and the elapsed time (DUR) between switch ON and switch OFF are defined. When a unit passes the CLOCK ON switch, its number is recorded in the UNIT column, and a check is placed in the ON column. When the unit passes the OFF switch, a check is placed in the OFF column, and the DUR is calculated and recorded. The DUR is simply the difference between the SIM TIME at which the unit trips the OFF switch minus the SIM TIME at which the ON switch was activated; that is:

$$DUR = E.E.T. \ (OFF, \ UNIT \ N) - E.E.T. \ (ON, \ UNIT \ N)$$

The first four lines of the CHRONOLOGICAL list of Table 7.5 have been transferred to illustrate the clocking procedure. The completion of work task 2 ($t = 3.0$) releases the laborer flow unit (labeled L1) to pass to QUEUE node 1. Since the ingredience requirements for COMBI 2 are met, the unit transits immediately to 2, tripping the OFF switch, To note this, the OFF column is checked and the DUR value in this case is 0. The same sequence of events occurs at $t = 7.9$, and the L1 unit STAT clock is turned ON and OFF at $t = 7.9$, yielding a zero DUR value.

The CHRONOLOGICAL list to include the CLOCK columns for the hand simulation of Table 7.5 is reproduced as Table 7.6. The table shows the ON and OFF switching for the L1 clock throughout the 29.0 minutes of the hand simulation. Examination of the CLOCK column verifies the same results obtained using the schedule pattern. The STAT clock for L1 is tripped ON and immediately OFF for all transits through QUEUE node 1 with two exceptions. These exceptions at $t = 12.3$ and $t = 14.4$ result in DUR values of 1.7 and 4.2.

Table 7.6 CHRONOLOGICAL LIST WITH CLOCK COLUMNS

		Chronological List										
ACT	SIM TIME	CLOCK_____				CLOCK_____				CLOCK_____		
		UNIT	ON	OFF	DUR	UNIT	ON	OFF	DUR	UNIT	ON	OFF	DUR
2	3.0	L1	√	√	0								
5	4.0												
2	7.9	L1	√	√	0								
6	8.5												
5	8.9												
5	9.0												
5	9.9												
2	10.3	L1	√	√	0								
2	11.3	L1	√	√	0								
5	11.3												
2	11.8	L1	√	√	0								
2	12.3	L1	√										
6	13.0												
5	14.0	L1		√	1.7								
2	14.4	L1	√										
6	17.6												
6	18.3												
5	18.6	L1		√	4.2								
5	19.3												
2	21.4	L1	√	√	0								
6	22.0												
6	22.1												
5	23.0												
5	23.1												
6	23.2												
2	24.0	L1	√	√	0								
5	25.0												
2	25.5	L1	√	√	0								
2	27.5	L1	√	√	0								
6	27.8												
5	28.8												
6	29.0												

Total laborer idleness = 5.9 minutes

7.6 HAND SIMULATION ALGORITHM WITH STATISTICS

Implementation of the STAT clock concept leads to the addition of some steps to the hand simulation algorithm presented previously in Table 7.1. Each time units move in the system, a check must be made to determine whether any unit clocks are switched ON or OFF. This means that it is possible that clocks are tripped ON or OFF at points in the GENERATION phase and the ADVANCE phase. Following step 3 in the procedure in Table 7.1, a check should be made to see if the movement of the flow unit causes the tripping of any STAT clock switches. Similarly, following step 12, at which time units are released to flow to following elements, a check of the unit movement to establish whether STAT clocks are turned ON or OFF should be made.

The revised algorithm for hand simulation now contains 14 steps and is shown in Table 7.7. A flow diagram for the revised procedure is shown as Figure 7.13. In order to understand the new algorithm, consider the first few steps of the original simulation as presented in Section 7.4 to see how its operation implements the functioning of STAT clock switches placed around the idle state 1 of the masonry problem.

The first activity to be commenced is work task 2, and the units are moved from QUEUE nodes 1 and 4. This would trip the laborer unit STAT clock to the OFF position, but this is not relevant, since it was never turned on. The time generated is 3.0 and the E.E.T. for COMBI 2 is 3.0. This is recorded, and the algorithm moves to the ADVANCE phase.

The E.E.T. of 3.0 is transferred to the CHRONOLOGICAL list and the SIM TIME is advanced to 3.0. This results in the termination of COMBI 2 and the release of the laborer and stack unit. In its return transit to QUEUE node 1, the laborer passes the CLOCK ON position, the unit number, L1, is recorded, and the ON column checked.

Returning to the GENERATE phase, the COMBI 2 work task can be started again. The laborer unit moves to the LOAD SCAFFOLD work task, passing the CLOCK OFF position and tripping L1 STAT clock OFF. A check is placed in the OFF column, and the DURATION value is calculated as zero. This happens because the SIM TIME value has not advanced since the L1 STAT clock was switched on.

7.7 HAND SIMULATION OF GEN-CON COMBINATIONS

The operation of the GEN and CON function during hand simulation introduces an added degree of complexity. In hand simulation, consider only GEN-CON combinations in which the N value associated with the GEN element (i.e., QUEUE Node) and the CON element (i.e., FUNCTION Node) are equal. In discussing GEN-CON structures in Chapter Four, a typical case of this type was presented. In this instance, a truck unit was subdivided

Table 7.7   REVISED HAND SIMULATION ALGORITHM

GENERATION PHASE	1.	CAN ANY WORK TASK START (i.e., are ingredience conditions met)?
	2.	IF YES, continue; if NO, go to step 9.
GENERATE PHASE	3.	MOVE units that can begin transit of a WORK TASK into WORK TASK element.
	4.	CHECK to see if movement of the flow units has caused any clocks to be switched ON, record the unit number (or label), and place a check mark on the ON subcolumn of the CLOCK column of the CHRONOLOGICAL list. If a unit STAT clock has been switched OFF, place a check in the off column, locate the SIM TIME at which the clock was turned on, compute and record the DUR value.
	5.	GENERATE duration of transit time (i.e., transit or processing delay).
	6.	CALCULATE scheduled termination of activity (E.E.T. = TNOW + DUR).
	7.	RECORD E.E.T. in EVENT List.
	8.	Return to step 1.
ADVANCE PHASE	9.	TRANSFER next earliest scheduled event on the EVENT list to the CHRONOLOGICAL list. (Cross out entry on EVENT LIST).
	10.	ADVANCE Sim Clock to time of transferred event.
	11.	TERMINATE work tasks with associated E.E.T.
	12.	RELEASE unit(s) to following elements.
	13.	CHECK STAT clock switches (Same as step 4).
	14.	Return to step 1.

into N load units (N = 5) and then reconsolidated into a truck unit for further transit. This situation is shown again in Figure 7.14.

Essentially, the hand simulation of a GEN-CON combination is similar to that of the STAT clocks. Two checkpoints are involved. At the point that a flow unit enters the GEN QUEUE node, a switch establishing an account of N items to be generated must be established. At a point just prior to the CON FUNCTION node, a second switch is established that decrements or reduces

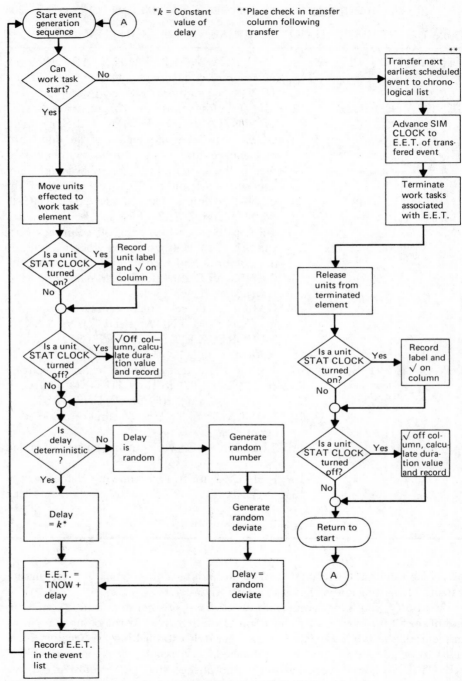

FIGURE 7.13 Algorithm with STAT CLOCKS.

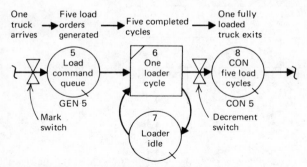

FIGURE 7.14   GEN-CON segment.

by one the account established at the first switch until the account is reduced to zero (i.e., $N - N = 0$). At this point, the FUNCTION node is realized, and a single flow unit is released. Typically, a switch establishing the GEN account is located between the GEN QUEUE node and the element preceding it. The decrementing switch is located between the processing work task (i.e., the COMBI) and the CON FUNCTION node. These are shown in Figure 7.14 as the MARK switch (establishing the account) and the DECREMENT switch.

In order to accommodate this switching scheme, one of the STAT CLOCK columns of the CHRONOLOGICAL LIST is relabeled and its subcolumns redesignated. This is shown in Figure 7.15. As with the STAT clocks, all actions must be associated with the commencement and completion of an active element. No actions are tied to a QUEUE node, since QUEUE nodes are not recorded as entries in the CHRONOLOGICAL LIST. As an active element (i.e., COMBI, NORMAL, FUNCTION, or ACCUMULATOR) preceding a GEN QUEUE node is completed and its E.E.T. is transferred to the CHRON LIST, an entry is made in the MARK subcolumn of Figure 7.15. This entry is of the form Nn, where:

N is the integer value associated with the GEN-CON combinations

and   n is a lowercase letter of the alphabet used for control purposes.

In Figure 7.15, at time 3.0 the entry "5a" is made in the MARK column; indicating that ACT 4 preceding GEN QUEUE node 5 is completed and five units (Group a) are generated. As units trip the DECREMENT switch between COMBI 6 and the FUNCTION node 8, decrement entries are made in the DEC column. This is also illustrated in Figure 7.15. The form of this entry is 1n, where the n again indicates the control account against which the decrement should be made. In this case the decrement is against 5a, so the entry is 1a. In the REM column, a running account of the number units remaining to

A C T	SIM TIME	GEN – CON I				Clock–			
		Mark	Dec	Off	Rem	Unit	On	Off	Dur
4	3.0	5a							
6			1a		4a				
6			1a		3a				
6			1a		2a				
6			1a		1a				
6			1a	√	0				

Chronological List

FIGURE 7.15   Modified CHRON list.

be decremented is kept. When this REM value reaches zero, a check is placed in the OFF subcolumn, and the FUNCTION node is realized. At this point, a single unit is released to the element following the FUNCTION node.

All actions relating to the GEN-CON combination take place at the completion of a work task. Therefore, modification of the hand simulation algorithm requires the addition of some actions in the ADVANCE phase. Step 11 in Table 7.7 is expanded to read as follows.

11. TERMINATE work tasks with associated E.E.T. If the work task is followed by a MARK switch, establish an account for the following GEN QUEUE node. If the work task is followed by a DECREMENT switch, make a decrement entry and update REM column. If REM value drops to zero, TERMINATE CON FUNCTION node.

The revised algorithm flowchart to include the GEN-CON function operation is shown in Figure 7.16.

Consider this extended algorithm within the context of a simple dual cycle system in which a truck arrives, generates five load commands to a loader, the loads are consolidated, and the truck departs to the fill area. This system is shown in Figure 7.17. For simplicity assume that the cumulative distribution associated with the "BACK CYCLE TIME" (element 5) is the same as

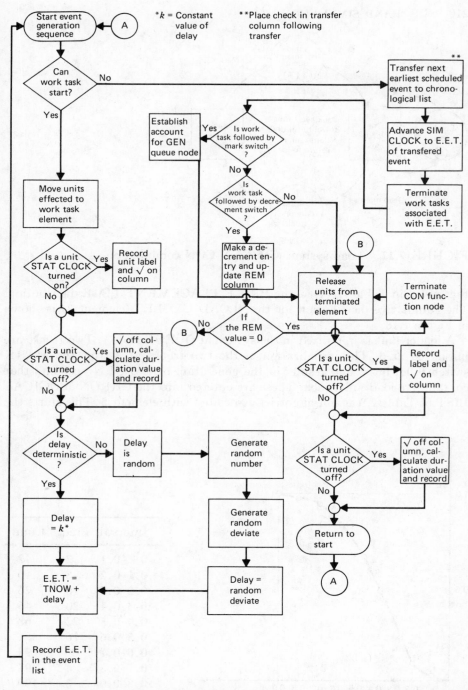

FIGURE 7.16   Algorithm with STATCLOCKS and GEN-CON FUNCTION.

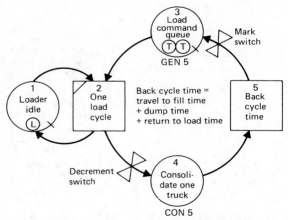

FIGURE 7.17.   Load system with GEN-CON combination.

that given in Table 7.2(a) as the BRICK PLACEMENT TIME distribution. The cumulative distribution for the "LOAD CYCLE" (element 2) is shown in Figure 7.18.

A loader unit is initialized in the system at QUEUE node 1. Two trucks are initialized at 3. The initialization of the two trucks at 3 trips the MARK switch following 5 and results in the generation of two GEN accounts that contain five load units each. These are entered into the CHRONOLOGICAL LIST of Table 7.8 as initial entries associated with element 5. This starts the

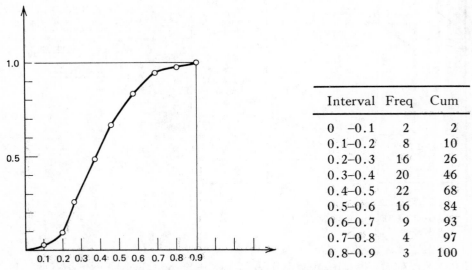

Interval	Freq	Cum
0 −0.1	2	2
0.1–0.2	8	10
0.2–0.3	16	26
0.3–0.4	20	46
0.4–0.5	22	68
0.5–0.6	16	84
0.6–0.7	9	93
0.7–0.8	4	97
0.8–0.9	3	100

FIGURE 7.18   "LOAD CYCLE" cumulative distribution.

Table 7.8   EVENT-CHRON LISTS FOR GEN-CON FUNCTION
            PROBLEMS

		Chronological List			
		GEN-CON I			
ACT	SIM TIME	MARK	DEC	OFF	REM
5	0.0	5a			
5	0.0	5b			
2	0.2		1a		4a
2	0.45		1a		3a
2	1.35		1a		2a
2	1.93		1a		1a
2	2.37		1a	√	o
2	3.00		1b		4b
2	3.38		1b		3b
2	3.91		1b		2b
2	4.39		1b		1b
2	5.00		1b	√	o
5	6.87	5a			
2	7.2		1a		4a
2	7.87		1a		3a
2	8.39		1a		2a

		Event List		
Transfer	ACT	TNOW	DUR	E.E.T.
√	2	0	0.2	0.2
√	2	0.2	0.25	0.45
√	2	0.45	0.90	1.35
√	2	1.35	0.58	1.93
√	2	1.93	0.44	2.37
√	5	2.37	4.5	6.87
√	2	2.37	0.63	3.00
√	2	3.00	0.38	3.38
√	2	3.38	0.53	3.91
√	2	3.91	0.48	4.39
√	2	4.39	0.61	5.00
	5	5.00	6.12	11.12
√	2	6.87	0.33	7.2
√	2	7.2	0.67	7.87
√	2	7.87	0.52	8.39

decrementing for accounts 5a and 5b. The hand simulation to include two GEN-CON sequences and the start of a third is shown in Table 7.8. By consulting the OFF column of the CHRONOLOGICAL LIST, it can be established that it takes 2.37 minutes to load the first truck and 2.63 minutes (5.00 − 2.37) to load the second. In each case, five cycles of the loader were required for each truck loading.

7.8 SYSTEM CHECKOUT USING HAND SIMULATION

As can be seen from the simple example in the previous section, the use of hand simulation in systems containing GEN-CON combinations can become very tedious. For this reason, these functions are normally introduced later during computer simulation, after the system has been checked out using hand simulation. Despite the fact that the system will ultimately require the use of GENERATE and CONSOLIDATE functions to establish compatibility between cycles or for some other reason, the system is initially checked without these functions. This procedure of checking logic by hand simulation is analogous to conducting a pressure check on the plumbing of a building before placing it into service.   If the logic of a system is not properly defined, the simulation of the system will break down at some point. This usually occurs when ingredience requirements for some element cannot be met because of the lack of a logical connection (i.e., ARC). Hand simulation reveals such logical inconsistencies as well as giving the process designer an insight into the interaction of flow units within the system.

In order to illustrate this check-out procedure for a multilink system, consider a building construction problem. The complete modeling of this system requires the use of GEN-CON combinations to achieve compatibility between links. Information regarding the project is as follows.

PROJECT TITLE: PEACHTREE SUMMIT BUILDING
PROJECT LOCATION: DOWNTOWN ATLANTA, PEACHTREE STREET
PROCESS DESCRIPTION: The observed process involves the repetitive installation of floor slabs in the Peachtree Summit Project. This repetitive process occurs from floors 4 to 30. A single form is used for each 34-foot square bay. This results in the use of prefabricated formwork that can be reused from floor to floor. The movement of these forms, the preparation of the forms for concrete placement, and the delivery of concrete to the forms are examined in this analysis.

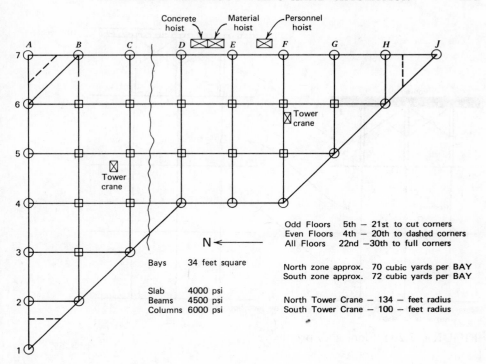

FIGURE 7.19   Site layout.

Drawings showing the site layout and appropriate elevations are given in Figure 7.19 and 7.20.

The steps involved in the forming and concreting process are as follows.

1. A 2 cubic yard concrete hoist is fed by a 10 cubic yard concrete truck. Hoist cycles are required to empty the truck.

2. The concrete hoist is lifted to the floor level under construction. A 2 cubic yard hopper is filled with concrete from the hoist. The hoist returns to the ground for another load from the truck.

3. A ¼ cubic yard buggy is filled from the hopper. Eight buggies are required to empty the hopper.

4. The buggies carry concrete to the formed bay, where it is poured and finished.

5. After curing has taken place, forms are "flown" to the next floor using the crane. Forms are placed for the next pour.

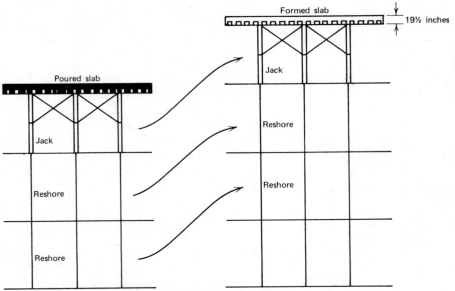

Note: Slab form keeps its same relative position on each floor, fits
      into slots on slab.

FIGURE 7.20   Floor slab profile.

This process is shown in CYCLONE system format in Figure 7.21. It introduces several interesting requirements for the use of GEN-CON combinations.

First, the arriving concrete truck must be broken into five hoist loads. This
is accomplished using a GEN-CON combination with N value of 5 at elements
4, 5, and 6. This results in five cycles of the hoist cycle for each single cycle
of the truck cycle.

Each hoist load (2 C.Y.) is lifted to a 2 C.Y. temporary storage hopper.
Here each 2 C.Y. hopper load is further broken into eight individual buggy
loads, which are carried to the pour site. This subdivision is also handled using
a GEN-CON combination. At this point we have converted the original truck
load of concrete into 40 (5 × 8) buggy loads.

Since each bay pour takes approximately 70 C.Y. of concrete, the total
bay pour requires 280 buggy loads (70 C.Y. ÷ ¼ C.Y. per buggy). Certain
actions must be initiated when the pour is complete. In order to note the completion of the CONCRETE POUR, a CON 280 element is inserted following
COMBI 21. Similarly, in order to generate the demand for 280 buggy loads
once the forms are ready for pour commencement, a GEN 280 is associated
with element 29. This insures that the buggy will cycle 280 times.

Certain control mechanisms have been inserted to insure that the concrete
delivery is coordinated with demand. A feedback loop is inserted from element

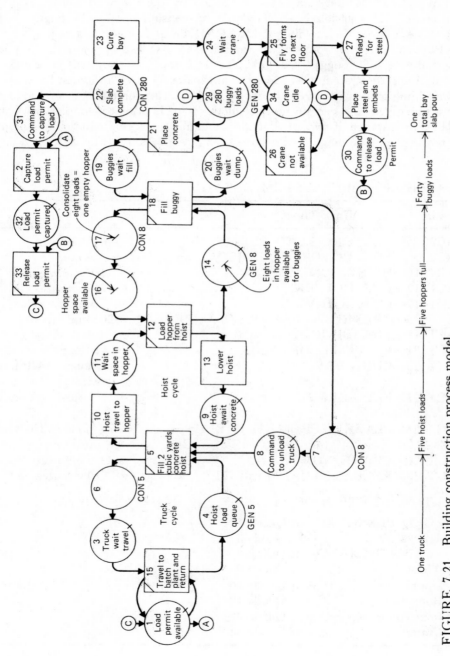

FIGURE 7.21   Building construction process model.

18 to COMBI 5. This insures that a concrete truck does not begin to unload until the 2. C.Y. (eight buggies) unloaded by the previous truck has been exhausted. This action is implemented by consolidating eight buggy loads at element 7 and then issuing an UNLOAD TRUCK command at 8.

Furthermore, the trucks on the haul cycle must be stopped once the pour is complete (280 buggy loads are placed). Otherwise they will continue to haul, filling the system with concrete that is not required and must be wasted. Depending on how many trucks are on the haul, this command to stop hauling must be given to a certain number of truck loads prior to completion of the slab. For instance, if two truck loads are hauling, the command can be issued after 64 C.Y. (256 buggy loads) have been placed. Assume that at least 2 C.Y.

TABLE 7.9   BASIC SYSTEM DATA

Estimated Work Task Times

ACT 2 —CAPTURE LOAD PERMIT	— 0.0 minutes
ACT 5 —FILL CONCRETE HOIST	— 2.0 minutes
ACT 10—HOIST CONCRETE	— 1.5 minutes
ACT 12—FILL HOPPER	— 2.5 minutes
ACT 13—LOWER HOIST	— 1.0 minutes
ACT 15—TRAVEL TO PLANT, LOAD, RETURN	—20.0 minutes
ACT 18—FILL BUGGIES (1 each)	— 0.5 minutes
ACT 21—PLACE FLOOR SLAB	— 1.0 minutes
ACT 23—CURE BAY	—24 hours (1440 minutes)
ACT 25—FLY FORMS	— 4 hours (240 minutes)
ACT 28—PLACE STEEL AND IMBEDS	— 4 hours (240 minutes)
ACT 26—CRANE NOT AVAIL	—30.0 minutes
ACT 33—RELEASE LOAD PERMIT	— 0.0 minutes

INITIAL POSITION OF SYSTEM

1 LOAD PERMIT	—at Qnode 1
2 Trucks	—at Qnode 3
1 LOAD COMMAND	—at Qnode 8
1 Hoist	—at Qnode 9
1 Hopper	—at Qnode 16
6 Buggies	—at Qnode 19
1 Pour command	—at Qnode 29
1 Crane	—at Qnode 34

are in the system and 4 C.Y. arrive loading at the batch plant. Therefore, a stop hauling command can be issued. For simplicity, assume that the command is released after 280 buggy loads.

The implementation of the STOP ORDER is achieved using a capture mechanism that withdraws the load permit associated with COMBI 15. Mechanisms in general and capture mechanisms in particular will be discussed in Chapter Nine.

Table 7.10   INITIAL SYSTEM SIMULATION OF FLOOR SLAB INSTALLATION

Transfer	Event List				Chronological List	
	ACT	TNOW	DUR	EET	ACT	SIM TIME
√	15	0.0	20.0	20.0	15	20.0
√	26	0.0	30.0	30.0	5	22.0
√	5	20.0	2.0	22.0	10	23.5
√	15	20.0	20.0	40.0	22	26.0
√	10	22.0	1.5	23.5	13	27.0
√	12	23.5	2.5	26.0	26	30.0
√	13	26.0	1.0	27.0	18	30.0
√	18	26.0	4.0†	30.0		
√	26	30.0	30.0	60.0		
√	21	30.0	240.0*	270.0		
					21	270.0
√	2	270.0	0.0	270.0	2	270.0
√	23	270.0	1440.0	1710.0		
					23	1710.0
√	26	1700.0	30.0	1730.0	26	1730.0
√	25	1730.0	240.0	1970.0	25	1970.0
	26	1970.0	30.0	1800.0		
√	28	1970.0	240.0	2210.0		
					28	2210.0
					33	2210.0
√	33	2210.0	0.0	2210.0		
	15	2210.0	20.0	2230.0		

*The placement time for the 280 buggies is taken for the check as 240 minutes.
†The load time for eight buggies (8 × 0.5) has been used for this checkout.

In order to check this system for logical integrity, a hand simulation will be conducted. Since this exercise is designed to check only the consistency of the logical linkages in the system, the GEN-CON combinations will not be implemented. The CON FUNCTION nodes that follow COMBI elements will be disregarded, and multiple units will not be defined at the GEN QUEUE nodes. Deterministic durations will be used for simplicity. Basic data for the systems simulation are given in Table 7.9. Table 7.10 gives the initial entries in the system simulation.

PROBLEMS

1. Conduct a hand simulation of the simple masonry model shown in Figure 5.9. Use the transit times given in Table 7.2. Assume that five masons are supported by two laborers. Allow the system to cycle so that 10 brick placement sequences are completed. How does the system production achieved with hand simulation compare with the value predicted by deterministic methods?

2. Hand simulate the first three cycles of the masonry system shown in Figure 7.5, assuming that there are six stack locations on the scaffold and masons supported by two laborers. Compare the productivity achieved with that found in Problem 1 above.

3. Compare the impact (using hand simulation) of having all bricks stacked on the scaffold at the beginning of shift versus having no bricks on the

Table P7.4.1   REPAIR TIMES

Interval (Hours)	Frequency
0 −0.4	1
0.41–0.8	5
0.81–1.2	17
1.21–1.6	37
1.61–2.00	61
2.01–2.40	78
2.41–2.80	62
2.81–3.20	93
3.21–3.60	42
3.61–4.00	28
4.01–4.40	36
4.41–4.80	27
4.81–5.20	13

Table P7.4.2 RUNNING TIMES

Interval (Hours)	Frequency
0–2	4
2–4	19
4–6	47
6–8	79
8–10	76
10–12	67
12–14	54
14–16	39
16–18	34
18–20	26
20–22	20
22–24	15
24–26	10
26–28	7
28–30	3

scaffold, as in the system of Figure 7.5. Plot the production transients for each system (see Figures 4.11 and 10.10).

4. A number of machines are repaired by a maintenance team when they break down. After repair the machines continue to run until they break down again. The repair times are given in Table P7.4.1. The running times are given in Table P7.4.2. The machines are repaired in sequence in which they break down; that is, the machine that breaks down first is repaired first. Do a hand simulation of five cycles of this system.

5. Conduct a hand simulation of the tunnel model developed in Chapter Five (Figure 5.19). Assume the following deterministic times for the active states.

State	Description	Time in Minutes
ACT 2	Position pipe	20
ACT 3	Replace jacks and collar	25
ACT 5	Jack 6 feet forward	115
ACT 8	Reset jack rams	45
ACT 12	Remove jacking collar	35
ACT 18	Mole maintenance	90

Allow the system to cycle three times. What is the idle time associated with the tunneling machine at QUEUE node 19?

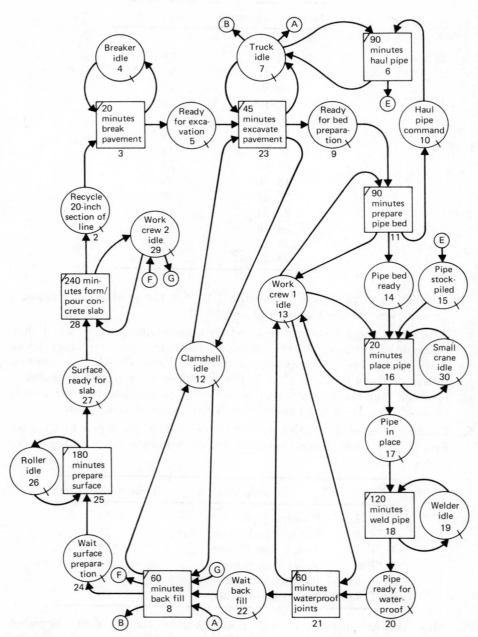

FIGURE P7.6 System Representation of sewerline job.

6. A solution of the sewer line problem (Problem 5.6) is given in Figure P7.6. Do a hand simulation of this model using deterministic times as shown in the figure and determine the percent time the crews at QUEUE nodes 13 and 29 are idle using STAT clocks.

7. The CYCLONE model for a material hoist that lifts concrete blocks to the upper floors of a hospital project is shown in Figure P7.7a. The hoist

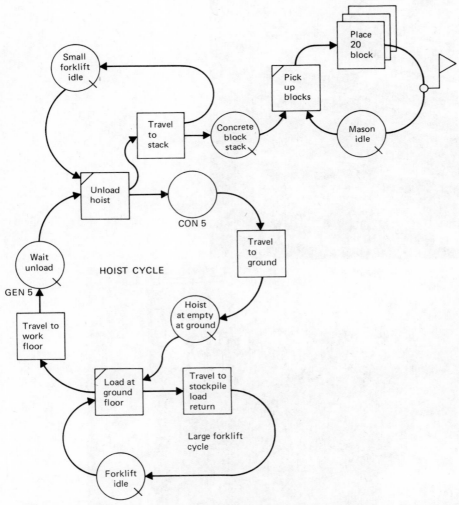

FIGURE P7.7   (*a*) Material hoist model. (*b*) Hoist travel. (*c*) Small forklift returns to hoist.

(b)

(c)

FIGURE P7.7   (*Continued*)

is loaded at the ground floor with a large fork lift and is unloaded at the work floor with a small hand-operated forklift. Five cycles of the small forklift are required to unload the material hoist. Figures 7.7b and 7.7c show pictures of the hoist and the small forklift in operation. Conduct a hand simulation to determine the hourly production of this system. Productivity is measured at the COUNTER in 20-block increments. The element transit times are as follows.

Large Forklift		Small Forklift	
Travel to Block Stockpile and Return			
Time in Minutes and Seconds		Travel to Stack	
3:30–3:45	2	0 –0:30	1
3:45–4:00	8	0:30–0:45	8
4:00–4:15	14	0:45–1:00	17
4:15–4:30	21	1:00–1:15	26
4:30–4:45	20	1:15–1:30	22
4:45–5:00	15	1:30–2:00	6
5:00–5:15	11		
5:15–5:30	9		

Hoist

Load Time

Ground Floor	Travel to Work Floor		Unload	Travel to Ground Floor	
Constant 1:30	0:15–0:20	4	Constant	0 –0:05	2
	0:20–0:25	10	1:00	0:05–0:10	9
	0:25–0:30	24		0:10–0:15	25
	0:30–0:35	26		0:15–0:20	28
	0:35–0:40	11		0:20–0:25	12
	0:40–0:45	5		0:25–0:30	4

Assume the pickup time for the mason is constant and equal to 1 minute. Assume the placement time per 20-block increment is twice the values given in Table 7.2.

CHAPTER EIGHT

Modeling Control Structures

A goal of good planning and field control in construction is the efficient use and smooth interaction of project resources. At the construction operation level this requires that the relevant man power, equipment, materials, and know-how are assembled in the right place at the right time. Any deficiency in supplying these resources will result in either a total hold-up delay in starting the operation or an inefficient limited progress of the operation. In either case, constraints arise affecting the productivity, duration, and other attributes of the construction operation involved.

Often the technology required for a construction operation is such that certain work tasks or processes must wait for other internal work tasks or for the release of specific resources assigned to the operation that are currently engaged elsewhere within the operation. In this way, the internal interaction of resources, work tasks, and processes becomes a controlling factor that influences the productivity and progression within the construction operation.

The modeling of these internal controlling logic structures is essential for any useful predictive model and is a requisite to a complete knowledge of the construction operation itself. This chapter introduces the reader to the modeling of control structures.

8.1 ORIGIN OF CONSTRAINTS IN CYCLONE

The very nature of the ingredience requirement has implicit in it the concept of constraints. The idea is that one or more flow units must wait for another unit before an activity can start. This constrains the flow on the waiting units until the awaited unit is available. This is typical of operations in the real world, as anyone who has ever used the expression "What's holding us up?" is well aware. It is common to have all units required for a certain operation except for one or two. Thus it can be said that the units not available, but required, are constraining or "holding up" the starting of some work task.

Two aspects of state network models assist in developing constraints in correspondence to the real-world situation. First, the flow units themselves, in conjunction with the ingredience constrained work task concept, operate to constrain the flow of other units. Second, the network topology or structure dictates to which state or work task a required flow unit can be assigned at a given time. This results in the capture of the flow unit for processing and effectively withholds it from other work task sequences until it is released to them. In this way the network topology adds an additional dimension to the constraining power provided by key flow units.

An example of a typical flow unit constraint situation is shown in Figure 8.1. The flow unit (unit A) is constrained to a single path in which it transits activities 1, 2, and 3 sequentially and then returns to begin the sequence again. This constrains the flow of units entering the waiting state X, since unit A must return before these incoming units can transit the sequence.

Thus, once unit A starts work task 1, it cannot further provide access to this stretch of network (for units at X) until it has passed work tasks 1, 2, and 3 and thus is released. The initial requirement for A at ingredience-constrained work task 1 makes unit A a controlling unit. The magnitude of this constraint is a function of the path that A must follow in returning to its idle state.

Another illustration of the topological influence of a network on control unit constraint is shown in Figure 8.2. Here a single control flow unit A is available at any one time for either the high-priority process (work task 1) or the low-priority process (work tasks 2 and 3). The behavior of the network for units arriving at QUEUE nodes X and Y depends on the state and location of the control flow unit A.

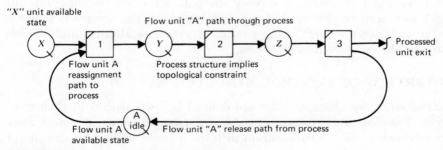

FIGURE 8.1   A control unit constraint.

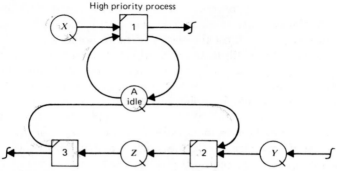

FIGURE 8.2 A control unit constraint with topological (i.e., process) priorities. Delay to high-priority process X units dependent on current location of control unit A.

Thus, for example, if the flow unit A is already at work task 2 in the low-priority process when a unit arrives at X, it must also complete 3 before returning to "A IDLE." Then, based on the QUEUE node routing rule (section 4.4), it will be routed into work task 1 to commence the high-priority process. If, however, flow unit A is idle at A IDLE, it will be captured by the first process that is initiated by a unit arriving at queue nodes X or Y. If simultaneous arrivals occur, the labeling rule will ensure that the high-priority process is serviced first.

A typical example of this occurs when a sequence of lower-priority tasks that must be done in sequence are started and thus delay the start of a higher-priority task. This happens when forming is not ready on a critical or progress-constraining building component such as a column or a beam and the pouring crew is diverted to start pouring floor deck. Even though the forming of the structural frame is completed before completion of the floor deck pour sequence, it must wait until the deck work is completed, since breaking that sequence would be inefficient and would lead to unfinished concrete. A model of this situation is shown in Figure 8.3. Since the path structure (topology) of the network does lead to this kind of constraining effect, the process designer is forced to consider the effect of the routing structure established and develop a policy on the movement of resources.

8.2 OVERVIEW OF CONTROL STRUCTURES

Control structure characteristics are defined by two aspects of their composition. The flow unit that acts to constrain system flow at one or more points is a *control unit*. In conjunction with the ingredience-constraint concept associated with COMBI activities, control units can cause delays in the

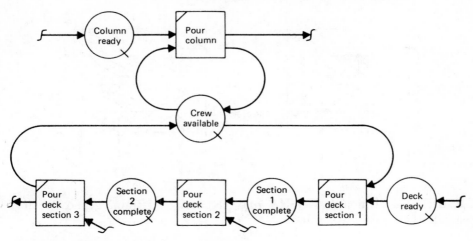

FIGURE 8.3   A concrete crew constraint.

system at various points.   In general, such units can be categorized (according to the real-world units or entities that they model) as:

1.   Physical or resource unit constraints (e.g., a crane or crew).
2.   Nonphysical constraints (e.g., an inspection or permit).

A special subcategory of nonphysical constraints develops because of the need to implement managerial decisions in a given model. For example, suppose a section of pipe in a nuclear power plant must be inspected using nondestructive testing methods before it can be incorporated into the construction. The requirement, although it is a managerial or safety consideration, should be included in the model. Other constraints of this type develop because of union rules and other considerations (e.g., safety and space constraints). In the erection of steel framing, for instance, the columns are often fabricated so that three tiers or floors of construction can be framed at one time. Union rules as well as OSHA* provisions, however, prohibit the framing work to proceed more than two tiers ahead of the decking or floor installation. This is so that a worker falling from the frame will not fall more than two tiers or floors. This is a nonphysical constraint that must be included in any model of the steel erection process. The rule is shown schematically in Figure 8.4. Numerous other examples of nonphysical constraints exist.

In general, the physical control units are obvious, since they are pieces of equipment or labor units that are clearly contributive to the system productivity and easily identified. They can be considered control units since, at

*Occupational Safety and Health Act.

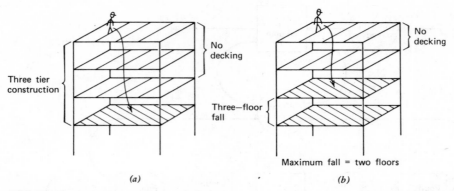

FIGURE 8.4   Union and OSHA rule on phasing of framing and decking. (*a*) Not allowed. (*b*) Corrected.

various points in their flow paths, they cause delays and thus constrain or control the movement of other units. Other physical units are less easily identified. Space units such as those referred to in Figure 5.3 when discussing the curing tunnel are not production units per se, and they tend to be over-looked sometimes. Access to a work site is a common constraint encountered in construction work. This is a form of space unit that controls how many productive units (e.g., carpenters, trucks, and scrapers,) can exit or enter a given work site. The work site itself may be small and may constrain the num-ber of workers active at any given time. This again leads to a space constraint that can be modeled using a space unit in a fashion similar to that used in the brick factory curing tunnel situation. Space is physical and therefore can be considered a physical flow unit. However, it is a nonproductive or passive flow unit in that it does not contribute to or accomplish work or placement of construction in the same sense that a carpenter or a bulldozer does. The nature of flow units, then, is one defining aspect of a control structure.

The second characteristic of a control structure that defines its impact on the system is the topology or paths associated with the flow of the unit through the system. Each control unit has a path or set of paths through which it must transit. This can be considered its associated path topology or path struc-ture. The complexity of the path structure associated with a given control unit defines the nature and complexity of the constraining influence of the control unit. The simplest path topology is that associated with the "slaved" unit discussed in Chapter Three. The slaved unit can be considered a control structure, since it constrains the passage of flow units through the COMBI activity with which it is associated. While a single slave unit is captured within the COMBI work task, arriving units at the processor are delayed until the slave unit is released.

In general the flow path topology of control units in work task sequences can be used as the basis for classifying control structures. Thus the following control structure classes are defined.

1. *Class I Structures: Sequential Loop.* The control unit traverses through a single sequential set of work tasks and then returns to complete its single loop.

2. *Class II Structures: Nonconsecutive Sequential Loop.* The control unit traverses through several independent sets of sequential work tasks before it returns to complete its single loop.

3. *Class III Structures: Butterfly or Multiloop.* The control unit can traverse repeatedly through independent sets of sequential work tasks and its path thus forms a butterfly or multiloop pattern with the control unit idle state as a focal point.

4. *Class IV Structures: Open Path.* The control unit is generated in one sequence of work tasks and then traverses along an open path to initialize another independent sequence of the work tasks. No looped control path exists.

5. *Class V Structures: Switches.* The control unit path depends on the status of a number of independent sequences of work tasks and it thus acts as a switch in the selection of possible paths.

6. *Class VI Structures: Parallel Processors.* Control units allow simultaneous processing of more than one processed unit.

The following sections will discuss the above class structures in detail.

8.3 CLASS I CONTROL STRUCTURES: THE SEQUENTIAL LOOP

Sequential Loops are paths associated with a given control unit in which the unit traverses a single closed cycle or loop and passes through every work task in a segment sequentially (i.e., one after another) without skipping or changing the order of any work task in the segment. The simplest example of a sequential loop is the slave unit structure. This Class I control structure consists of a single waiting state in which the control unit is idle and a single COMBI work task that constitutes the segment to be traversed and controlled. In general the segment to be controlled consists of several COMBI, NORMAL, or QUEUE node elements that are traversed in sequence. Following transit, the control unit returns to its idle state QUEUE node element and awaits another incoming unit to be accompanied through the cycle.

The structure in a general form is shown in Figure 8.5. The NODE elements can be divided into three subsets. One subset contains the idle state QUEUE node element reserved for the control unit. The second subset contains those QUEUE node elements associated with COMBI work tasks included in the

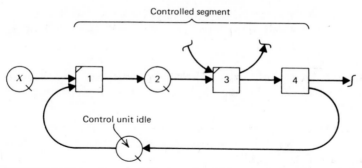

FIGURE 8.5   General Form of a Class I Control Structure. The sequential loop.

controlled segment. The final subset consists of the COMBI, NORMAL, and other elements in the controlled segment.

The controlled segment shown in Figure 8.5 (i.e., elements 1–4) consists of two COMBI work tasks (1 and 3), the QUEUE node element (2) associated with COMBI work task 3, and NORMAL work task 4. The minimum and maximum values for the sets defining a sequential loop structure are as follows.

Element Type or Category	Minimum Number of Elements	Maximum Number of Elements
COMBI work task	1	Unlimited (n)
Control unit idle state (QUEUE node)	1	1
QUEUE nodes associated with COMBI in controlled Segment	0	Unlimited ($n - 1$)
NORMAL work task	0	Unlimited

A Class I control structure constrains the number of units that may be in transit in the controlled segment at any given time. This is a function of the number of controlled units initially defined at the idle state QUEUE node when the system is initialized. In the case of the sequential loop in Figure 8.5, if one control unit is defined, then only one unit (other than the control unit) entering the system at QUEUE node X can be in transit in the controlled segment (elements 1–4) at a time. If 20 control units are initially defined, then the number of units at any one time in the controlled segment is constrained to be 20 or less.

The class I control structure is particularly handy in real-world systems in which units in a particular area are constrained because of space or other considerations. Again referring to the brick factory curing tunnel, the number of brick carts in the tunnel at any time is constrained by the number of cart spaces available. This is a space constrained version of the steam tunnel example used in Section 4.8. If the number of carts that can be positioned simultaneously in the tunnel when full is 50, then the constraint can be modeled with a sequential loop, as shown in Figure 8.6c. In the curing process as depicted, each brick cart (containing bricks for curing) is pulled into the curing tunnel and posi-

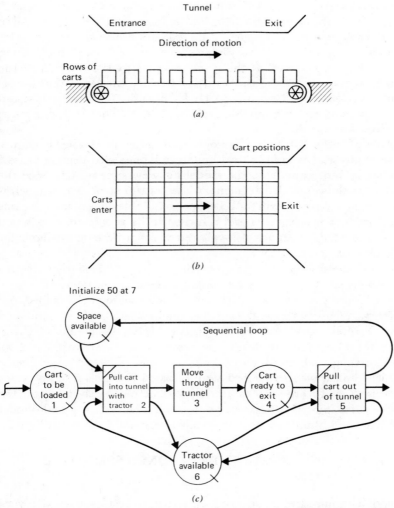

FIGURE 8.6   Brick curing tunnel model. (a) Elevation view of tunnel. (b) Plan view of tunnel. (c) CYCLONE model.

tioned using a small tractor. The cart then advances from the entrance to the exit of the tunnel on a large conveyor until it arrives at the exit. At this point, the tractor pulls it out of the tunnel. In the plan and elevation views shown in Figures 8.6a and 8.6b, the direction of motion is from left to right.

The Class I structure used to model this situation is shown in Figure 8.6c. The space control units cycle in the loop 7–2–3–4–5. To model the 50 tunnel positions available, 50 control units at 7 are initialized. The loop controls access to the tunnel segment (2–3–4–5) and insures that no more than 50 carts can be in the sequence 2–3–4–5 simultaneously. Each cart arriving at 1 must wait until both the tractor unit (6) and a space unit (7) are available before proceeding. When both supporting ingredients are available, the cart enters the tunnel, "capturing" the available space unit and carrying it along until it exists at 5. This insures that the space occupied by the entering cart is blocked until such time as it exits the tunnel. After the cart exits the tunnel, the space unit is released and returns to QUEUE node 7, where it is again available to accompany another unit through the tunnel. That is, the blocked space now becomes available again.

Any sequence of work tasks that is controlled through its length by the requirements for physical or nonphysical units can be modeled using a Class I structure. What is important from the modeler's viewpoint is that once the control unit enters the controlled segment of the control structure, it is withdrawn from the system and will not become available again until it has transited the length of the segment and returned to its idle state. It is like a boat that has entered the "tunnel of love" and will not become available for passengers again until it exits.

In some cases, it may be possible to reduce a multielement controlled segment to a single COMBI work task, thus reducing the sequential loop to a slave unit structure. For instance, if the cart tractor were not detached and reutilized after each load and unload, tunnel elements 2 to 5 could be combined into a single COMBI work task (e.g., "MOVE THROUGH TUNNEL"). However, this would completely delete the system constraint introduced by the tractor's shuttling back and forth to load and unload carts and produce a system in which the carts essentially load and unload themselves. This is not desirable in this case. In order to retain the interaction of the tractor, the tunnel, and the carts in correct perspective, the work tasks 2, 3, 4 and 5 must remain separated, and a multielement controlled segment must be utilized.

8.4 CLASS II CONTROL STRUCTURES: NONCONSECUTIVE SEQUENTIAL LOOPS

A slight modification of the sequential loop structure results when the work tasks in the controlled segment do not follow one another consecutively. This results when other work tasks that do not require the controlled unit intervene in the segment. This situation results in a single loop structure in which the work tasks linked by the control unit loop do not follow one another consecutively in the total systems model. This type of loop is called a Class II control structure. Figure 8.7 shows a typical nonconsecutive sequential loop structure. Such structures are useful when a unit controls flows at two or more nonconsecutive COMBI work tasks, but must itself maintain a sequentially of flow.

A common example of this situation occurs in the control of haul road systems in heavy construction when a certain stretch of the system is constricted to one-lane traffic. On such stretches, traffic can move only in one direction at a time. A flag system is used to control such systems. Traffic queueing at one end of the construction must wait until a flag (normally carried by the last truck or vehicle in a column) is passed from traffic moving in the opposite direction to the vehicles of the waiting column. In more modern systems, a temporary traffic light system is used in which the "flag" is replaced by the flash of a green light. Such systems can be modeled using a nonconsecutive sequential loop, as shown in Figure 8.8. Traffic arriving at QUEUE node 1 is consolidated into columns that await the arrival of the flag from the last vehicle in the column moving to the west. The flag cycles along path 2–5–4–6 in a sequential path that traverses nonconsecutive work tasks. This insures that the sequentiality of traffic movement from east to west to east, and so on, is maintained. That is, two columns cannot move to the east while nothing moves west. After work task 4, "TRANSIT TO THE WEST," is completed, the control entity (i.e., the flag unit) passes to 6, where it releases the column at 1. If no column is available at one, it waits until a column develops. Then work task 2, "TRANSIT CONSTRICTION," commences, capturing the flag unit. Following completion of 2, the flag passes to 5, where it releases the column waiting to go west (if any) or waits until a column develops.

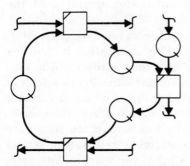

FIGURE 8.7   The nonconsecutive sequential loop.   General form of a Class II structure.

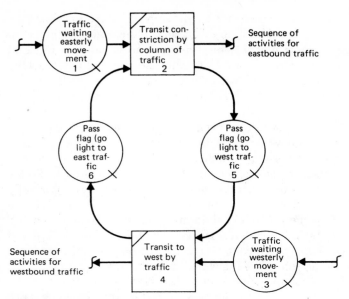

FIGURE 8.8   Single lane traffic control.

The flag system has its drawbacks, since quite a few vehicles can accumulate at 3, waiting to go west, while the flag remains idle at 6, waiting for an easterly moving column to develop. Since the passing of the flag depends on the accumulation and movement of a column, the system is nonautomatic and can overload in one direction or another. The operation of the green light as shown in this model is also not automatic. The assumption required to make this model correct is that the last vehicle of the moving column activates the green light for the standing or waiting column. In the next chapter, a capture mechanism will be developed that models the automatic operation of the traffic light on a preset or fixed transit time basis.

Another situation that can be modeled using nonconsecutive sequential loops is the movement of a ferry across a river. The assumption is that the ferry moves to one side of the river, unloads, and then waits until a load accumulates before returning to the far shore. A similar situation develops in a tunneling situation in the movement of mucking trains. The muck cars are loaded at the working face by conveyors that exhaust material from the tunneling machine. When the muck cars pulled by the locomotive are full, they travel out of the tunnel to a dump location where they release the tunnel material. If the dump operation requires either a human or machine resource, the locomotive must wait until that resource becomes available. Once the muck cars have been emptied, the locomotive and its train of cars return to the

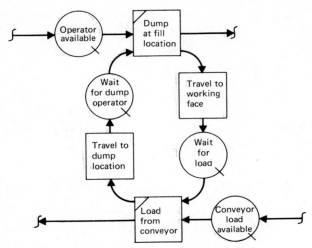

FIGURE 8.9   Operation of muck train.

working face to "muck out" again. A model describing this process might appear as shown in Figure 8.9. This model could also be used to show the operation of a man-hoist operating between levels of a high-rise building. As shown, the assumption would be that management policy requires the hoist to operate only when loaded. That is, no empty trips are allowed. The model structure would be the same as that shown for the muck train. Definition of the flow units required and their paths is left as an exercise for the student.

A crew that works various nonconsecutive stations in a sequence can also be modeled using a Class II structure. It must be noted the term nonconsecutive must be defined in terms of the mainline process and flow unit. That is, the stations are not consecutive in the mainline process, although when viewed from the level of the crew, they are both sequential and consecutive.

8.5 CLASS III CONTROL STRUCTURES: BUTTERFLY OR MULTILOOP

The first two classes of control structures are similar in that the path topology consists of a single closed loop. Class III structures are multiloop configurations in which two or more loops radiate from a central idle state QUEUE node. Upon completing a traverse of any loop, the control unit returns to the idle state for reassignment. Because of the winged appearance of the radiating loops, this type of structure is called a butterfly structure. The number of "wings" that a butterfly pattern may have, however, is not limited to two but is a function of the number of work tasks or work task sequences with which the control unit is associated. A typical butterfly structure consisting

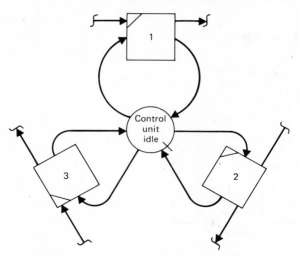

FIGURE 8.10   General form of a Class III structure: butterfly or multiloop.

of three loops is shown in Figure 8.10. The individual loops associated with a central idle state can also be categorized as simple loops, single-element loops or multielement loops. The loops shown in Figure 8.10 are single-element loops, since they pass through only one COMBI work task and return. Multielement loops traverse sequences of work tasks either consecutively or nonconsecutively. Thus the multielement loops of Class III structures can be further classified as either Class I or Class II loop structures. Figure 8.11 shows Class I and Class II multielement loops within a butterfly structure.

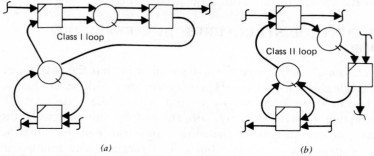

FIGURE 8.11   Class III structures with Class I and Class II multielement loops. (a) Butterfly with Class I multielement loop. (b) Butterfly with Class II multielement loop.

Butterfly structures are useful for modeling situations where a resource is shared between several work tasks or task sequences. Viewing this from the constraint or control point of view, the control unit constrains the flow of units through several work tasks or work task sequences. The multiloop topology of the butterfly structure leads to the need for the definition of a routing priority system since, if the control unit returns to the idle state and is required along two or more of its loops simultaneously, a method for determining which loop receives priority must be developed.

The rule used to establish routing priority was previously presented in Section 4.4 and is now restated as follows:

In multiloop (Class III) structures, the control unit is always routed to the loop in which the ingredience requirements of the COMBI work task (or of the lead COMBI in the event of a multielement loop) are satisfied. In the event the ingredience requirements for commencement of several loops are satisfied simultaneously, the unit is routed to the loop with the COMBI labeled with the lowest number (i.e., ties are broken by a labeling rule). Therefore, if loops with COMBI work tasks 3 and 7 both can commence (based on the ingredience criterion), the control unit is routed to the loop containing COMBI work task 3, since it has the lower numerical label.

The effect of this rule from the control structure standpoint is to delay units passing certain loops for a lesser period of time than those passing other loops. That is, the control unit gives priority to certain work tasks. In the next chapter this characteristic of Class III structures is used to develop various routing mechanisms.

In order to illustrate the use of Class III structures, consider the following cases. The most common instance in which butterfly structures are used is the modeling of shared resources. Most processes have certain resources that are involved in several tasks that are not sequential or related. Cranes, for instance, are often required to operate between several nonrelated tasks or processes, because lifting may be common to several processes on the job site. The crane on a given site may be required to (1) lift steel to the framing area, (2) unload steel from arriving low-bed trucks, (3) "shake out" or order the steel in lift sequence for later incorporation into the structure, and (4) lift concrete for floor decking. The crane controls progress on four separate tasks, and it is a constraining resource. Moreover, the priority that the crane gives to the tasks to which it is common also affects the progress of the project. Figure 8.12 shows a Class III structure that is capable of modeling the work task pattern of the crane on this project.

It is now necessary to determine which work tasks have priority if a conflict arises. One possible solution to the problem is as follows. Priority for the crane must go to framing the building, since this is the critical process in the project. Therefore the lift steel for framing work task should have the lowest

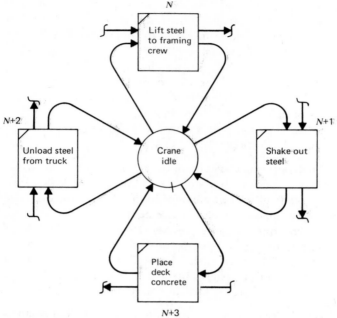

N

Lift steel
to framing
crew

$N+2$

Unload steel
from truck

Crane
idle

Shake out
steel

$N+1$

Place
deck
concrete

$N+3$

FIGURE 8.12   Shared resource (crane) control structure.

work task number of the four work tasks served by the crane. The next most important work task affecting the progress of construction is the ordering or shaking out of the steel, since it is required before the steel can be lifted for framing. Next in order of priority is the unloading of steel, since the steel cannot be ordered until it is unloaded. The lowest priority goes to the concreting operation, which can be done when steel work is not available. Therefore, in agreement with this priority system, the COMBI work tasks in Figure 8.12 would be labeled as follows.

Work Task	Label
LIFT STEEL TO FRAMING CREW	N
"SHAKE OUT" STEEL	$N+1$
UNLOAD STEEL FROM TRUCK	$N+2$
PLACE DECK CONCRETE	$N+3$

This solution presumes that when the crane cannot lift steel for framing, it moves back and starts sorting or shaking out steel. This may not, in fact, be the most efficient policy from the standpoint of total equipment utilization, since trucks carrying steel to the site will wait idle until all steel is sorted before they are unloaded. A more reasonable priority sequence might reverse the positions of the shake out and unload work tasks. The correct sequence is a function of the actual situation and is tied to the rate at which steel arrives on site from the fabricator.

In some systems the priority routing may vary across the life of the project. For example, Figure 8.12 indicates that priority should go to the framing of steel. However, as has been mentioned previously, union rules and safety provisions often require that the decking operation may not lag behind framing more than two floors. This means that at the point when framing exceeds the two-floor limit, priority for the cranes operation must be given to the concreting operation. A control structure to implement this change in priority must be developed. This can be achieved by designing a triggering control unit that is released at such time the framing tier exceeds the floor to which the structure is decked by more than two levels. Such a triggering structure will be discussed in the section on open path control structures (Section 8.7).

8.6 CLASS III CONTROL STRUCTURES: MULTIELEMENT LOOP

In the crane example, a more realistic presentation of the work sequence might require the crane to shake out and lift the steel without returning to the idle state for routing to another work task. This would lead to a multielement loop, replacing two of the single-element loops shown in Figure 8.12. The graphical model for this process would be as shown in Figure 8.13. Some additional detail to the basic Class III structure has been added to indicate that both work tasks incorporated into the multielement loop maintain their COMBI work task status. Since the ground crew is required for the shake out work task and the framing crew on the building structure itself is required for framing, these work tasks remain separate. If this were not the case, the shake out and lift work tasks could be combined into a single work task and a multi-element loop would not be required. As things stand, the crane is required in sequence for both the shake out and lift work tasks, while the ground crew is released after shake out. The framing crew is captured upon lifting of the steel and remains combined through a multielement sequence not shown in the figure.

A slight modification of the situation presented in Figure 8.13 results in a Class II multielement loop being included in the crane routing pattern. It is assumed that the crane must systematically work the shake out and lift work tasks as shown above. However, a welding crew welds the members following shake out, but prior to the lift work task. Since the crane is not required for the welding operation, it is effectively placed on standby until the welding operation is completed. This is an inefficient usage of the crane, since it must

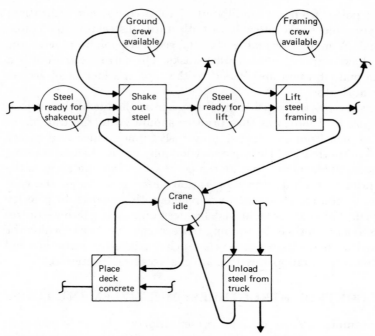

FIGURE 8.13   Multielement loop structure for steel erection model.

undergo an enforced idleness during the welding operation. It could be unloading steel during this period. However, in the actual situation, management may not want to mobilize the crane on another work task that will
capture the crane and possibly delay lifting of the steel (i.e., delay a critical
task). In order to insure availability of the crane for lifting following the welding work task, management accepts this slight inefficiency in the utilization
of the crane. The new system providing for a sequential hold on the crane
during the welding operation results in a Class II multielement loop within
the crane utilization Class III control structure. This is depicted in Figure 8.14.

The enforced idleness of the crane following the shake-out activity is obvious,
since it enters an idle state pending completion of the welding work task.
The careful reader will be able to identify an alternate solution to the Class II
loop model. It is possible to maintain a Class I loop including the three work
tasks of shake out, welding, and lifting in sequence without diverting the crane
to an idle state. This would, however, sacrifice the delay information on the
idle state of the crane by deleting the QUEUE node element "CRANE IDLE
(WELDING)." The saving of an element may compensate for the loss of this
system information.

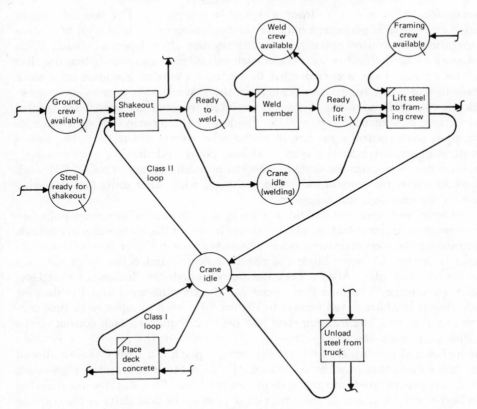

FIGURE 8.14   Crane pattern with Class II multielement loop.

This modification points up the potential use of segments of the Class II loop as a monitoring device to measure travel time between two points in the network. As shown in Figure 8.14, the crane unit delay in the "CRANE IDLE (WELDING)" state is the same as the time required for the steel units between completion of shake out and the beginning of the lift work task. This is a so-called BETWEEN statistic. The system feature that measures the idleness of the crane unit in the QUEUE node provides the mechanism of measuring transit time between the separate points in the network. This will be developed in greater detail in Chapter Ten when discussing system monitoring concepts.

8.7 CLASS IV CONTROL STRUCTURES: OPEN PATH

The Class I to Class III structures have in one form or another relied on the use of closed cycles or loops through which the control unit passes, always returning to an idle state from which it is redeployed. The control unit in such structures is generated once at the beginning or initialization of system operation. The control of one section of the system often depends on completion of work tasks in another section. In such instances, a message indicating that certain events have occurred must be sent. This acts as a release on a work task or work task sequence elsewhere in the system. This type of linkage between two sections of the system is achieved by using an open path control structure, as shown in Figure 8.15. The key characteristic that differentiates an open path control structure from the other structures is that the control unit is generated into the system at one point and, having performed its release function, exits the system. The exit normally occurs at a COMBI work task at which the control unit is consolidated with other units and following which the unit does not reappear.

In order to better understand how an open path control structure links two processes or triggers start-up of one process based on the completion of another, reconsider the steel erection problem. Essentially each tier or floor of construction is framed by assembling the required beams and columns as specified in the erection plan. After a floor has been framed, the decking of that level may commence. The unit flow cycles for the framing steel and the decking are shown in abbreviated format in Figure 8.16, with an open path link connecting the two. The framing steel and decking sequences each consist of the same four work tasks: (1) off-load, (2) shake out (i.e., order in erection sequence and prepare for lift), (3) lift and (4) place. To reduce the number of mainline units that must be initialized, the units, after exiting the place work task, are regenerated into the system and recycled. This was discussed earlier in Sections 5.3 and 5.4. In the situation depicted, the flow units in the framing steel cycle are "packets" of framing members (e.g., beams and columns). For simplicity, it is assumed that five packets of framing are sufficient to release a section of the structure for decking with one decking packet. The

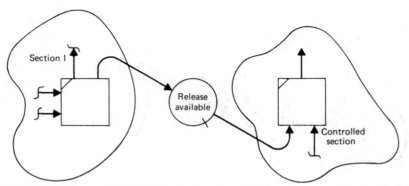

FIGURE 8.15 General form of Class IV control structure: open path.

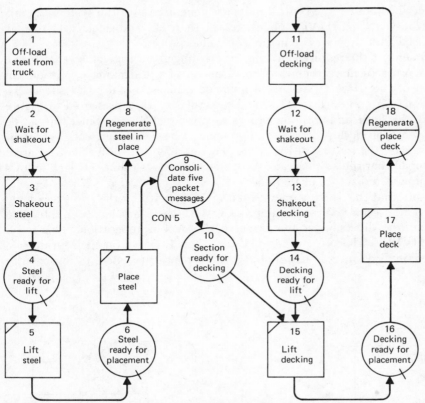

FIGURE 8.16   Class IV structure linking framing and decking sequences.

control link must "notify" the decking cycle when five packets of framing steel have been erected so that the decking cycle can commence.

This release of the decking cycle for start-up is achieved by using a Class IV structure, which includes a FUNCTION node that consolidates units and emits a single unit. The CONSOLIDATE node receives message units from the "PLACE STEEL" activity in the framing cycle, and having received five placement messages, generates a single message to the "SECTION READY FOR DECKING" QUEUE node. This message is an ingredient common to the "LIFT DECKING" operation. Therefore, decking cannot be lifted for inclusion into structure until a message unit is available at QUEUE node 10. This controls the cycling of the decking sequence so that it occurs once each time the framing sequence cycles five times. This is a typical use of the Class IV structure and illustrates how various processes can be triggered or phased through its operation.

As described above, the control units are generated into the system (in this case at the CONSOLIDATE node), travel the link, and effectively exit at the COMBI that they service in the controlled process.

To implement the requirement that priority be given to deck so that frame does not progress more than two floors ahead of the decking operation, an extension of the Class IV structure must be utilized. Figure 8.17 shows the integration of a crane Class III structure similar to that presented in Figure 8.12 and the deck and framing processes presented above. Since the crane controls work on both processes, the problem of insuring sufficient deck progress focuses on the routing of the crane. Therefore, a control mechanism must be developed that routes the crane to the deck process so that the deck follows the framing at a distance of two floors.

To insure that the decking operation is given priority by the crane following the erection of each floor of framing, a COMBI work task has been added to implement the diversion of the crane to the decking operation. This work task, "DIVERT CRANE," is labeled work task 1. Based on the routing rule presented in Section 8.5, it will be given top priority once its ingredience

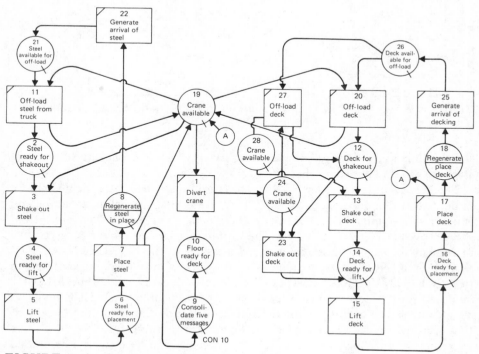

FIGURE 8.17   Steel erection system.

requirements are satisfied. The two required units are the crane from QUEUE node 19 and the completion message from QUEUE node 10. With the completion of the placement of five packets of framing, one floor of framing is completed and the crane will be diverted to the "SHAKE-OUT" work task. In order to prevent the situation in which the crane is diverted to 24 and then cannot commence the shake-out work task because of lack of off-loaded deck, the crane can be routed alternatively to activity 27 to off-load the deck. The addition of the parallel COMBI work task ("OFF-LOAD DECK") at 27 insures that conflicts in the system will not occur.

8.8 CLASS V CONTROL STRUCTURES: SWITCHES

A slight modification of the Class III multiloop structure leads to the formation of another very useful control structure. It can be described as a *switch*, since it allows routing of flow units along one of a set of paths based on the realization of certain requirements. This switching type pattern is referred to as a Class V structure. A typical Class V routing structure is shown in Figure 8.18. The structure consists of an idle state QUEUE node into which the unit to be routed is directed. The QUEUE node is followed by COMBI work tasks that represent the alternate paths that the routed unit may follow. In the figure, the two paths possible are either COMBI 1 or COMBI 2. The number of following COMBI work tasks and paths available is not constrained. The actual path selected by the flow unit depends on where the COMBI work task ingredience requirements are first realized. The ingredience requirement for COMBI 1 is:

$$\text{IF A AND B THEN 1 (i.e., } A \cap B \rightarrow 1)$$

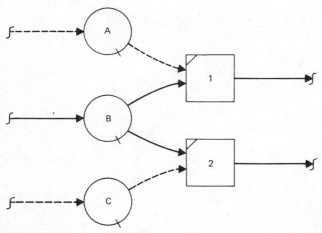

FIGURE 8.18   General form of Class V control structure: switches.

For COMBI 2 the requirements are:

IF B AND C THEN 2 (i.e., B ∩ C → 2)

This means that if a unit arrives at A while the flow unit to be routed is at B, then the unit at B is routed to 1. Another possibility is that a unit is already at A when the unit to be routed arrives at B. Again, the unit arriving at B would immediately take the path to 1. Similar logic applies to a flow unit routing to COMBI 2.

If units are available at both A and C when a unit to be routed arrives at B, then the priority routing rule applies. In this case, the unit arriving at B would be switched to 1, since COMBI 1 has a lower numerical label (i.e., lower work task number) than COMBI 2. It now becomes clearer that a Class V structure acts as a switch or routing structure that diverts flow units to points within the network, much as switches in a railroad yard divert cars from one track to another for assembly into a train.

Consider more closely the functioning of the switching structure used in conjunction with the Class IV (open path) structure in the steel erection problem of the previous chapter. Figure 8.19 shows this structure projected out of the overall network. Utilizing the rules developed above, this structure establishes:

IF 26 AND 24 THEN 27 (26 ∩ 24 → 27)

or

IF 12 AND 24 THEN 23 (12 ∩ 24 → 23)

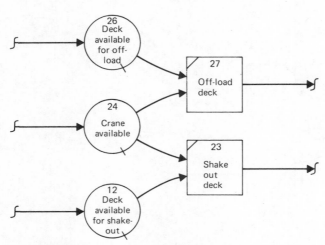

FIGURE 8.19 Crane routing switch.

If units are located both at 12 and 26 when the crane unit arrives at 24, there results

IF 26 AND 12 AND 24 THEN 23 (26 ∩ 12 ∩ 24 → 23)

The set notation expressions for the logical conditions described are given in parentheses next to the textual descriptions.

Therefore, according to the switching logic defined, if the crane unit arrives at 24 and deck is available for off-loading, but no deck is ready for shake out (i.e., 26 ∩ 24 → 27), the crane proceeds to the off-loading activity (27). If the crane finds deck available for shake-out, but none available for off-loading (i.e., 12 ∩ 24 → 23), then it is routed to the shake out work task (23). If deck is both available for shake out and off-loading (i.e., 26 ∩ 12 ∩ 24 → 23), then preference is given to the shake-out work task and the crane is routed to COMBI 23. This policy is consistent with giving priority to the work tasks that result in progress on the main structure of the building.

Another Class V structure is embedded in the steel erection problem. It is also part of the open path structure used to route the crane to the decking process. It is desired to divert the crane to decking operation following construction of one tier or floor of structural frame. This segment of the steel erection network is shown in Figure 8.20. This switch operates to pull the

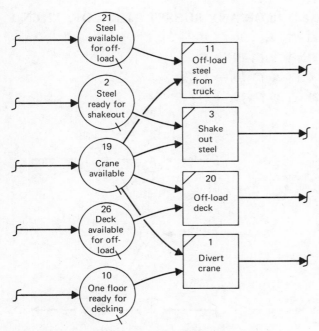

FIGURE 8.20   Crane diversion switch.

crane out of its Class III (butterfly) structure and divert it to a separate path. In fact, a combination of Class III and Class V structures is operative. The crane idle state at 19 is part of the butterfly structure radiating to operations 3, 11, and 20. Elements 19 and 10 are part of the Class (open path) structure created to divert the crane to the decking operation sequence following framing. Finally, the Class V (switching) structure ties together the butterfly and the open path structure linking them through COMBI work task 1, "DIVERT CRANE."

Another item of interest regarding the interrelationship between Class III and Class V structures now becomes apparent. The switching structure can be viewed as the front half of the butterfly pattern. Therefore taking the butterfly structure and deleting the return arcs creates a switching structure. This is illustrated in Figure 8.21, which shows that the switching structure is nothing more than an open butterfly pattern.

Returning to the crane diversion switch depicted in Figure 8.20, the following logical relationships are operative.

IF 21 AND 19 THEN 11 (21 ∩ 19 → 11)

IF 2 AND 19 THEN 3 (2 ∩ 19 → 3)

IF 26 AND 19 THEN 20 (26 ∩ 29 → 20)

IF 10 AND 19 THEN 1 (10 ∩ 19 → 1)

IF 21 AND 2 AND 26 AND 10 OR ANY SUBSET THEREOF, THEN 1

$$(21 \cap 2 \cap 26 \cap 10 \to 1)$$
$$(21 \cap 2 \cap 10 \to 1)$$
$$(2 \cap 26 \cap 10 \to 1)$$
$$(21 \cap 26 \cap 10 \to 1)$$

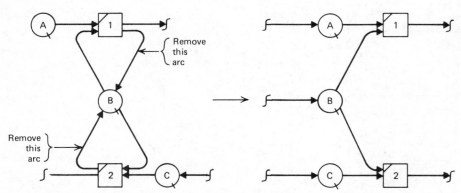

FIGURE 8.21   Conversion of Class III to Class V structure.

Therefore, the priorities on movement of the crane out of its idle state at 19 are:

1. First priority is given to the diversion of the crane to the decking operation. If a unit is available at 10 when the crane returns to 19, the crane is routed through 1 to the decking operation. A unit can only be available at 10 following completion of the framing. Therefore decking proceeds only after framing is completed. The lowest number (i.e., one) has been assigned to the divert work task to insure that it receives top priority once a tier of framing is completed. Note that the crane returns to 19 from 17 following the decking sequence. Since the unit at 10 is removed when the crane is diverted through 1 from 19, three options confront the crane upon return to 19: (a) routing to 3, (b) routing to 11, or (c) routing to 20.

2. Second priority goes to the lowest numbered of the remaining COMBI work tasks tied to 19. Therefore, the shake-out work task at 3 receives second priority on the allocation of the crane's effort.

3. Third priority on the crane's effort is assigned to the off-loading of structural steel at 11.

4. If work is not available at 1, 3, or 11 (i.e., ingredience requirements are not satisfied at these COMBI work tasks), the crane is routed to 20 to off-load the deck.

The versatility of the switching structure in handling routing problems should now be obvious. Because of its utility in diverting flows based on sets of conditions being realized, it is particularly helpful in the construction of mechanisms. Mechanisms are combinations of control structures designed to perform functions that are not intrinsic to the basic modeling elements.

Before leaving the subject of switching control structures, the special case of probabilistic switches will be considered. The idea of probabilistic exit arcs was introduced in Chapter Four. Using this feature allows the development of probabilistic routing structures. In such structures, the switching path is not predetermined but is selected randomly. Any element other than a QUEUE node can have probabilistic ARCs associated with it. Therefore, any element (other than the QUEUE node) can be used to introduce a probabilistic switch into process design. In some cases, the switch is associated with a system element with a primary function other than switching. That is, the switching function is an addition to a primary function such as COMBINATION or CONSOLIDATION. In other instances, a separate FUNCTION node is defined with associated probabilistic ARCs for the primary purpose of introducing a random routing pattern into the system.

Returning again to the steel erection system, if it is more realistic to assume that the crane is routed randomly to either the steel rigging or deck placement sequences, there is a requirement for a probabilistic switch. One solution to this random routing problem is shown in Figure 8.22. In this configuration,

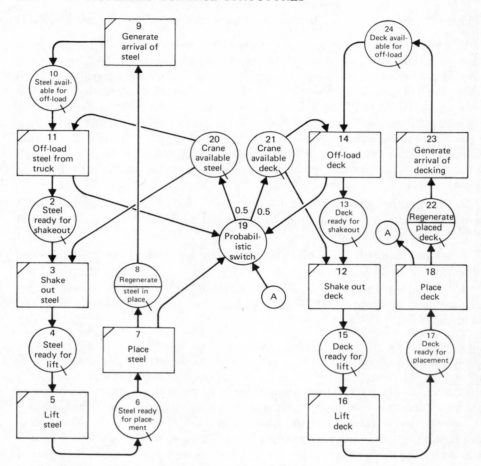

FIGURE 8.22   Steel erection system with probabilistic switch.

the crane, instead of returning to the "CRANE IDLE" node, is returned to a FUNCTION node (19) from which it is randomly routed to an availability position for steel rigging or deck placement. The availability positions are defined by breaking the "CRANE IDLE" QUEUE node into two nodes (20, 21). The chance of a crane transiting to either one of these newly defined QUEUE nodes is 50% (as specified on the ARCs in the diagram). The movement of the crane flow unit to steel rigging and deck placement is now realized on a random basis. If, after arriving at FUNCTION node 19, a random number ≥ 0.50 is generated, the crane moves to QUEUE node 20. Otherwise, the crane is made available to the deck installation sequence through QUEUE node 21.

Probabilistic switches used in combination with other control structures will be further developed in Chapter Nine.

8.9 CLASS VI CONTROL STRUCTURES: PARALLEL PROCESSORS

One final control structure must be discussed before proceeding to the topic of mechanisms. This structure differs from those presented so far in this chapter. It is not defined in terms of network topology but, instead, is a function of the ingredience requirements associated with a given COMBI activity.

A control entity and its associated cycle introduced solely to control the number of units that can be processing a COMBI activity simultaneously are defined as a parallel processor structure. It consists of a single QUEUE node, its associated COMBI processor, the arcs connecting the queue node to the COMBI processor, and the control units that define the number of processed units that can be processing the COMBI simultaneously. It can be viewed as a simple case of the Class I structure designed to control the volume of unit flow through a single processing station. In this sense, it controls the number of processing channels available at its associated COMBI element.

The number of processor locations associated with a given COMBI processor can be established by the number of entities initialized in the processor control structure. That is, it is possible to model a parallel processor situation by defining the appropriate number of control entities in the associated control queue. This is illustrated in Figure 8.23. In Figure 8.23a, two control entities are initialized at 2, while in Figure 8.23b, five control entities are initialized at 4. The number of parallel processor actions possible is illustrated in CYCLONE format by showing shadow work tasks at COMBI elements 1 (two processor work tasks) and 3 (five processor work tasks).

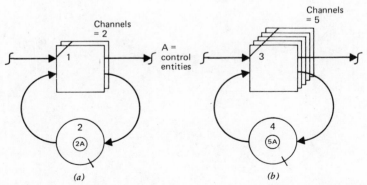

FIGURE 8.23   General form of Class VI control structure: parallel processors.

In general, the number of processor locations associated with a COMBI processor is defined by the relationship:

$$C = \min_{i=1}^{m} (n_i)$$

where

C = maximum number of processor locations associated with a given COMBI processor

m = number of preceding queue nodes (i.e., ingredients in the ingredience set)

n = the maximum number of flow units that can occupy queue i at any time (i.e., the sum of all entities flowing in cycles associated with queue i).

In this way, a single COMBI work task can be used to represent several processors based on how many ingredience sets it is possible to have at a given time. For example, in Figure 8.24, the ingredience requirement for work task 4 (load truck) calls for a truck, a loader, and a load of material from the stockpile. The maximum number of "truck load" processing work tasks that can take place is constrained by the number of units in the smallest ingredience set category (i.e., the loaders). Therefore, in this case, the maximum number of truck load operations that can occur simultaneously is constrained to two, since there are only two loaders. This is illustrated by showing the work task squares for LOAD TRUCK backed by the appropriate number of shadow work tasks such that the maximum number of possible simultaneous processes, C, is indicated. This feature, in conjunction with the number of units in each

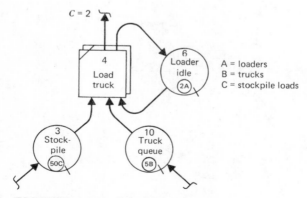

FIGURE 8.24   COMBI processor in shadow format.

ingredience set category, provides the capability of modeling the multichannel situation by simply varying the number of entities in each ingredience set category.

8.10 DEVELOPMENT OF MECHANISMS

This chapter has presented six control structures that are basic to the development of flow unit networks and the modeling of construction processes. These six structures provide the fundamental constructs in an algebra for graphical systems definition and model building. They can be used to build mechanisms capable of handling a wide range of modeling functions not intrinsic to the systems elements themselves. Mechanisms are combinations of control structures that interface to achieve a given function not realizable with simpler constructs or elements. The crane routing subsystem (discussed in Section 8.8), which combined Class III, Class IV, and Class V structures to divert the crane at the proper time to the deck sequence is an example of a systems mechanism. Mechanisms are usually designed to meet specific conditions or achieve specific results. They are comparable to algebraic equations. In a graphical sense they are unique expressions designed to reflect a given logical situation. Some of them are encountered more often than others. In chapter Nine the subject of mechanisms and their development will be presented and some general classes of mechanisms will be developed.

PROBLEMS

1. Extend the brick curing operation model discussed in the text and illustrated in Figures 8.6a and 8.6b to include Class I, Class II and Class III control structures based on the cart, tractor, and space resource units.

2. Identify and classify the various types of control structures used in the concrete delivery models illustrated in Figures 13.1 to 13.4.

3. Extend the precast decking operation discussed in the text and illustrated in Figure 4.14 to incorporate the management of an on site storage bay for 10 precast elements, as an initial start permit trigger for the decking operation.

4. A carpenter foreman is in charge of a slab formwork erection crew consisting of three carpenter-laborer pairs. The foreman instructs each work pair team on the sequencing and locating of forms, and inspects erected forms for stability and level. If faulty erection occurs the foreman assists in whatever corrective work tasks are required. If correct erection occurs the foreman indicates the location of the required next formwork section and monitors the other work pairs.

Develop a basic Class III control structure for the foreman's management assignment and incorporate probabilistic exit arcs to Class II work task loops to model his work sequence whenever faulty erection of formwork occurs. How would you establish the probabilities to be assigned to the probabilistic exit arcs?

5. By field observation, establish the various work task sequences performed during the working of special equipment operations such as pile driving and drilling. Then establish field working priorities and develop a suitable CYCLONE model, incorporating the relevant priority control structures.

6. An earthmoving operation involves a fleet of scrapers and a single pusher dozer for the loading operation. Scraper operators have been instructed to bypass the pusher dozer and self-load, to minimize bunching effects, whenever the dozer is engaged with a scraper in the load operation and a second scraper is already available in a queue position awaiting the pusher dozer.

Develop a suitable control structure to model the earthmoving operation.

7. Comment on the following statements:
   (a) The work task breakdown of a construction operation is useful for the estimating function and as a means of describing the work content of the construction operation.
   (b) The construction technology of the construction operation is shown in the basic CYCLONE model static structure and are readily developed if control structures representing field expertise and management are ignored.
   (c) The mere inclusion of resource idle state QUEUE nodes in the basic model static structure is sufficient to sensitize, or alert, field management to observing the impact of nonuse of field resources.
   (d) The inclusion of control structures to represent management policies and priorities or to model field expertise is only necessary when attempts are being made to design the field operations or to explore ways of improving field productivity.

8. The process under consideration involves the movement of precast double T parking decks from a storage yard to the job site where a parking garage is being constructed. The double T precast elements are moved using low-bed trucks. Two elements can be loaded on each truck. Each section is lifted into position at the job site using a tower crane. Because of the congested nature of the job site trucks must wait in the street while being off-loaded. Only two trucks can wait at a time. When a truck has been unloaded, a signal is sent to the storage yard to release the next truck. Using a Class I structure, model the precast element transport system described.

9. In Problem 8, assume that trucks are not held at the storage yard, but move to the job site directly. At the job site, there are two off-loading positions. Incoming trucks are delayed at a holding area until one of the two off-loading positions becomes available. As an off-loading position becomes available, the next truck moves forward, off-loads, and departs. Model this slightly modified situation using a Class II structure.

10. Using a Class II structure, model the operation of a "muck train" in a tunneling operation in which the train can be recalled from the work face on demand before it is completely filled.

11. The 900-foot 6000-ton center span of the Fremont Bridge was lifted 160 feet into position above the Willamette River in Portland, Oregon.* The unique lifting system consisted of the following components: 200-ton hydraulic jacks were used to lift 32 4-inch diameter threaded rods, eight fastened to each corner of the bridge. Fig. P8.11a shows elevation views of the span during and after completion of the lift. Details of the lifting sets at each corner of the bridge are shown in Figures P8.11b and P8.11c. In a typical lifting cycle all 32 jacks pull rods through a 24-inch (610 millimeter) stroke, with the rams pushing against nuts on the rods. As the rams rise, iron workers keep a second set of nuts screwed down snug against bearing seats below the jacks (see the figures). The stroke completed, rams are retracted and the load transferred to the lower nuts until the next cycle began. The iron workers run the top nuts down with the rams as they are retracted. Both sets of nuts are split so they can be opened to allow couplers to pass through. The two halves of each nut are held together by a slip-on collar.

    Model the jacking cycle using a Class IV structure to divert control to the sequence which allows passage of the coupler through the center hole jack.

12. Trucks off-load at a warehouse that has four unloading bays. The nature of the location is such that trucks waiting to off-load must wait on the main street, since the access road would be blocked otherwise. A site plan is shown in Figure P8.12. Develop a model of this process and conduct a hand simulation to determine the time trucks must wait to off-load. The model should be constructed so as to indicate periods during which the access road is blocked to other traffic because of the maneuvering of trucks that are docking. Determine the average duration of such blockages. The durations of relevant activities are as follows.

    (a) Truck entry time (from waiting line): 5 minutes.

*See Alpo J. Tokola and E. J. Wortman "Erecting the Center Span of the Fremont Bridge," *Civil Engineering*, July 1973.

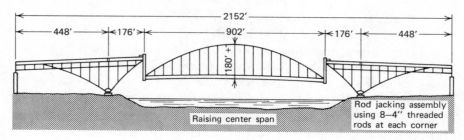

(a)

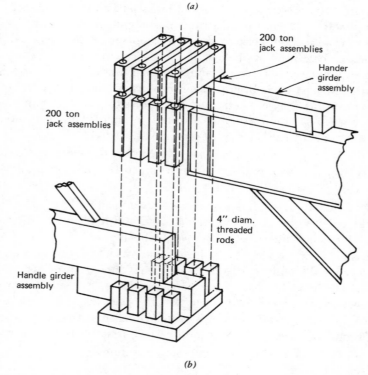

(b)

FIGURE P8.11   (a) After bridge is jacked into position. (b) Jacking system.
(c) Details of rod jacking device.

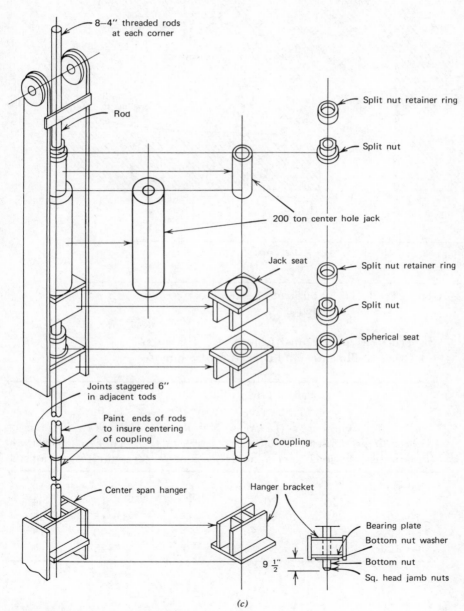

8–4″ threaded rods at each corner

Rod

Split nut retainer ring

Split nut

200 ton center hole jack

Jack seat

Split nut retainer ring

Split nut

Spherical seat

Joints staggered 6″ in adjacent tods

Paint ends of rods to insure centering of coupling

Coupling

Center span hanger

Hanger bracket

Bearing plate
Bottom nut washer

$9\frac{1}{2}''$

Bottom nut

Sq. head jamb nuts

(c)

FIGURE P8.11   (Continued).

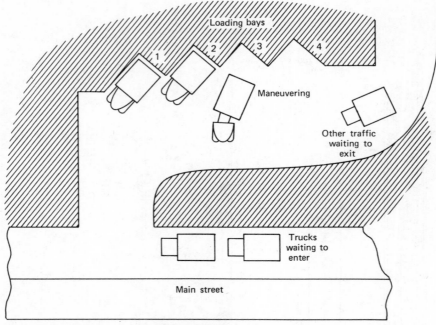

FIGURE P8.12   Truck off-loading site plan.

(b) Truck departure time (from dock): 4 minutes.
(c) Other traffic entry and departure: 1.5 minutes.

Off-loading (in minutes)	Number of Observations	Truck Arrivals (Between Times) (in minutes)	Number of Observations	Other Traffic (Between Times) (in minutes)	Number of Observations
50–60	12	0–20	28	0–5	16
60–70	36	20–40	39	5–10	34
70–80	21	40–60	22	10–15	25
80–90	17	60–80	8	15–20	15
90–100	14	80–100	3	20–25	10

Modeling Mechanisms

In the development of networks to model construction processes there is often the need for implementation of logical system characteristics that cannot be developed easily from the intrinsic characteristics of elements or the control structures presented in Chapter Eight. In order to achieve the required functions, new constructs must be developed by combining elements and control structures. *Mechanism* is the general term used to describe these graphical combinations or expressions. A mechanism can be thought of as a kind of graphical equation stating a relationship between various flow units within the system. In this chapter, categories of mechanisms will be investigated and the method of developing mechanisms will be discussed.

9.1 COMPLEX FUNCTIONS AND MECHANISMS

As noted in Section 8.10, the control structures identify some of the basic operations that can be modeled using existing system logic. The systems elements themselves have certain elemental functions that they are capable of performing. The elements and the control structures form the basic components and set of operations that define a graphical process modeling algebra. By properly ordering the elements either singly or within the control structures, graphical expressions that result in a transformation of, or functional relationship between, various flow units can be developed. The resulting system segment synthesized from control structures is a mechanism. The mechanism performs a specific function that may be unique to one model, or very general and applicable to many systems. The combination of Class III, Class IV, and Class V structures presented in Section 8.8 to achieve routing of the crane from the framing to decking sequence is a typical example of a system mechanism. In this case, the mechanism developed was more or less unique to the particular system developed. Other types of mechanisms are more general in their application.

The development of mechanisms to solve a given problem is similar to the solving of an identity in algebra. The intuition and skill of the modeler in seeing relationships between structures and elements that will lead to economic representation of the required function or relationship is of prime importance. Once the elements and structures are completely understood, the development of mechanisms becomes an acquired talent that is best learned by practice. Therefore, the best way to approach the study of mechanism development is to define some modeling functions or operations that are of interest and attempt to develop a mechanism to handle the desired function.

A number of categories of mechanisms of a general nature can be identified. Some fairly characteristic categories are:

1. *Switching or Routing Mechanisms.* Switching mechanisms route flow units through networks to a variety of destinations dependent on current system state conditions and logic formulation. In some instances, the switching pattern is random and probabilistic exit ARCs are used. Gating mechanisms are a special case of the general switching mechanism that are used to separate portions of networks and to deny or facilitate access to the various subnetworks through priority functions.

2. *Capture Mechanisms.* Capture mechanisms enable various flow units and subnetworks to be captured or withheld for specific periods of time or until logical conditions are met.

3. *Mechanisms for Nonstationary Processes.* These mechanisms enable time-dependent variations to be incorporated into the characteristics of model components so that nonstationary processes can be modeled.

4. *Unit Generation Mechanisms.* These mechanisms enable flow units to be generated or emitted into the network model from time to time to suit specific time or logical functions.

The need for these mechanisms occurs frequently in the modeling of construction processes and will be developed later in this chapter.

9.2 SWITCHING OR ROUTING MECHANISMS

One of the most commonly used modeling functions is that of switching or rerouting units through the network based on conditions that obtain across time. Development of switching mechanisms is possible using various configurations of the Class V structure. The versatility of this structure in handling situations in which flow units must be rerouted has already been discussed and illustrated in Section 8.8. The use of switches is central to the functioning of logical systems. The computer, for instance, can be thought of as a gigantic box of switches that are set and reset according to a *program* developed to solve a particular problem. The writing of such a program leads to the requirement to switch back and forth to access certain actions. The programming language used. to communicate with the computer defines a function that provides generally IF a certain criterion is met, THEN some action is to occur. This IF–THEN relationship is nothing more than a switching function that says in everyday language:

IF (this and this and this happen), THEN go do that

The above expression can be symbolically modeled as follows.

Let the symbol X represent "this and this and this"

and the symbol Y represent "that";

then the expression becomes

$$\text{IF } (X) \text{ THEN } Y \tag{9.1}$$

if the definition of the symbol X is generalized as

X represents ingredients "A and B and C"

then Equation 9.1 becomes

$$\text{IF } (A \text{ AND } B \text{ AND } C) \text{ THEN } Y \tag{9.2}$$

or, using set notation,

$$\text{IF } (A \cap B \cap C) \rightarrow Y \tag{9.3}$$

A graphical form of statement (9.3) is achieved in Figure 9.1 using a Class V structure. Figure 9.1 indicates that units at D are routed through Y if A and B and C are present, since the routing rule (see Section 8.5) gives the priority to the node labeled N (i.e., $N < N + 1 < N + 2$).

Other expressions represented by this construct are:

$$
\begin{array}{llll}
\text{IF } A \rightarrow X & \text{IF} & (A \cap B) \rightarrow Y & \\
\text{IF } B \rightarrow Y & \text{IF} & (A \cap C) \rightarrow X & \\
\text{IF } C \rightarrow Z & \text{IF} & (B \cap C) \rightarrow Y & (9.4)
\end{array}
$$

The basic functioning of the Class V structure has been illustrated in Section 8.8. To better understand its use in the development of a switching mechanism, consider a more complex routing problem presented within the context

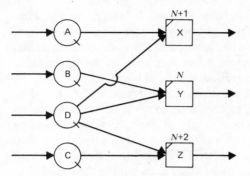

FIGURE 9.1   Graphical statement of expression (9–3).

(a)

Condition STUB	Condition ENTRY
Action STUB	Action ENTRY

(b)

		A	B	C	A∩B	B∩C	A∩C	A∩B∩C
If	A	X			X		X	X
	B		X		X	X		X
	C			X		X	X	X
Then	X	X	X				X	
	Y		X		X	X		X
	Z		X					

① ② ③ ④ ⑤ ⑥ ⑦

FIGURE 9.2   Decision table layout. (a) General layout of decision table. (b) Decision table described in Figure 9.1 structure.

of a decision table. A decision table is a device used to consolidate multiple switching requirements into a compact format. As shown in Figure 9.2a, the decision table consists of four labeled subcomponents: (1) condition stub, (2) condition entry, (3) action stub, and (4) action entry.* Figure 9.2b shows the decision table describing expressions (9.3) and (9.4) and that is modeled by the Class V structure shown in Figure 9.1. The condition entry area indicates the IF aspect of the IF–THEN relationship, detailing in register format the possible conditions that are to be considered (i.e., in this case A, B, and C). The condition entry contains columns indicating the possible combinations of the conditions considered. The action stub register indicates the action (or, in this case, the path) to be taken in the event certain combinations of conditions obtain. This corresponds to the THEN aspect of the IF–THEN relationship.

*For more information regarding decision tables, consult references (McDaniel, 1968).

Finally, the action entry contains columns corresponding to those in the condition entry containing markers (X) designating the action appropriate to a given set of conditions. Therefore, if conditions B and C obtain, combination column (5) in the condition stub is realized. The action entry associated with column 5 in the action stub is Y. Hence the decision table gives:

$$\text{IF } (B \cap C) \rightarrow Y$$

The operation of a switching mechanism in modeling complex decision table situations becomes clearer when descriptive titles are associated with the conditions and actions. As an example, consider a situation in which ducting for the heating, ventilation, and air-conditioning system of a high-rise building is being lifted to be placed at various floors. Assume that the site is serviced by a tower crane, a roof hoist, and a man hoist. The conditions relevant for the duct lift problem might be defined as follows.

Let A = Tower crane available for lift

   B = Roof hoist available for lift

   C = Man hoist (i.e., lift cage) available for lift

The actions would be similarly defined as:

$$\text{Lift duct using} \begin{cases} X = \text{Tower crane} \\ Y = \text{Roof hoist} \\ Z = \text{Man hoist} \end{cases}$$

The management decision table and graphical structure formulated for the duct lift problem would appear as shown in Figure 9.3. As presented, the ducting would normally be lifted with the roof hoist if available. Second priority for lifting would go to the tower crane; only if the tower crane and the roof hoist are unavailable would the man hoist be used (if it is available). Although in this example the number of actions was equal to the number of conditions, this is not required. In some cases, there may be as many actions as there are combinations of the conditions. In such a case, the graphical expression would be expanded to appear as shown in Figure 9.4.

The ingredience requirements for each of the action activities (i.e., R through Z) are the same as the column entries in the condition entry. Connectors Ⓐ, Ⓑ, and Ⓒ have been used to avoid cluttering the figure with arrows. For action Y (corresponding to column 6), the ingredients are the routed unit D and the conditions A and C. By checking column 6, the conditions required for this action are, in fact, A and C.

The number of possible actions can also be expanded by the number of unit types to be routed through the switching mechanism. If, for instance, three commodities are to be routed, the number of possible actions is expanded. Assume that the three lifting devices not only transport ducting, but also are

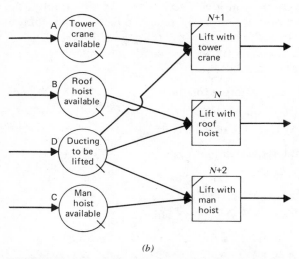

		①	②	③	④	⑤	⑥	⑦
If	Tower crane available	X			X		X	X
	Roof hoist available		X		X	X		X
	Man hoist available			X		X	X	X
Then lift with	Tower crane	X					X	
	Roof hoist		X		X	X		X
	Man hoist			X				

(a)

(b)

FIGURE 9.3 Switching mechanism for ducting lift problem. (a) Decision table. (b) Graphical expression.

required to lift rebar (reinforcing steel) and brick pallets. In this case the number of actions is defined in Table 9.1. The table also shows the priorities to be observed in lifting in parentheses. Eight commodity lift activities are possible. The only combination excluded is the lifting of rebar with the man hoist. The decision table for this situation is given in Table 9.2 and illustrated in Figure 9.5.

The above examples illustrate that once a decision problem involving priority policies can be formulated as a decision table, then switching mechanisms can be developed for the decision table and incorporated into the network model. The decision table logic can be developed independently of network model considerations. Once the relevant conditions, policies, and actions have been defined, it is a fairly simple problem to identify the relevant flow and control units in the network model. Finally, the network model must be modified by

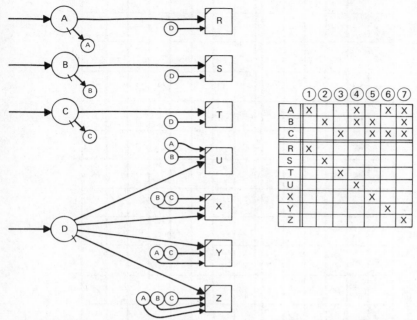

FIGURE 9.4   Graphical structure for all combinations of A, B, C.

incorporating or "wiring in" the mechanism for the decision table according to the flow units and relevant states involved in the mechanism.

In certain instances, it is necessary to make transit along two or more paths mutually exclusive. As discussed in Section 8.4, this can be accomplished using a Class II structure. However, if it is desired to maintain mutual exclusiveness and give priority to one path over all others, a routing structure develops

Table 9.1   POSSIBLE ACTIONS WITH THREE COMMODITY LIFT PROBLEM

Commodity \ Lifting Device	Tower Crane	Roof Hoist	Man Hoist
Reinforcing steel	X (1)	X (4)	Not feasible
Ducting	X (6)	X (5)	X (7)
Brick pallets	X (3)	X (2)	X (8)

Table 9.2 DECISION TABLE FOR THREE COMMODITY PROBLEM

	1	2	3	4	5	6	7	8
Rebar available			X			X	X	X
Ducting available	X			X			X	X
Bricks available		X			X		X	X
Tower crane available	X	X	X		X		X	
Roof hoist available				X	X	X		X
Man hoist available						X	X	X
Rebar with tower crane			X				X	
Rebar with roof hoist						X		X
Ducting with tower crane		X			X			
Ducting with roof hoist				X				
Ducting with man hoist							X	
Bricks with tower crane	X						X	
Bricks with roof hoist				X				X
Bricks with man hoist					X	X		X

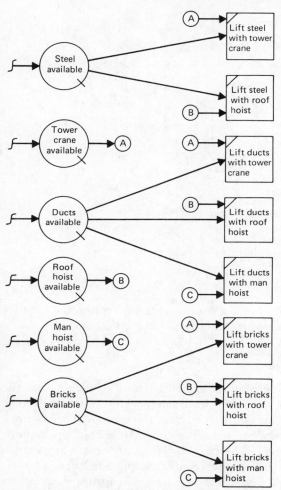

FIGURE 9.5   Logical method for three commodity problem.

that differs from the basic switching. In this application and extensions of it, the Class III structure is used and referred to as a "gating" mechanism. The COMBI work tasks in such a mechanism are called "gates" and operate to provide entry to a section of the system. The control unit in the idle state of the Class III structure and the COMBI numbering sequence used determine the priority of passage through the "gating" mechanism. In Figure 9.6, a four-gate mechanism is shown giving priority to gates A, B, C, and D in that sequence. Complex gating mechanisms can be developed by expanding the

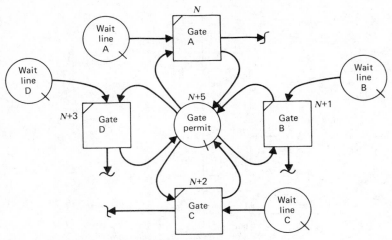

FIGURE 9.6   Typical gating mechanism.

QUEUE NODE idle state ($N + 5$) into a control structure consisting of other mechanisms. This will be illustrated in an example in Section 9.5.

9.3 CAPTURE MECHANISMS

Very often a modeling situation requires the withdrawal of a unit from the system for a certain period of time and its reentry. This function can be handled by using a capture mechanism that diverts the unit to be withdrawn into a holding activity for a prescribed period of time. After holding the unit, it is released back into the system. The breakdown or periodic maintenance of a resource can be modeled in this manner. Similarly, the 8-hour on, 16-hour off shift cycle of a crew can also be represented using a capture mechanism.

The basic capture mechanism consists of two Class I control structures interfaced at a common COMBI work task that holds or retains the unit to be withdrawn for a predefined period and then releases it back into the system. The capture mechanism in its simplest form is shown in Figure 9.7a in conjunction with a slave cycle structure.

The diverting of the unit to be withdrawn takes place at an idle state in its flow cycle. In this instance, the resource to be diverted is "captured" from QUEUE node 2. Upon being withdrawn, it is routed to the HOLD TIME WORK TASK. The withdrawal cycle 1–2 is ancillary to the resource's primary cycle 2–3. The withdrawal action is initiated by a withdrawal command that circles between 1 and 4. The frequency with which the resource unit is withdrawn is tied to the release time activity (4) in the withdraw command cycle. The time for which the unit is held in its withdrawn status is defined by the duration of the hold activity (1).

In order to illustrate this function, let us examine the capture mechanism for withdrawing a crew unit for 16 hours (off-shift time) and releasing it for a work period of 8 hours (on-shift time) once every 24 hours. This situation is shown in Figure 9.7*b*.

The crew unit to be shifted is in a primary Class I structure (slave cycle) between activities 1 and 2. During the time the withdraw command unit is in the release activity at 4, the crew unit is free to cycle between 1 and 2. Upon exit of the withdrawal command from 4, it moves to 5, where it establishes a capture situation. The crew unit, upon returning to QUEUE node 2, will now be diverted to 1, since its ingredience requirements are now satisfied (i.e., units at 2 and 5) and it is the lowest number COMBI work task to which the crew unit can be routed (i.e., 1 < 3). Therefore the crew unit is routed to 1, where it is held for 16 hours. During this time the crew is not available at 2 and the work task 3 is shut down. After the 16 hours of off-shift time is expired, the crew unit returns to its available position at 2 and the withdrawal command moves to 4, where it is held for 8 hours. This allows the crew unit to remain actively in service for 8 hours until the withdrawal command again moves to 5. The crew unit is alternatively "captured" and then released on a 24-hour timetable. The cyclic withdrawal and insertion are governed by the individual and additive durations of the work tasks in the withdrawal command cycle.

In general, a capture mechanism consists of:

1. An access point (QUEUE node) in the primary cycle of the unit to be withdrawn.

2. An ancillary cycle into which the unit to be withdrawn is directed and in which it is held for a prescribed length of time until reinserted.

3. A withdrawal command cycle that contains a unit that captures and releases the withdrawn unit on a predefined time table.

The cycle time of the withdrawal command that governs the cyclic movement of the withdrawn unit is:

RELEASE TIME + HOLD TIME = TOTAL COMMAND CYCLE TIME

The duration of the WORK TASK is not normally related to the RELEASE or HOLD times other than it is in most situations considerably smaller in duration. In the crew shift example, for instance, the work task may be the rigging of a steel member with an average duration of 15 minutes. This work task will take place many times during the release period. It is however, generally independent of the hold and release times.

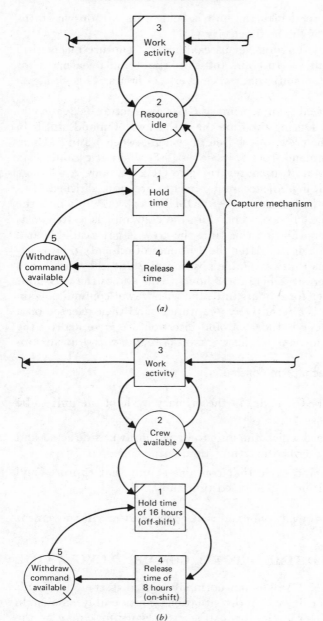

FIGURE 9.7 Capture mechanisms. (*a*) Basic capture mechanism. (*b*) 8 ON/16 OFF shift mechanism. (*c*) Command initialization preceding release hold.

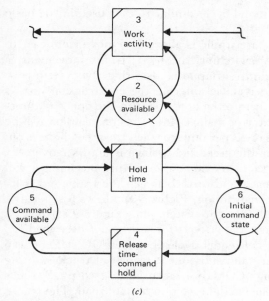

(c)

FIGURE 9.7   (*Continued*).

One other item of importance relative to the basic capture mechanism is the initial position of the withdrawal command. This determines whether the resource to be withdrawn is in a withdrawn (i.e., hold) state at time $t = 0$. In the model of Figure 9.7*b*, there is only one QUEUE node (at 5) in the withdrawal command cycle. Therefore, the command unit must be initialized at this point and the resource unit (i.e., the crew) will be withdrawn to 1 at time $t = 0$. In order to insure that the crew resource unit is active at time $t = 0$, another QUEUE NODE must be inserted in the command cycle preceding release activity 4. This allows the command to be held for the release period prior to capturing the resource unit at 1. This modified structure is shown in Figure 9.7*c*.

9.4 EXTENSIONS OF CAPTURE MECHANISMS

The capture mechanism can be used in conjunction with other constructs to model a variety of commonly used functions. In Section 8.4, a Class II control structure is used to model the flag control system on a constructed single-lane roadway with haul traffic moving east and west. If the traffic movement problem is to be modeled with an automatic light system in operation instead of the control flag, then the problem requires the combination of a Class II structure with a command cycle similar to that used in the basic capture mechanism.

Using a green light system, suppose traffic is allowed to move from east to west for 3 minutes and then, by switching the light, traffic movement is allowed from west to east for 3 minutes. Suppose, also, that the system is to allow an extra half minute for any vehicle entering the construction just as the green light changes, to transit before the green light in the opposite direction is switched on. The rate of entry of vehicles into the construction will be assumed to be one every 5 seconds or 12 per minute when the green light is on. The unit that must be captured is the green light, since it allows movement and plays the role of the flag in this situation. The capture mechanism will act to move the green light unit back and forth between the traffic lights located at the entry points to the construction. Figure 9.8a shows a schematic view of the traffic constriction with its light system, and Figure 9.8b shows a logical model designed to implement the criteria for movement described.

The green light unit in the logical model cycles between QUEUE node 6 and 11. When at 6, it permits entry and subsequent transit of trucks waiting to move to the east. It cycles through COMBI 4, essentially "escorting" trucks waiting in the QUEUE at 3 into the one-lane section of haul road. The trucks are escorted into the construction once every 5 seconds. Having entered the construction, they take an average of 30 seconds to pass through the one-lane portion of road.

During the time the green light is at 6, the withdrawal command unit is in a delay work task at 7. The withdrawal command cycles between two divert activities at 1 and 2. After exiting the delay at 7, the withdrawal command unit moves to 8 and captures the green light moving it to 1. The green light unit is delayed in 1 for $\frac{1}{2}$ minute, allowing any truck entering the constriction at 4 just before the light changed to pass before releasing traffic in the opposite direction. For this half minute, both control lights are red. Following the delay at 1, the green light moves to 11 and becomes available to escort traffic from east to west again on a 5-second entry cycle. The command unit exits and moves to a 3-minute delay work task, releasing the movement of traffic to the west for this period of time. Following the delay, it moves to 9 again, capturing the green light and causing red lights to go on at both control signals. The movement of the green light to 6 and the command unit to 7 is similar to the sequence required to release traffic in the westbound direction. Figure 9.8 incorporates a structure that can also be referred to as a gating mechanism. If the subnetwork associated with the green light availability and timing (i.e., including nodes 1, 2, 6, 7, 8, 9, 10, and 11) is envisaged as collapsing into a single node then the node becomes the gate permit queue node of Figure 9.6, gate A corresponds to COMBI 4, and gate B to COMBI 13.

A more complex mechanism results if there is a three-way train haul system such as the one shown in Figure 9.9a. Such systems are typical in earthmoving

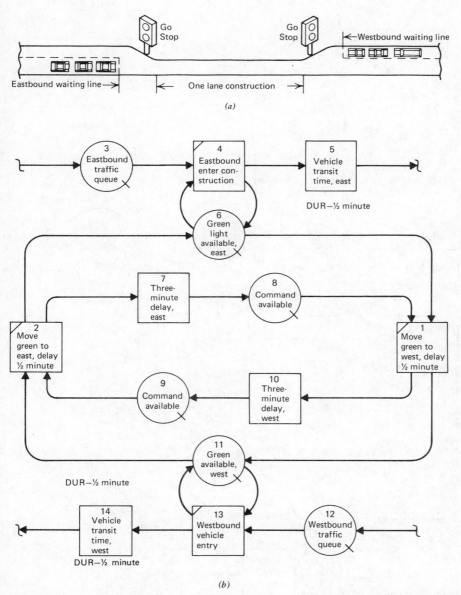

FIGURE 9.8   Haul road control with automatic light. (*a*) Schematic of haul traffic construction. (*b*) Capture mechanism for haul traffic construction situation.

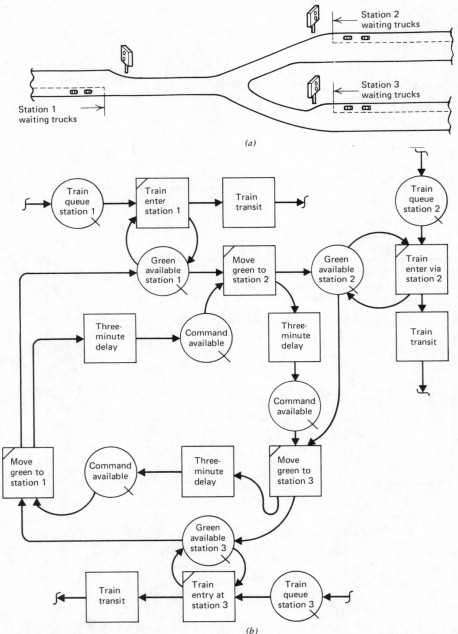

FIGURE 9.9 Three-way control system. (a) Schematic of three-way haul problem. (b) Logical model.

situations as well as in tunneling and mining operations. The solution to this control problem is possible using a capture mechanism that is an extension of that developed above. Again, as was the case in the two-way system, the command unit cycle is nested within a Class II structure that defines the green light area. The extension of the problem leads to a butterfly pattern radiating from the nested capture mechanism and Class II structure. The COMBI work tasks implement train entry from the individual loops of the butterfly structure.

Other functional constructs can be developed by combining the capture mechanism with other control structures. Some typical combinations are shown in Figure 9.10. In some situations, a resource is constrained by two factors. In this case, a double capture structure might be used. In a beach reclamation project, a barge-mounted crane is driving pile for a retaining wall. The work can only be pursued during high tide, since at other times the water depth is not sufficient to handle the draft of the barge. Because of union rules, the crane and driving rig cannot be operated at night. The tide is in for 9 hours and out for 18 hours. There are 13 hours of light and 11 hours of darkness. The mechanism of Figure 9.10c can be used to model the availability of the crane by setting hold 1 and release 1 to 18 and 9 hours durations to account for the tidal action. The hold 2 and release 2 durations would be defined as 13 and 11 hours, respectively, to implement the constraint on work after dark.

9.5 MECHANISMS FOR NONSTATIONARY PROCESSES

Often, in modeling processes, situations are encountered where the delay or travel times associated with the process activities do not remain the same, but change or vary as the process progresses. A typical example of a construction process in which this is the case is that of trucks hauling asphalt from a batch plant that is located alongside a linear construction site such as a roadway. Such a situation is shown schematically in Figure 9.11.

The diagram shows the stretch to be paved divided into five separate sections. The batch plant is located at the center of the stretch to be paved, and work progresses from left to right. The average travel times (T.T.) to sections 1 and 5 are the same. The loops indicate the haul cycle to these sections and T.T. 1 as the average duration of the haul cycle. The travel time values can be considered either as deterministic or stochastic (i.e., random about a mean). The average travel time to sections 2 and 4 is given as T.T. 2 and the section 3 travel time is T.T. 3. Assume, for the sake of illustration, that the values of the travel times are as follows.

$$\text{T.T. 1} = 15 \text{ minutes}$$
$$\text{T.T. 2} = 10 \text{ minutes}$$
$$\text{T.T. 3} = 5 \text{ minutes}$$

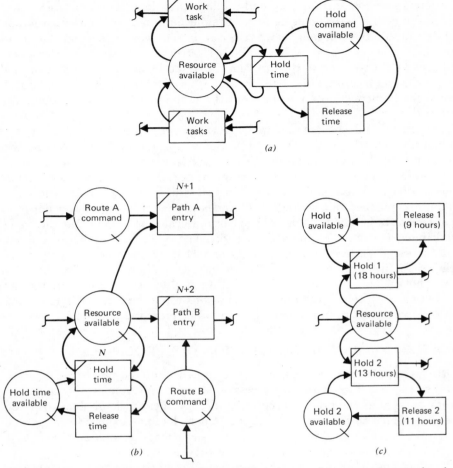

FIGURE 9.10 Typical structure combinations with capture mechanism.
(a) Capture mechanism with Class III structure. (b) Capture mechanism with
Class V structure. (c) Double capture mechanism.

These are average travel times. Figure 9.12 shows a graphical presentation of
the distributions of these travel times, assuming a deterministic analysis and
a probabilistic analysis. In the deterministic analysis, the probability of T.T. 1
during work on sections 1 and 5 is 1.0. That is, the travel time for sections
1 and 5 is exactly 15 minutes. In the probabilistic analysis, the probability of
a given travel time is given by a distribution with mean T.T. (i) ($i = 1, 2, 3$).
The distribution mean changes across the life of the project starting with
$i = 1$. The means vary with time in the following sequence: (i, T.T.) 1, 15;
2, 10; 3, 5; 2, 10; 1, 15. Therefore, the travel times vary from 15 down to 5

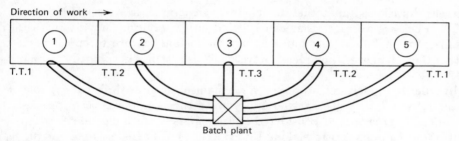

FIGURE 9.11 Schematic diagram of roadway paving project.

and back to 15 minutes. Variation of this type leads to a nonstationary process. That is, when time parameters vary across the life of a process, it is said to be nonstationary. If, however, time parameters stay the same across the life of a process, it is said to be a stationary process.

For this illustration, it is not important whether the travel times are deterministic or probabilistic, since this attribute is defined when specifying the

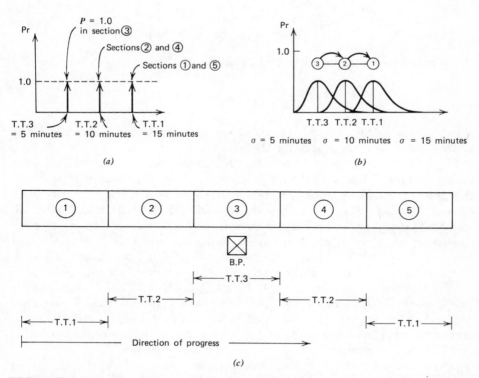

FIGURE 9.12 Travel time distributions. (a) Deterministic values. (b) Probabilistic distributions. (c) Nonstationary travel times.

element duration in an actual model. In this example, it is assumed that the
T.T. values are averages that can be used either deterministically or as the
means of probabilistic distributions. The important feature to be considered
is that they change as the process proceeds. This is shown graphically in Figure
9.12c.

In order to model such a process, a mechanism must be designed capable of
changing the T.T. values at the appropriate times as the process proceeds.
An initial approach is to consider the problem as a routing problem that requires routing of haul trucks along PATH 1 when T.T. 1 is required and along
paths 2 and 3 when T.T. 2 and 3 are required. A routing mechanism is needed
that switches the travel time parameter as required. A Class V control structure will be needed, since it is capable of performing the switching function.
A nonstationary process mechanism can be developed by combining a Class V
structure with a series of capture mechanisms.

Assume that 50 haul cycles are associated with each section and that work
proceeds from left to right; this means that after the first 50 haul cycles,
section 1 is completed, after the second 50 cycles, section 2 is completed, and
so forth. The section S being worked is simply the integer division:

$$S = \left| \frac{\text{Haul cycles to date}}{50} \right| + 1 \qquad (9.5)$$

If the number of cycles worked to date is 35, then Equation 9.5 becomes:

$$S = \left| \frac{35}{50} \right| + 1 = 0 + 1 = 1$$

Based on this calculation, it is possible to determine which T.T. is appropriate
and to select the correct path for the haul truck.

The logical model to implement this switching function is given in Figure
9.13. The basic Class V structure radiates from the truck queue at 4. From
this point, the asphalt trucks can be routed by either PATH 5, 6, or 7. Each
path activity has a control unit associated with it at QUEUE nodes 12, 18,
and 22. The capture and release of these control units determines which path
and the travel time to which the truck units will be routed. Capture mechanisms are associated with each control unit at COMBI work tasks 1, 2, and 3.
The activation of these capture mechanisms is controlled by the CONSOLI-
DATE function node located at 9. This CONSOLIDATE function node is
linked to the WORK PLACEMENT work task 8 through the off-page con-
nector A. After the WORK PLACEMENT work task is completed, the truck
unit is returned to 4 for routing and a pulse is sent to 9 to note that one haul
cycle has been completed.

When the first 50 haul cycles have been completed, node 9 releases a pulse
to 10 and 11. The pulse to 11 activates the capture work task 1 and results in

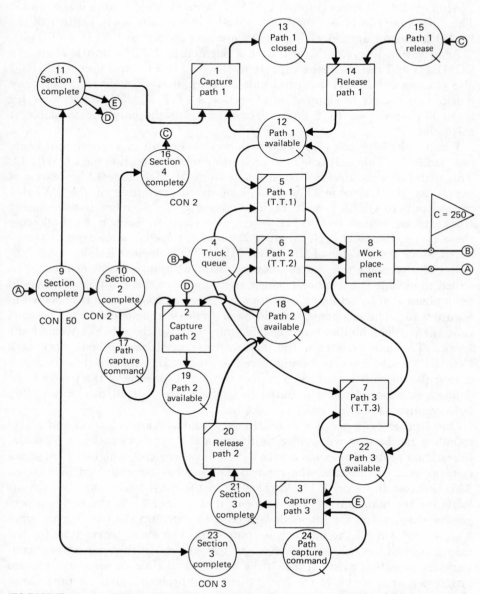

FIGURE 9.13 Nonstationary process mechanism (asphalt haul process).

the diversion of the PATH 1 control unit to 1. The path work tasks have been numbered in ascending sequence so that the next lowest-numbered COMBI work task available for truck routing is work task 6. Therefore, truck units are now routed to PATH 2 with an average T.T. of 10 minutes.

After cycling 50 times through PATH 2 (work task 6), a pulse is sent to 17. The pulse (unit) to 17 is routed to COMBI 2 and results in the capture of the PATH 2 control unit. Arriving trucks are now routed to PATH 3 (T.T. 3).

Again, after 50 cycles, a third unit is released to 24. The unit is routed to COMBI 3 and results in the capture of the PATH 3 control unit. COMBI 3 also releases the PATH 2 control unit back to QUEUE node 18. Fifty more truck cycles result in completion of section 4 (T.T. 2). The travel time for these 50 cycles is set to T.T.(2), since PATH 2 is the only one available at this point.

Finally, the 50 cycles associated with section 4 result in a fourth unit being released to 16. This unit results in realization of the condition for COMBI 14. This withdraws the PATH 2 control unit from QUEUE node 13 by release of a unit to QUEUE node 17 and implements the rerouting of the PATH 1 control unit to QUEUE node 12. The system cycles 50 more times, causing trucks to be routed by PATH 1 (corresponding to Section 5). Following these cycles, the total cycle count is 250 and the system simulation is terminated, since the C value of the ACCUMULATOR element is 250.

In order to make this system more realistic, a capture mechanism could be added to include the work shift effect as it applies to this example. This could be implemented by adding a capture mechanism at 4 similar to that shown in Figure 9.10c. The mechanism would have a release time equal to the on-shift time (i.e., trucks hauling) and a hold time equal to the off-shift time of the trucks. To implement this, the shift capture mechanism "hold work task COMBI" would have to be numbered 4 (i.e., <5) to insure priority for it during the off-shift period. Also, a hold command for each of the trucks to be defined would need to be initiated to insure capture of all trucks (i.e., "n" hold commands for "n" trucks on the haul).

One further example of a nonstationary construction situation and corresponding mechanism will suffice to illustrate this type of construct. Consider the routing problem associated with a man hoist servicing a high-rise construction project. In this case, the vertical building has been divided into zones. This is shown in Figure 9.14. At the start of the construction, most of the lift loads for the man hoist will be concentrated in zone 1. As the building progresses the demand for lift loads will tend to move up the building to zones 2, 3, 4, and 5 until the building is completed. The mean travel time for the man hoist will increase across the life of the project. In this case a linear construction situation exists in which the mean travel time varies as the process progresses. In the case of the batch plant haul problem, the haul time variation reversed itself, starting at 15 minutes, reducing itself to 5 minutes, and

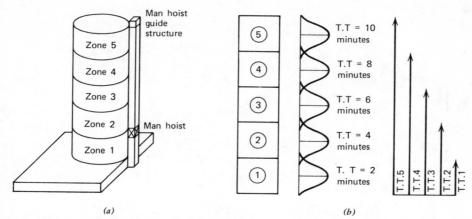

FIGURE 9.14 High-rise building with nonstationary process. (a) Schematic of high-rise project. (b) Travel time distributions.

then returning to 15 minutes. In this example, the mean travel time continues to increase as construction progresses. Assuming that five mean travel times are enough to characterize the problem environment, a mechanism can be developed to model this situation. It is assumed that each zone will require 1000 lift loads. It is important to know how long it will take to move the required 5000 lift loads, since this will constrain the duration of the project.

The development of the logical model for this problem is similar to that used for the asphalt haul problem. The logical model is shown in Figure 9.15. A Class V distributive structure is combined with appropriately placed capture mechanisms in sequence. Paths are withdrawn in sequence from 1 to 4 until only one PATH 5 is available. To implement the withdrawal sequence, control units are associated with each of the path activities. PATH i is associated with travel time i (T.T. 1) which, is associated with zone i.

Priority on path routing is changed following every 1000 cycles of the system, since a 1000 lift load is associated with each zone. Therefore, following the first 1000 cycles of the system, a pulse unit is generated at 13 and sent to QUEUE node 14, which triggers withdrawal of the PATH 1 control unit. The PATH 1 unit is routed through COMBI 1 to a sink node at 25. Since the paths are numbered in ascending sequence, the path priority now goes to PATH 2 and COMBI work task 8. Checking the rest of the system is left as an exercise for the reader.

9.6 UNIT GENERATION MECHANISMS

To this point, only closed systems have been considered in which the exiting production unit is recycled and enters the system as a new unit. This method

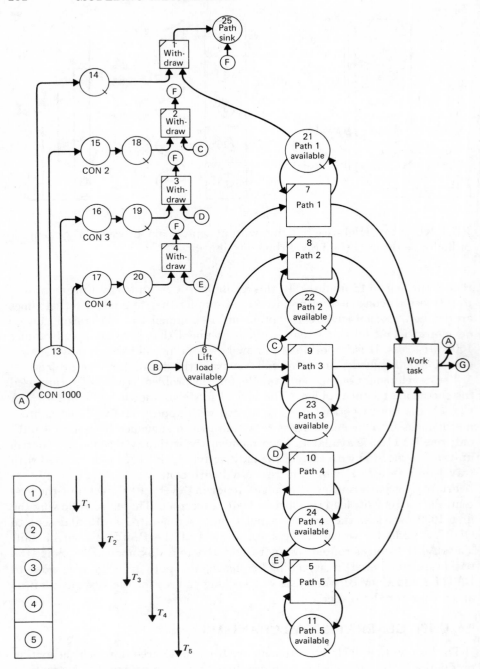

FIGURE 9.15   High-rise man hoist model.

was used in the tunneling problem in Chapter Five and the steel erection problem in Chapter Eight. In the tunneling system, sufficient pipe sections were defined at the "PIPE STOCKPILE" QUEUE node to insure that the system is not constrained due to lack of pipe sections in the stockpile. The rate of arrival of new pipe sections in the stockpile is equal to the productive release rate of COMBI 5 ("JACK 6 FEET FORWARD").

In some cases, it may be necessary to introduce units into the system at a specified rate. In closed systems, this function can be achieved by adding an element to handle the function of releasing units from the recycle QUEUE node at a predetermined rate. The QUEUE node to which units are returned after "exiting" the system acts as a reservoir of units available to reenter the system as "new" or "arriving" units. As such, it can be referred to as the recycle unit reservoir. By adding a COMBI work task between the recycled unit reservoir and the arriving unit location, units can be drawn from the reservoir and placed in the arriving unit QUEUE node at a specified rate and according to a predefined input process. This is shown schematically in Figure 9.16. The rate at which units are drawn from the reservoir and are reentered into the system is defined by the distribution and duration parameters associated with the COMBI work task. The COMBI work task acts as a buffer between the reservoir and the reentry QUEUE nodes. The control unit that is connected with the entry rate COMBI in a slave cycle insures that units are passed individually from reservoir to reentry QUEUE nodes at the rate defined by the COMBI work task.

Consider the tunneling model. In order to release pipe section units into the system at a specified rate of arrival, a unit reentry mechanism must be defined preceding the "PIPE SECTION STOCKPILE" QUEUE node. This modification is shown in Figure 9.17a. Assume that the system requires arrival at the stockpile at a Poisson input rate. This is achieved by associating an exponential delay distribution with the inserted COMBI unit. The COMBI element, with its associated control unit, insures that recycled units are passed from the "RECYCLED UNIT RESERVOIR" QUEUE node to the "PIPE SECTION IN STOCKPILE" work task in conformance with a Poisson

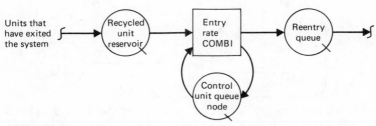

FIGURE 9.16   Unit reentry mechanism.

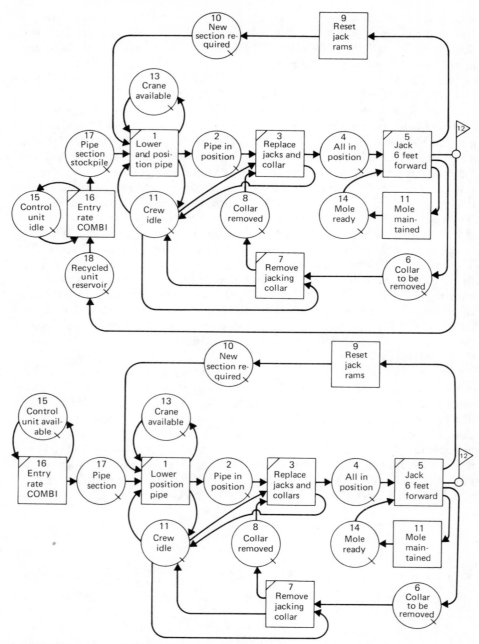

FIGURE 9.17 (a) Tunnel system with unit generation mechanism (closed system). (b) Tunnel system with unit generation mechanism (open system).

process. This assumes, of course, that the recycling reservoir never becomes empty. The COMBI element acts as a unit generation mechanism, implementing the desired arrival rate.

To insure that the system never becomes constrained by the recycling process, an open-ended system can be utilized in which units enter from an infinite source and exit into a sink node (normally an ACCUMULATOR element). The tunneling model in this configuration is shown in Figure 9.17b. The appearance of the open ended model is essentially the same as the closed production cycle model. The only difference is that the recycling reservoir element has been deleted. In this configuration, a unit is initialized at QUEUE node 15 and cycles continuously at COMBI 16, resulting in generation of units into the stockpile QUEUE node 17 at the specified rate. Units exit the system through a sink node at ACCUMULATOR (12). The rate of units arriving in the system both in the open and closed systems depends on the distribution type and transit time duration or duration parameters associated with the buffering COMBI element (i.e., COMBI 16 in this example).

It is sometimes desirable to temporarily break in on or interrupt the defined arrival rate. In order to interrupt the specified input rate of certain frequencies, a capture mechanism can be defined in conjunction with the unit generation mechanism. Such a mechanism is used to withdraw the control unit at certain points in time. For example, assuming that the delivery of pipe to the stockpile location can occur only during the first 4 hours of any work day, a release time of 4 hours and a hold time of 20 hours can be associated with the control unit. This is depicted in Figure 9.18. As shown, the control unit is initialized at QUEUE node 17 and cycles normally between 16 and 17. The capture command is initialized at 18 and enters the "RELEASE TIME" COMBI. This QUEUE node (18) is provided so that the system will not immediately capture the unit generation control unit. If the capture command unit had been initialized at 20, the generation of incoming pipe sections to the STOCKPILE QUEUE node would be delayed because COMBI node 15 would have priority and would be executed before COMBI node 16.

A detailed discussion of arrival patterns is beyond the scope of this text. However, some arrival patterns are so commonly encountered in real-world situations that they are worth brief comment. The arrival rate most commonly encountered is one approaching total randomness. In this type of arrival rate, the probability of a unit arriving in any given interval, t, is proportional to the interval. The actual distribution of such random arrivals can be shown to be exponentially distributed. The mean interarrival time is referred to as T, and the mean interarrival rate is

$$\lambda = \frac{1}{T} \qquad (9.5)$$

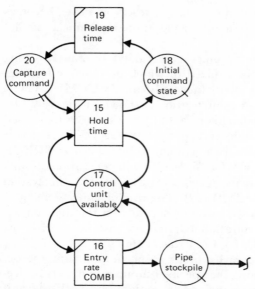

FIGURE 9.18   Unit generation mechanism with interrupt.

Based on this fact, the probability of an arrival in the period Δt is $\lambda\,\Delta t$, since λ is the average rate of arrival per unit time. The probability density function of the exponentially distributed interarrival times is given as:

$$f(t) = \lambda e^{-\lambda t} \quad \text{for} \quad t \geq 0 \tag{9.6}$$

This distribution is shown in Figure 9.19a. The cumulative density function $F(t)$ is the probability that an interarrival time of duration less than t will occur.
That is,

$$F(t) = \int_0^t \lambda e^{-\lambda t}\,dt$$

$$= \left| -e^{-\lambda t} \right|_0^t$$

$$= -e^{-\lambda t} - (-1) = 1 - e^{-\lambda t}$$

The distribution of the number of arrivals in the interval of duration t is a random variable. For a system with exponential interarrival times, it can be shown that the probability of n arrivals occurring in an interval of length t is

$$P(n) = \frac{(\lambda t)^n e^{-\lambda t}}{n!} \quad (n = 0, 1, 2 \ldots) \tag{9.7}$$

This is a discrete probability distribution and appears as shown in Figure 9.19b. This distribution is called a Poisson distribution, and processes that exhibit exponential interarrival patterns are commonly called Poisson input processes.

By taking the inverse of the cumulative function of exponential distribution, the random variate techniques presented in Chapter Six can be used to

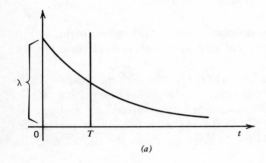

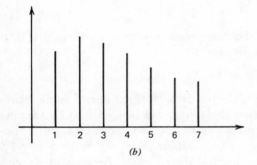

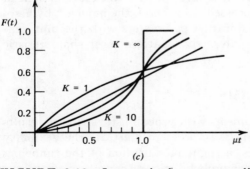

FIGURE 9.19 Interarrival pattern distributions. (a) Exponential distribution (P.D.F.). (b) Poisson P.D.F. (c) Cumulative density functions for Erlang distributions with various values.

generate exponential interarrival times. The inverse is given as:

$$\lambda t = -\log (1 - y)$$

If y is generated as a random number from a uniform distribution defined on the interval (0, 1), then $(1 - y)$ is also a uniformly distributed random number between 0 and 1. Solving the t in terms of y, the relationship is:

$$t = \frac{-\log (y)}{\lambda} = -T \log (y)$$

Therefore, by generating a random number (0–1.0) and substituting it into the above expression, an exponentially distributed interarrival time is computed.

The exponential distribution is a special case of the Erlang class of distributions named after A. K. Erlang, who found them to be descriptive of telephone call arrivals. This set of distributions is characterized by a shape parameter, k. By varying the parameter, k, the set of distributions varies from exponential to normal. The probability density function for the Erlang distribution is given as:

$$f(t) = (k\lambda)^k \left[\frac{e^{-k\lambda t}}{(k-1)!} \right] t^{k-1} \tag{9.8}$$

By setting the value of $K = 1$, the above expression reduces to the probability density function of the exponential distribution

$$f(t) = \lambda e^{-\lambda t}$$

Figure 9.19c shows a plot of the cumulative density functions of some representative k values ($1 \leq k \leq \infty$). From this plot it can be seen that by varying the K value, the Erlang distribution set can be used to approximate distributions varying from exponential ($k = 1$) to constant ($k = \infty$). With a $k = 10$, the Erlang reduces very nearly to a normal distribution. Because of its versatility, the Erlang distribution often is helpful in describing the arrival of units arriving in the system to be processed. In general, the modeler, by associating the correct distribution and duration parameters with the unit generation mechanism, can achieve the interarrival pattern most characteristic of the system under study.

9.7 PROBABILISTIC MECHANISMS

The combination of various mechanisms with probabilistic switches results in a wide variety of modeling *constructs*. In order to illustrate the flexibility offered by such composite structures, a slight modification of the tunneling problem will be considered incorporating probabilistic and capture mechanisms. The extension of this problem relates to the tunneling speed that can be achieved because of the resistance of the material being penetrated. Assume that generally three types of material are encountered:

1. Soft clay: Penetration rate = 6 feet per hour.
2. Hard clay: Penetration rate = 4 feet per hour.
3. Hard clay with fragmented rock: Penetration rate = 2 feet per hour.

The differing penetration rates associated with the three types of material cause the time required for the "JACK 6 FEET FORWARD" activity to vary from 1 to 3 hours. The type of material encountered is, however, a random variable. Records show that the material is encountered with the following expected probability:

1. Soft clay: 30% of the time.
2. Hard clay: 30% of the time.
3. Hard clay with rock: 40% of the time.

That is, there is a 30% chance that the material encountered in the next 6 feet of construction will be soft clay, a 30% chance that it will be hard clay, and the remaining 40% of the time it will be hard clay with fragmented rock. In order to interject this randomness of the material encountered, a probabilistic switch is used, as shown in Figure 9.20a. This model is relatively simple and does not consider the constraint of the maintenance required on the tunneling machine (i.e., "MOLE MAINTENANCE"). The assumption is that the jacking activity commences as soon as the jacking collar and jacks are in place. Introduction of the tunneling machine maintenance cycle adds a considerable degree of complexity. This occurs since two flow units must now be properly routed from preceding QUEUE nodes to the COMBI associated with the randomly selected penetration rate.

A model achieving this routing pattern is shown in Figure 9.20b. The movement of the processed unit (at QUEUE node 3) through the system is controlled by the single control unit, which recycles after each 6 feet forward motion of the tunnel liner returning to the probabilistic switch at FUNCTION node 2. From this point the control unit is routed randomly to one of the QUEUE node locations 41, 51, 61. Following 6 feet of penetration, the tunneling machine is routed to the maintenance activity at 8 and then enters the QUEUE node at 9. When the jacks and collar are in place and the maintenance is complete on the tunneling mole, the processed unit is routed from QUEUE node 9 to one of the COMBIs 4, 5, 6, depending on the location of the randomly routed control unit. In effect, the type of material encountered and the rate of penetration vary after each penetration cycle. Although in this case the variability in rate is associated with the rate of tunneling progress,

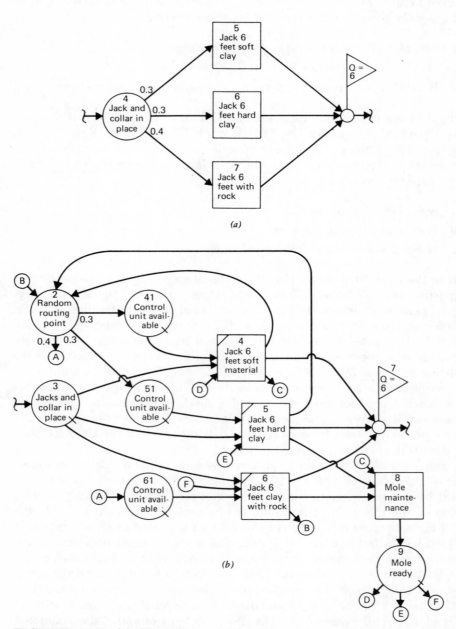

FIGURE 9.20 (a) Simplified model with random penetration rates. (b) Probabilistic mechanism.

the mechanism could just as easily have been used to model the variable rates of progress per story in a high-rise apartment building. In such an application, the rate for good conditions, fair conditions, and poor conditions could be associated with the COMBI work tasks 4, 5, and 6. For instance, the time required to complete pouring floor slabs and columns or shear walls (i.e., floor cycle time) with good conditions might be 3 working days with fair conditions, 5 days and, with poor conditions, up to 7 days. These rates would be associated with the appropriate COMBI work tasks to model the variability. In this context, the cycle time for the completion of a building floor can be thought of as a rate of penetration or progress and the building construction as a linear process in a vertical direction.

One further extension using the probabilistic switch will be considered. Suppose information is available regarding the seam widths of various strata encountered in tunneling. It may be possible to predict that soft clay when encountered will normally continue for 60 feet or 10 penetration cycles. Similarly, information regarding hard clay strata may indicate that when encountered, hard clay seams are 25 feet or approximately 4 penetration cycles in width and fragmented rock seams are 12 feet in width. By inserting appropriate CONSOLIDATE functions in the present model, this aspect of the operation can be reflected. The incorporation of this feature into the tunneling process model is shown in Figure 9.21. Once the control unit is routed to one of the QUEUE nodes 41, 51, or 61, it is not returned immediately to the probabilistic switch at FUNCTION node 4. It continues to cycle until the CONSOLIDATE function node (i.e., 44, 54, or 64) following the associated COMBI is realized. Realization of the associated function node causes activation of capture mechanism COMBI work tasks at 40, 50, or 60. The action of the capture mechanism returns the control unit to the random switch at 4. As defined above, the appropriate numbers of consolidations for each CONSOLIDATE function are shown at nodes 44, 54, and 64. With this structure, 10 soft clay cycles will be processed before releasing the control unit back to the random switch. Similarly, 4 hard clay cycles and 2 fragmented rock cycles will be completed before random rerouting of the control unit occurs.

9.8 SUMMARY

This presentation of mechanisms has not been intended to be comprehensive. Instead, it is intended to illustrate the construction of some typical mechanisms and to indicate the wide variety of constructs that can be developed from the basic CYCLONE elements and control structures. These constructs provide the laboratory apparatus in terms of which process experiments can be developed and analyzed. As such, they give the modeler adequate latitude in defining the process to be investigated at the level of detail appro-

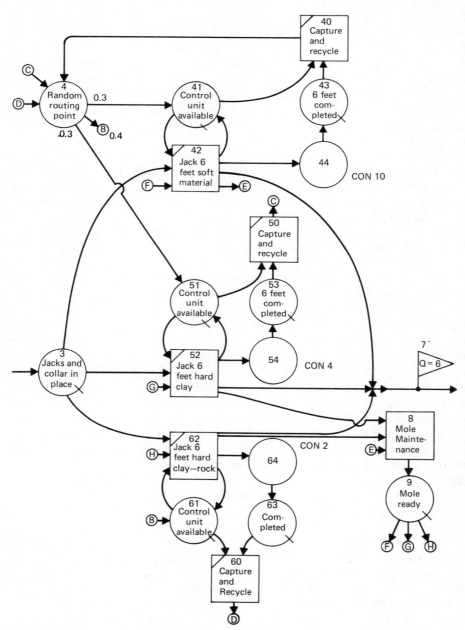

FIGURE 9.21   Composite mechanism—probabilistic switches and capture mechanisms.

priate to the analysis. A certain amount of innovative skill on the part of the modeler is required to arrive at the most efficient and representative model structure. This skill is easily acquired once a basic understanding of the logical relationships between the various CYCLONE elements and control structures has been achieved.

PROBLEMS

1. Develop decision table formulations for the steel erection and crane operations discussed in Section 8.8.

2. Extend the precast decking operation (see Figures 4.14 and Problem P8.3) to include the delivery of precast units at the loading dock and subsequent unloading to storage area or to immediate installation in the decking operation as a function of the current status of the number of elements held in the storage area and the work decision to initiate, proceed with, or terminate the precast decking operation. Develop a suitable decision table formulation of the precast unit delivery, storage, and decking operation and then develop a suitable CYCLONE mechanism to portray the construction management situation.

3. Assume that a high-rise city building site is serviced by a tower crane, a climbing roof hoist, and a progressively heightened man hoist as frame construction proceeds. Identify the major material flow mixes at the various stages of construction (i.e., initial frame construction, both frame construction and lower-floor finishing construction, and final cleanup and crane and hoist dismantling). Then develop suitable material handling decision tables for each construction stage.

4. A three-man labor crew is involved in trench excavation wherein one man uses an air spade to loosen material and, by working two faces, keeps two men occupied in shoveling material out of the trench. Every hour the air spade operator is relieved by one of the shovelers so that in any 3-hour period each man shovels for 2 hours and uses the air spade for 1 hour.
    If 15-minutes operation of the air spade generates 25 minutes of shoveling work, and the changeover of the work tasks takes 3 minutes, develop a capture mechanism model for the trench excavation operation. Develop a crew balance chart for the operation.

5. Consider an extensive fast-moving pipeline construction project located in a remote area that requires camp facilities for construction labor. Assuming that the normal construction operation of easement clearance, excavation, pipe welding, pipe laying, backfill, and so forth are to be modeled as separate crew work tasks, develop a nonstationary process model for the pipeline construction, with travel to and from camp included in the labor workday. How would you go about developing a

model to assist in the decision process for camp relocation as a factor in maximizing production?

6. In the road mix haulage operation discussed in Section 9.5, truck haulage distances vary, depending on the distance between the batch plant and the constantly moving road paver. A CYCLONE model formulation of the nonstationary process is shown in Figure 9.13. How would you go about developing the model to:

   (a) Determine the variation in truck fleet size to ensure constant production at the paver?

   (b) Determine the varying paver production for a constant fleet size?

   (c) Establish the maximum haulage distance before batch plant relocation becomes necessary?

7. A tunnel project uses a rail haulage mucking system with individual cars loaded by the mucker. The rail system is organized with a California switch at the loading face, with suitable provision of additional California switches along the length of the tunnel to allow locomotive trains to pass each other.

Develop a CYCLONE mechanism for the rail haulage operation for N locomotives each with M muck cars with California switches at I-foot intervals in a tunnel of length L feet.

8. (a) Given the model shown (Figure P9.8) for the transport of beam and deck sections to the site of a precast building project, utilize an unit generation structure to insert deck and beam sections into the model and a capture mechanism to withdraw the trucks from the transport sequences for 16 hours during every 24-hour period.

   (b) Modify the model to maintain the features required in part (a) above and provide for a break period of 5 minutes during each hour for each truck.

   (c) Half of the time, deck and beam sections on trucks arriving to off-load can be lifted directly into position. Otherwise, the sections are off-loaded into a stockpile location. Thirty-three percent of the time sections (if any) in stockpile are released for lift into the structure when the crane becomes idle due to lack of trucks to be off-loaded. Modify the model of Problem 2 to reflect this feature of the crane's operation.

9. Modify the paving model developed in Problem 5.3 into a nonstationary model covering five sections (See Figure 9.12).

10. Using a probabilistic switching mechanism, develop a model for the operation of the man hoist shown in Figure 9.14. Assume that the work has progressed to a stage such that the lifts demanded are as follows.

   (a) To zone I: 10%.

   (b) To zone II: 20%.

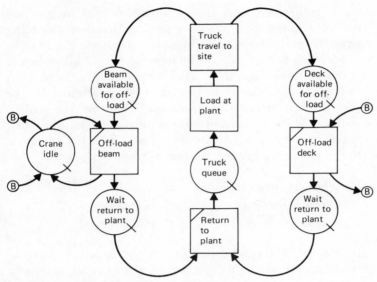

FIGURE P9.8   Transport model.

(c) To zone III: 15%.
(d) To zone IV: 20%.
(e) To zone V: 35%.

11. Using a probabilistic switch, modify the three-commodity system of Figure 9.5 to provide for the following arrival process.

(a) Steel lift-loads: 30% of the time.
(b) Brick lift-loads: 50% of the time.
(c) Duct lift-loads: 20% of the time.

Using the following lift times, do a hand simulation of the system to determine the system production in lift loads per hour.

Lift	Time (Deterministic)
Steel with tower crane	15
Steel with roof hoist	22
Ducts with tower crane	10
Ducts with roof hoist	16
Ducts with man hoist	20
Bricks with tower crane	15
Bricks with roof hoist	20
Bricks with man hoist	20

12. A river estuary is subject to tidal effects. Flat-bottomed freight and passenger ferryboats can negotiate the river to city Alpha for unloading regardless of the tide level. Normal freighters can negotiate the river to and from the off-loading berths only in the interval from 2 hours before high tide to 2 hours after high tide. The docking berths have been deepened so that once a freighter has docked, it is not affected by the level of the tide. Because of crowded conditions, ships are required to wait at the mouth of the river until a berth becomes free. This is signaled from the unloading area by radio. Movement to and from the mouth of the estuary (to the unloading berths) requires the following times:

> Flat-bottomed freighter: 25 minutes
> Normal freighter: 40 minutes
> Passenger ferry: 20 minutes

Five berths are available and can be used for off-loading either the ferry or the freighter. Because of the narrowness of the channel, only one ship at a time can enter or leave. Priority on use of the channel is as follows:

> Highest priority: Normal freighter
> Next highest: Passenger ferry
> Lowest: Flat-Bottomed freighter

Arrival of both types of freighters at the mouth of the river is random. The ferry arrives every 120 minutes ($\pm$20 minutes). The arrival patterns of the freight carriers have been observed as follows.

Interval Between Arrivals (in Hours)	Number of Observations, Freighter	Number of Observations, Flat-Bottomed Freighter
0–10	11	27
10–20	20	31
20–30	29	19
30–40	15	13
40–50	10	10
50–60	9	0
60–70	6	0

Berth time for the passenger ferry is fixed at 15 minutes. Unloading times for the freighters and flat-bottomed freight boats are as follows.

Time (in Hours)	Freighter	Flat-Bottomed Freighter
0–1	0	22
1–2	0	33
2–3	17	25
3–4	33	14
4–5	19	6
5–6	18	0
6–7	13	0

The interval between high tides is 13 hours. Using a capture mechanism to handle the action of the tide, develop a model that describes the operation of this system. Using 5-minute time units (1 unit = 5 minutes), hand simulate the operations of the system and determine the average time the normal freighters are delayed before entering to unload.

CHAPTER TEN

System Monitoring Concepts

Construction managers must continuously monitor and reevaluate progress to date in relation to planned progress, uncontrollable external influences, and the commitment and productivity of resources at their disposal. The management process amounts to an iterative decision cycle involving the planning, monitoring, and control functions.

A flowchart of the real-world management decision cycle is shown in Figure 10.1 in a process design format. The decision cycle, however, is essentially the same, whether the manager is concerned with flying a plane, driving a car, managing a football team, or running a construction job (see Table 10.1). A plan of action is developed. The planned progress is affected by the environment in which the activity takes place. The interaction of the plan with the environment results in an existing status that must be detected or monitored. The manager must then interpret the monitored status information and make decisions that affect the future status of the process.

CYCLONE provides a graphical modeling format for construction operation and process definition, which the manager can use to investigate and monitor system response. Ideally, the manager in the planning stage would like to have a laboratory in which system experimentation with various decision patterns would give indication of better operations and policies. In this way decisions relative to a model result in a system response that indicates potential reactions of the real-world situation.

Table 10.1   MANAGEMENT DECISIONS FUNCTIONS

Activity	Piloting an Airplane	Coaching a Football Team	Managing a Construction Project
(1) Planning	Flight plan	Game plan	Time and cost plan Site layout C.P.M. Process plans
(2) Environment	Head winds Weather patterns	Injuries Opposing team Tactics	Weather, labor situation (strikes, workouts) Materials supply Design changes
(3) Status	Location Altitude Air speed	Score Time left in game Yardage position	Percent complete Cost to date Time to completion
(4) Monitoring	Aircraft Instruments	Spotters Player reports	Field reports Scheduling updates Progress payment
(5) Decisions	Corrections Using aircraft controls	Substitutions New plays Alteration of game plan	Reallocation of labor Rescheduling Use of different technology (e.g., pump concrete)

The CYCLONE system element set provides a notation in terms of which the manager can define a laboratory model of a construction operation or process requiring investigation. The monitoring system in CYCLONE allows the manager to observe the response of the laboratory model and therefore gain insights into the possible response and interaction of the real-world system.

The concept of statistics collection in conjunction with CYCLONE system elements was introduced in Chapter Seven. Statistics, counting, and monitoring functions can be associated with the QUEUE and FUNCTION nodes and

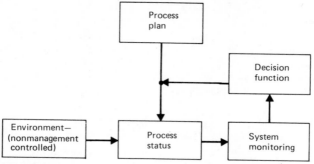

FIGURE 10.1   Management decision cycle.

aid in the detection of system performance. In addition, the accumulator element is used to determine system production. As described in Chapter Seven, the use of STAT clocks in hand simulation enables transit time information to be developed between various points in the system.

This chapter considers CYCLONE system monitoring functions and their use in investigating the response of CYCLONE operations or process models. The structure of the chapter results from the consideration of the following questions.

1.   What types of information regarding systems status and operation are of interest to the manager?

2.   In what form or format can this information best be presented?

3.   What are the mechanics of collecting this information?

10.1 INFORMATION AND STATISTICS COLLECTION

Several types of information regarding systems operation are of interest to the construction manager at the field level. The management information of interest in process design and operation relates to the following topics.

1.   Estimated system productivity.

2.   Delays in processing.

3.   Trace information.

4.   Resource utilization.

5.   Flow rates and station interarrival times.

Certain CYCLONE elements act as monitoring devices that allow this information, as developed by the model, to be assessed.

Management's main interest is the rate of production that can be expected from the process or system as configured. Production is directly tied to the delays in processing experienced by the processed units. In simple dual-cycle systems such as the shovel-truck system, the production delays are associated with the balance of productive capacity between the processor and processed units and all resources are fully utilized. In other systems such as the tunneling project (Figure 10.2), slaved units must be made available to insure expediting of the processed unit. In this process, for instance, a crane is slaved to the processed pipe cycle. The crane may not be highly utilized, but it must be available at all times so that delays experienced by the pipe sections processed are held to a minimum. If projected system productivity is not as high as expected or desired, delays being experienced by the processed units must be examined. The QUEUE nodes along the processed unit cycle act as monitoring stations that immediately indicate the length of delay as well as number of units delayed preceding each COMBI activity station. They operate as

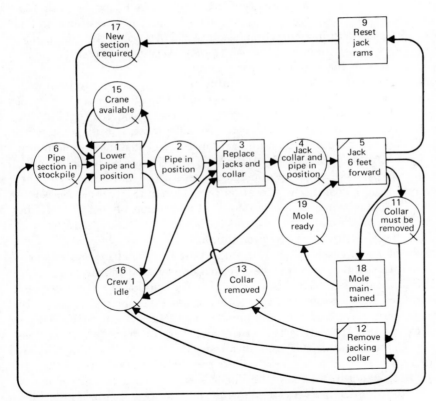

FIGURE 10.2 Total system with informational unit.

"windows" at which information regarding systems delays can be read. If long delays at various points in the system are causing bottlenecks for the processed units, the indication is that resource production rates at one point in the system may greatly exceed or be out of balance with resource rates at a later point. For instance, the pavement breaker in a sewer line problem may be considerably faster in producing sections available for excavation than the excavator is at processing them. This imbalance leads to a growing stockpile of sections waiting to be processed.

In determining the reasons for bottlenecks along the processed unit cycle, it is interesting to know something about the transit times of the required support units between various points in the system. This information can be generated using trace or interval statistics, which measure the time required for a unit to transit between two specified points in the system. This is the type of information developed using STAT clocks in Chapter Seven. For instance, in the tunneling problem, it may be interesting to know what the average transit time of the jacking collar is from the completion of jacking work task (5) to the commencement of the "REPLACE" work task (3). Such statistics can be developed using a trace on the flow unit. The unit is traced from its position following completion of COMBI 5 to the commencement of 3. If this transit time is excessive, the REMOVE work task (12) may be delayed because of the requirement for a crew.

Not only is system production important, but the cost of production is even more critical. The production cost is a function of the hourly rate of production and the hourly cost of the complement of resources (equipment, crews, etc.) required to achieve the rate. In most cases, it is possible to improve the rate of production by adding resources. For instance, the amount of delay at work task 3 REPLACE JACK & COLLAR can be reduced in the tunneling model by increasing the number of crews at QUEUE node 16. This adds, however, to the cost of the operation by an amount equaling the wages of the crew. This question is one of economical utilization of the crew. If the crew is idle 90% of the time, the utilization is poor and increased cost will probably not be offset by increased production.

Utilization of resources is, therefore, important in evaluating system performance and determining whether best return on invested resources is being achieved. Utilization is a term associated with the support or nonprocessed flow units, since these normally constitute the resources in the process. In many cases, these resources are incorporated into the system in a Class I (slave cycle) or Class III (butterfly) control structure format. As will be illustrated later in this chapter, this simplifies the process of determining utilization, since occupancy statistics relating to the QUEUE node(s) of the associated control structure indicate the idleness of the resource unit or units. Conversely, knowing the percent of the time that the resource units are idle, the percentage of the time that they are involved in production (i.e., the utilization) can be determined. For instance, in the tunneling example, by knowing the percent of the time a crew is idle at QUEUE node 16, the percent of the time the unit is active at 1, 3, or 12 can be calculated.

Rates of flow can also be established utilizing a statistic other than the trace or interval method. It may be of interest to keep track of the inter-arrival times of units arriving at a specified element. That is, it may be desirable to record the times elapsing between arrivals at a given element. In a typical haul type of problem it is relevant to audit the interarrival times of trucks arriving at the loader station. This type of statistic is also referred to as a "between" statistic. In the tunneling problem, measurement of the total jacking collar cycle time can be developed by specifying a "between" statistic at some point along its cycle, such as QUEUE node 13. Since only a single flow unit representing the jacking collar is cycling along this path (i.e., 3–4–5–11–12–13), the interarrival time of that unit at 13 will also be its total cycle time.

In some systems, a simple count of the number of units passing a particular station may provide important information. A counter placed somewhere within an open or closed path can be used to indicate the number of units passing a given point in the system. In complex systems, such information may be difficult to forecast and yet may be of interest to the manager. In a system such as the three-commodity problem shown in Figure 9.5, it may be relevant in a preplan analysis of a project to know the actual number of times the bricks are lifted by the tower crane, the man hoist, and the roof hoist, respectively. This can be achieved by inserting counters following each of the lift COMBI elements associated with the "BRICKS AVAILABLE" QUEUE node.

10.2 INFORMATIONAL FORMATS

Several modes of presenting systems information can be utilized. The method of presentation is significant, since it prescribes the method of data collection. Two types of presentation formats are used commonly.

1. Summary presentation.
2. Histogramatic or distribution presentation.

Counter, utilization, and some types of distributed data are commonly presented in summary format. Distributed types of information can be summarized in terms of minimum and maximum values, average or mean values, and standard deviations. Occupancy and utilization information is usually reduced to a percentage that indicates the part of the total observation period during which a unit is utilized in an active state or a QUEUE node is occupied. Counts are presented as integer values that indicate the accumulated observations.

Counts are basic to the reduction of all kinds of observed information to a summary format. In calculating both the mean and standard deviation of a set of distributed data, the formulas used require the maintenance of a cumulative record of the number of observations, n. The mean value of a set of N observation x_i is given as:

$$\bar{x} = \frac{1}{n} \sum_{i=1}^{n} x_i \tag{10-1}$$

and the standard deviation is:

$$s = \left\{ \frac{\sum_{i=1}^{n} (x_i - \bar{x})^2}{n} \right\}^{1/2} \tag{10-2}$$

Maintenance of utilization and occupancy statistics requires the calculation of percentages. The calculation is indicated graphically in Figure 10.3.

The fraction of the time a state is occupied is given simply as:

$$\text{O.F.} = \frac{1}{T} \sum^{N} (t_f - t_c) \tag{10-3}$$

where O.F. = the occupancy factor
T = the duration of the time period observed
N = the count or number of busy periods observed
t_c = the time of commencement of a busy period
t_f = the finish time of the busy period

With this type of statistic, a count of the number of occupancies as well as a cumulative total of the time the idle state is occupied must be maintained. This cumulative value of occupancy time is divided at the end of the monitoring period by the total elapsed time T.

It is desirable to know not only what percent of the time an idle state (QUEUE node) is occupied but also some indication of the number of units

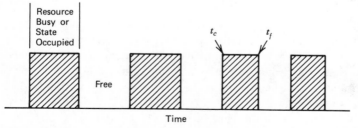

FIGURE 10.3   Utilization and occupancy statistics.

blocked or delayed in the state. One relatively simple approach to summarizing this information is the use of a time-integrated average statistic. This statistic, as well as the percent occupancy, is maintained as an intrinsic function of all QUEUE nodes defined when using the computerized CYCLONE system. A simplified diagram of the time integrated average statistic is given in Figure 10.4. This statistic is defined by the expression:

$$A_j = \frac{\sum_i^m N_{ij} t_{ij}}{T} \text{ (for all } j) \tag{10-4}$$

where

N_{ij} = The number of units waiting in QUEUE node j during time interval t_i

t_{ij} = the length of the time interval i during which the number of units N_{ij} in the queue j remains unchanged

T = the total time of the simulation

m = the number of time intervals during which QUEUE node j contains an unchanged level of delayed units

A_j = the time integrated average number of units in queue j

Values of the variables in the above equation are given in Figure 10.4 for a time space of eight time units. The y axis indicates the number of units delayed in the QUEUE node during time interval i. Substituting into Equation 10–4, there results:

$$A_j = \frac{1(1) + 2(1.5) + 3(1.5) + 0 + 2(1) + 1(1)}{8} = \frac{1 + 3 + 4.5 + 2 + 1}{8}$$

$$= \frac{11.5}{8} = 1.44$$

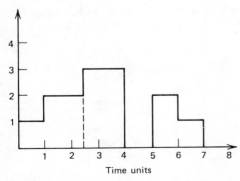

Observation	t_i	x_{ij}	$x_{ij} t_i$
1	1.0	1.0	1.0
2	1.5	2.0	3.0
3	1.5	3.0	4.5
4	1.0	0	0
5	1.0	2	2.0
6	1.0	1.0	1.0
7	1.0	0	0

Time units

FIGURE 10.4 Time-integrated statistics.

The percent of time units are delayed in QUEUE node j is given by a modification of expression 10–3:

$$D_j = \frac{\sum\limits_{i}^{m} t_{ij}}{T} \times 100 \qquad (10\text{–}5)$$

where

D_j = The percent of the time units are delayed in QUEUE node j

t_{ij} = the time periods during which units were delayed at QUEUE node j

and the other parameters are as defined above. Substituting values in Equation 10–5 from Figure 10.4, there results

$$D_j = \frac{6}{8} \times 100 = 75$$

This indicates that units were waiting in QUEUE node j 75% of the time. In Chapter Twelve the CYCLONE computer program output that gives this information will be described.

It is often advantageous to be able to examine observed data in their distributed state in order to gain more information regarding the characteristics of the distribution. Maintenance of this type of statistic is more complex, since a range of observation and cells into which the data are sorted (i.e., distributed) must be defined. Counts must be maintained for each cell or class interval defined and must also include values above and below the range defined. The upper and lower limits of the defined range and the width of the class intervals must, of course, be defined.

The observed data are sorted and assigned to the appropriate cell resulting in an incrementing of the associated cell counter. This is shown diagramatically in Figure 10.5. This data formating leads to the development of a histogram representing the distribution of the observations. Defined statistics associated with elements when using the CYCLONE system computer implementation are displayed in this fashion. This method of presentation can be illustrated

FIGURE 10.5   Distributed (histogram) format.

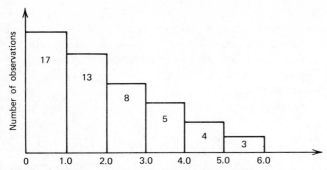

FIGURE 10.6   Histogram of helper idle times.

within the context of the masonry problem, which was hand simulated in Chapter Seven. The helper idle time was measured by inserting STAT clocks before and after the helper QUEUE node. The first two idle time durations were 1.7 and 4.2. If the simulation were to be continued, further observations would be accumulated. These would be sorted and the counters of appropriate histogram cells incremented. The observed distribution for the helper idle time is shown in Figure 10.6. In this instance the lower limit has been defined as 1.0. Since negative values are not considered (i.e., no negative idle times are possible), all underflow values are $0 \le x_i < 1.0$. The cell width is defined as 1.0 and the upper limit is 6.0. The number of cells defined is $n = 5$.

10.3 CYCLONE MONITORING FUNCTIONS

CYCLONE system elements have two types of associated attributes.

1. User-defined (input) attributes.
2. System-generated (output) attributes.

Those characteristics of an element that are specified by the modeler at the time of initializing the model are user-defined attributes. These items can be thought of as *input* establishing the structure of the system to be investigated. Some typical user-defined attributes are:

1. Element type (COMBI, QUEUE node, etc.).
2. Work task duration (to include parameters for probabilistic durations).
3. System logic (which elements precede and follow other elements).
4. Initial conditions (points in the system at which flow units are initialized).

The characteristics associated with certain elements that develop as a result of the system's actual operation are referred to as system-generated attributes. Observation of these attributes allows the modeler to take information out of the system and determine the level of system performance. These attributes are referred to as *output* attributes. The output characteristics of the CYCLONE system elements allow the development of the type of statistical information (described above) that is of interest to the construction manager. In a laboratory sense, the elements act as probes that capture information on system operation and performance.

Certain elements, by virtue of their structural functions in the topology of the system, provide information take-off points in the model. The QUEUE nodes represent states in which units are delayed. It is obvious that QUEUE nodes provide an ideal medium for measuring:

1. The percent of the time units are delayed along a given track pending realization of COMBI ingredience requirements.

2. The time integrated average of units delayed at a delay position pending realization of ingredience requirements.

3. The delay times experienced by individual flow units at an ingredience delay position at QUEUE node.

Measurement of these characteristics immediately provides information regarding bottlenecks and apparent imbalance in the resources committed to system operation. The QUEUE nodes can be thought of as gauging stations indicating the backup of flow units at various points in the process under investigation.

In Figure 10.7a the COUNTER is placed to measure the productivity following the dump operation, and the productivity measure could be in cubic units of material placed at the DUMP location. Figure 10.7b the element is placed in the loader cycle and indicates the number of bucket loads loaded by the loader. In some cases, cycles with dissimilar increments of production must be accumulated and reported in common units. This leads to the use of the CONSOLIDATE function to obtain compatibility. Such a situation will be illustrated in the next section.

These output attributes associated with the QUEUE nodes and ACCUMULATOR element are referred to as intrinsic monitoring functions, since the collection of information regarding these performances indicators is automatically established when the element is defined within the system. That is, the modeler gets information on these attributes as a by-product of the element's incorporation into the network. No further defining functions must be associated with these elements. This is in contrast to FUNCTION nodes, which must

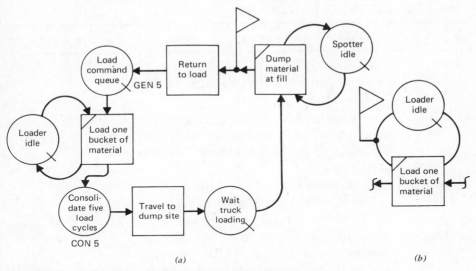

FIGURE 10.7   Placement of accumulator element.

have user-defined functions associated with them. Otherwise, their presence within the process model is meaningless. They have no function in the network model other than to provide take-off points at which statistics can be collected. The type of statistic to be measured or operation to be performed at the FUNCTION node must be defined by the user. Therefore, statistics such as those associated with FUNCTION nodes are called defined statistics. The FUNCTION node provides a special element that can be used to develop management information regarding:

1.   Unit counts.
2.   Interval of trace statistics.
3.   Interarrival or between statistics.

The types of defined statistics which are commonly associated with FUNC-TION nodes are given in Table 10.2.

10.4 HAND SIMULATION STATISTICS COLLECTION

Extensions to the hand simulation algorithms can be developed to handle the collection of both defined and intrinsic statistics. These procedures provide insights into the operation of statistics collection routines and aid in understanding the extent of the information available for model analysis.

Table 10.2   DEFINED FUNCTION

Name	Description	Numerical Code*
First	This statistic records the time at which the first entity arrives at the associated FUNCTION node.	1
All	This statistic records the time at which each entity arrives at the associated FUNCTION node. An average value is printed.	2
Between	This statistic records the time between arrival of entities at the associated FUNCTION node.	3
Interval	This statistic records the time between passage of a preceding MARK node and arrival at the associated function node.	4
Delay	This statistic is used in conjunction with the CONSOLIDATE function. It records the time between the first entity of the group to be consolidated and the last of the group.	5
Count	This statistic is a simple count of the number of units that transit the element with which it is associated.	6

*This code will be discussed later in Chapter Twelve in describing the CYCLONE system computer implementation.

In order to accumulate occupancy statistics at the QUEUE nodes defined in the system, a method must be devised to determine the percent of the total time period, T, that units are delayed in the QUEUE node. As noted above, this reduces to accumulating the t_i values in Equation 10–3. A method for determining the t_i values using STAT clocks is demonstrated in Figures 7.11 and 7.12. The t_i values observed allowed determination of the percent of the time the laborer is idle in the masonry problem. If more than one laborer is defined in the system, the procedure for determining t_i as defined in Equation 10–3 would be slightly modified. In this case, t_c in Equation 10–3 is marked as soon as a unit enters the empty QUEUE node, and t_f is not marked until the last unit in the QUEUE node exits, leaving it totally empty. With only one laborer defined in the masonry system, the model simulation indicates that the laborer is idle for a total of 5.9 minutes during the 29.0 minutes of simulated time. In such situations where a resource or resources are associated with a single COMBI in a slave cycle, the percent utilization of the resource(s) is given as:

$$U_\% = \{1 - \text{O.F.} \times 100\} \tag{10-6}$$

where O.F. (occupancy factor) is defined by Equation 10–3. In the case of the single laborer system this results in a percent utilization of

$$U_\% = \left(1 - \frac{5.9}{29.0}\right) \times 100 = 79.66\%$$

If two laborers are defined in the system, the occupancy factor would increase and the $U_\%$ would consequently decrease. Table 10.3 shows the CHRONOLOG-ICAL list for a system in which two laborers are defined. The two laborer system model is shown in Figure 10.8. By initializing two laborers in QUEUE node one, a two-channel station at COMBI 2 is generated. The flow unit markers as positioned in Figure 10.8 show the system state at $T = 29.0$ following the ADVANCE step. In this system, STAT clocks (L1-A and L2-A) have been associated with each laborer. An additional clock, T_1, has been introduced to measure the periods during which at least one unit is in the QUEUE node 1. An N column has been added to the CHRONOLOGICAL list to register how many flow units (i.e., laborers) are delayed at QUEUE node 1 at each time step. The steplike plot next to the N column can be used

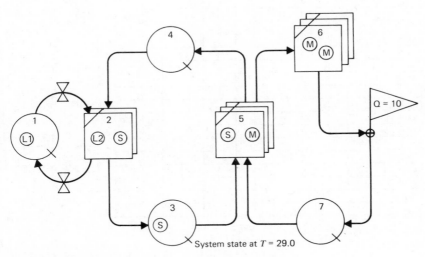

FIGURE 10.8   Two-laborer masonry system.

Table 10.3 CHRONOLOGICAL LIST FOR TWO-LABORER MASONRY SYSTEM

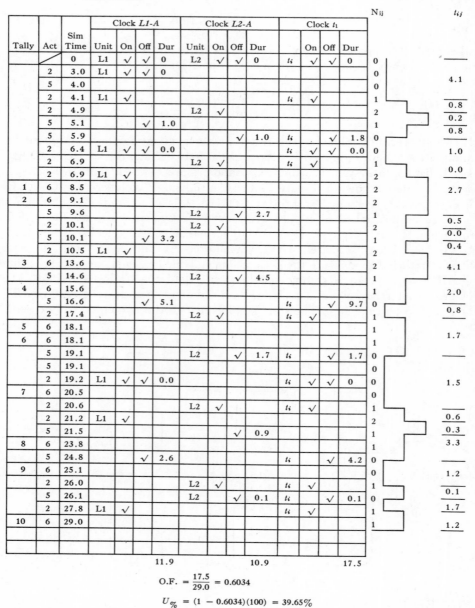

Tally	Act	Sim Time	Unit (Clock L1-A)	On	Off	Dur	Unit (Clock L2-A)	On	Off	Dur	Unit (Clock t₁)	On	Off	Dur	N_{ij}	t_{ij}
		0	L1	✓	✓	0	L2	✓	✓	0	t_i	✓	✓	0	0	—
	2	3.0	L1	✓	✓	0									0	4.1
	5	4.0													0	
	2	4.1	L1	✓							t_i	✓			1	0.8
	2	4.9					L2	✓							2	0.2
	5	5.1			✓	1.0									1	0.8
	5	5.9							✓	1.0	t_i		✓	1.8	0	
	2	6.4	L1	✓	✓	0.0					t_i	✓	✓	0.0	0	1.0
	2	6.9					L2	✓			t_i	✓			1	0.0
	2	6.9	L1	✓											2	
1	6	8.5													2	2.7
2	6	9.1													2	
	5	9.6					L2		✓	2.7					1	0.5
	2	10.1					L2	✓							2	0.0
	5	10.1			✓	3.2									1	0.4
	2	10.5	L1	✓											2	
3	6	13.6													2	4.1
	5	14.6					L2		✓	4.5					1	
4	6	15.6													1	2.0
	5	16.6			✓	5.1					t_i		✓	9.7	0	0.8
	2	17.4					L2	✓			t_i	✓			1	
5	6	18.1													1	1.7
6	6	18.1													1	
	5	19.1					L2		✓	1.7	t_i		✓	1.7	0	
	5	19.1													0	
	2	19.2	L1	✓	✓	0.0					t_i	✓	✓	0	0	1.5
7	6	20.5													0	
	2	20.6					L2	✓			t_i	✓			1	
	2	21.2	L1	✓											2	0.6
	5	21.5							✓	0.9					1	0.3
8	6	23.8													1	3.3
	5	24.8			✓	2.6					t_i		✓	4.2	0	
9	6	25.1													0	1.2
	2	26.0					L2	✓			t_i	✓			1	
	5	26.1					L2		✓	0.1	t_i		✓	0.1	0	0.1
	2	27.8	L1	✓							t_i	✓			1	1.7
10	6	29.0													1	1.2
						11.9				10.9				17.5		

$$\text{O.F.} = \frac{17.5}{29.0} = 0.6034$$

$$U_\% = (1 - 0.6034)(100) = 39.65\%$$

to develop the integrated time statistic for QUEUE node 1. By applying Equation 10–5, an occupancy factor can be calculated as follows:

$$\text{O.F.} = \frac{\sum t_1}{T} = \frac{17.5}{29} = 0.6034$$

This leads to a percent utilization for the two laborers of

$$U_\% = (1 - 0.6034)\,(100) = 39.65\%$$

The time integrated statistic calculated using Equation 10–4 reduces the following expression:

$$A_j = \frac{\sum\limits_{i=1}^{22} N_{ij}t_{ij}}{T} = \frac{26.0}{29.0} = 0.896$$

Accumulating the information for the t_1 clock leads to a departure from the procedure normally used in tripping the clock on and off. The t_1 clock is tripped on whenever the QUEUE node is empty and a flow unit enters. It is tripped off only when the unit departing causes the QUEUE node to be empty. This satisfies the requirements for t_1 established by Equations 10–3 and 10–5.

The t_1 clock as used in the development of the occupancy factor can be best thought of as associated with the QUEUE node for which statistics are collected instead of with a particular flow unit. Therefore, it can be considered as an *element* STAT clock instead of a *flow unit* STAT clock.

Accumulation of the N-column values required for the calculation of time-integrated statistics leads to a further modification of the basic hand simulation algorithm. In addition to the N (Number of units delayed) value, the time interval, t_{ij}, during which the QUEUE node contains the N units, is needed to substitute into Equation 10–4. This leads to two registers being defined for each QUEUE node.

In these registers, the N value and interval during which N has remained unchanged are maintained. Whenever the N value changes (i.e., a unit arrives or departs), the value $N_{ij}t_{ij}$ is calculated and added to the previous time-integrated balance for the QUEUE node. In the two-laborer QUEUE node, for example, the QUEUE node is empty until $T = 4.1$, at which time one laborer becomes idle. The $N_{ij}t_{ij}$ calculation results in $(0).(4.1) = 0$, which is added to zero (the initial time integrated value). The next change in the N_{ij} value occurs at $T = 4.9$, at which time the second laborer becomes idle. At this point, $N_{ij}t_{ij}$ is again calculated and added to the previous balance. At time 4.9, the $\sum N_{ij}t_{ij}$ is $0 + (1)(4.9 - 4.1) = 0.8$. At $T = 5.1$, one laborer reenters activity 2, and the $\sum N_{ij}t_{ij}$ becomes $0.8 + (2)(5.1 - 4.9) = 1.2$. The cumulative value of $N_{ij}t_{ij}$ is maintained in this way throughout the simulation. At the end of the simulation run, the cumulative $\sum N_{ij}t_{ij}$ is divided by T, the total duration of the simulated period, in accordance with Equation 10–4. The QUEUE node 1 changed its state (i.e., N_{ij} value) 22 times during the 29.0 minutes simulated in Table 10.3. Therefore, the aggregate value of $\sum N_{ij}t_i$ consists of 22 components as follows:

CHANGE

1	$0 \times 4.1 = 0$	6	$1 \times 0 = 0$	11	$2 \times 4.1 = 8.2$	16	$1 \times 0.6 = 0.6$
2	$1 \times 0.8 = 0.8$	7	$2 \times 2.7 = 5.4$	12	$1 \times 2.0 = 2.0$	17	$2 \times 0.3 = 0.6$
3	$2 \times 0.2 = 0.4$	8	$1 \times 0.5 = 0.5$	13	$0 \times 0.8 = 0.0$	18	$1 \times 3.3 = 3.3$
4	$1 \times 0.8 = 0.8$	9	$2 \times 0.0 = 0.0$	14	$1 \times 1.7 = 1.7$	19	$0 \times 1.2 = 0.0$
5	$0 \times 1.0 = 0.0$	10	$1 \times 0.4 = 0.4$	15	$0 \times 1.5 = 0.0$	20	$1 \times 0.1 = 0.1$
	$\sum = 2.0$		$\sum = 6.3$		$\sum = 11.9$		$\sum = 4.6$

21    $0 \times 1.7 = 0$

22    $1 \times 1.2 = 1.2$

$$A_j = \sum_{i=1}^{22} \frac{N_{ij}t_{ij}}{T} = \frac{2.0 + 6.3 + 11.9 + 4.6 + 1.2}{29} = \frac{26.0}{29.0} = 0.896$$

Collection of these statistics can be achieved by modifying the hand simulation algorithm as given in Chapter Seven. Following the STAT clock at the end of the ADVANCE phase and during the GENERATE phase (Figure 7.14), additional checks are included to update the QUEUE node STAT clocks and accumulate the $N_{ij}t_{ij}$ values as required. The additional flow graph segment that implements these checks is shown in Figure 10.9. Addition of this segment causes an updating of the cumulative $N_{ij}t_{ij}$ statistic for the period (just ended) during which the N_{ij} level remained constant. In addition, it implements the switching ON and OFF of the QUEUE node occupancy clocks.

Both the occupancy factor and the time-integrated average statistics are summary indications of QUEUE node characteristics. It may be of interest to the manager to have information regarding the distribution of delay times at the QUEUE node. The distribution of delays can be summarized using the mean and standard deviations of the observations for the period and the minimum and maximum values observed. The shape of the distributed values also can be shown using a histogram.

During the 29 minutes of simulated system operation, the two laborers experienced a total of 10 nonzero delays. The mean value, standard deviation, and minimum and maximum values are readily developed by consulting the DUR columns for Table 10.3 and calculating the values using appropriate equations (i.e., Section 6.6, Equation 6–9). For computer processing, it is advantageous not to have to remember or keep a record of each observation individually. Therefore, the information required for calculation of the mean and standard deviation is kept in a cumulative fashion (i.e., as for the time-

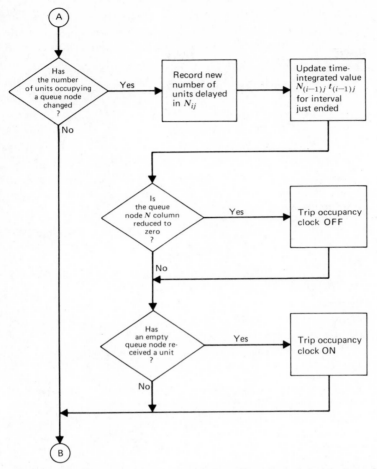

FIGURE 10.9   Occupancy and time-integrated statistic collection segment.

integrated statistic), thus reducing the number of values that must be updated at any particular point in the simulation for summary distribution
statistics. Five values must be maintained. Maximum, minimum, and number
of observations must be recorded. In addition, two values must be accumulated.
Accumulation of the value $\sum\limits_{i=1}^{n} x_i$, where x_i is the value of observed delay i,
allows the equation

$$\bar{x} = \frac{\sum\limits_{i=1}^{n} x_i}{n} \qquad (10\text{--}7)$$

to be used at the end of the simulated period to calculate the arithmetic mean of the observed delays, $\bar{x}$. A modified equation for the calculation of the standard deviation (s) of the observed data requires only the accumulation of the value x_i^2 in addition to the n and $\sum_{i=1}^{n} x_i$ values. The equation is:

$$s = \left[\left(\frac{\sum_{i=1}^{n} x_i^2}{n} \right) - \left(\frac{\sum_{i=1}^{n} x_i}{n} \right)^2 \right]^{1/2} \tag{10-8}$$

By maintaining the value of $\sum x_i^2$ in addition to those mentioned above, Equation 10–8 can be used to calculate the standard deviation of the observed data at the end of the simulated period. The five registers required to maintain the required summary information are:

Register 1	Register 2	Register 3	Register 4	Register 5
n	$\sum_{i=1}^{n} x_i$	$\sum_{i=1}^{n} x_i^2$	$\min_{x_i}$	$\max_{x_i}$

Updating of these registers as the simulation proceeds provides the necessary information to calculate all summary statistics and relieves the necessity of maintaining a record of each observation. Accumulating values in these registers from the DUR column of Table 10.3, the final values are:

Register 1	Register 1	Register 3	Register 4	Register 5
10	22.8	76.26	0.1	5.1

Using the above equations:

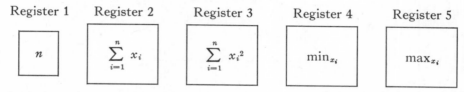

$$\bar{x} = 2.28 \text{ and } s = \left[\frac{76.26}{10} - \left(\frac{22.8}{10} \right)^2 \right]^{1/2} = 1.558$$

If, in the hand simulation algorithm, it is specified that the registers defined above are updated each time a STAT clock (either flow unit or element) is switched OFF, the necessary information for the calculation of summary statistics on the distributed data recorded by that clock will be available at the end of the simulation.

The accumulation of information required for the presentation of observed data in a histogram requires maintenance of a register for each interval defined in the histogram. An N_i register indicating the number of observations recorded in each interval, i, must be updated for each class interval. This form of presentation is, therefore, much more demanding in terms of values that must be stored. If, for instance, four intervals are defined between 1.0 and 5.0 with cell widths of 1.0 for the laborer QUEUE node 1, the number of pieces of information required is doubled (assuming the summary statistic registers are kept): that is, four registers for summary statistics and six registers for the histogram cell count. The total N register for summary statistics can be deleted, since total N ($\sum N_i$) can be developed from the cell register N_i counts. This results in a savings of one summary register.* If a histogram with 26 class intervals is defined, the number of registers would expand to 30 for each STAT clock (i.e., if distributed data presentation justified the additional complexity and expense of record keeping.)

Employing six registers to record the distribution of the observed idleness of the two laborers in Table 10.3, the following value would have been accumulated after 29.0 minutes of simulation.

Interval 1	Interval 2	Interval 3	Interval 4	Interval 5	Interval 6
2	3	2	1	1	1
<1.0	1.0–1.99	2.0–2.99	3.0–3.99	4.0–5.0	>5.0

In fact, definition of the four class intervals between 1.0 and 5.0 with a cell width of 1.0 and a lower bound of 1.0 is sufficient to develop the values shown, since $N + 2$ registers must be defined for every N class intervals specified. The two extra registers record the number of values less than the lower limit and greater than the upper limit.

Updating of the production information recorded by the ACCUMULATOR element is relatively straight forward. A register is associated with the AC-CUMULATOR, which is incremented each time a unit passes the element. As discussed in Chapter Four, the ACCUMULATOR can have a QUANTITY parameter associated with it. This is a scalar value indicating the unit of production. The ACCUMULATOR is essentially a counter. At the end of the simulation, the integer value in the ACCUMULATOR register indicates the number of production units (of size Q) that have passed the ACCUMULA-TOR.

The ACCUMULATOR is inserted in the masonry operation of Figure 10.8 following NORMAL work task 6 to measure the completion of brick placement. Each unit (in this case, each mason) passing this element represents a completed production of 10 placed bricks. Therefore, the QUANTITY scalar 10.0 has been specified on the ACCUMULATOR element symbol. In order to keep a tally of the number of units passing the ACCUMULATOR, an additional column is added to the CHRONOLOGICAL list to the left of the

---

*If calculation of the arithmetic mean and standard deviation using grouped data (i.e., Equations 6-10 add 6-11) is acceptable, two more registers can be deleted.

ACT column. This column allows a running tally to be kept on the number of units passing the ACCUMULATOR. The value recorded in this column is the same as that which would be in the register at the time step indicated. Examination of this tally for the 29.0 minute simulation indicates that 10 production units (representing 10 bricks each) passed the ACCUMULATOR. Therefore the production of the system for the first 29 minutes is 100 bricks placed.

Using a linear projection of this value to estimate hourly production, as is in Chapter Seven, an expected hourly production results of:

$$\text{Production} = \frac{60 \text{ minutes per hour}}{29.0 \text{ minutes}} \times 100 = 206.9 = 207 \text{ bricks placed per hour}$$

This value is an increase of approximately 20 bricks per hour over the production calculated for a system using only one laborer. The additional production, however, would have to be weighed against the added cost of the second laborer. This production figure is conservative, since the system is still a transient and has not yet reached steady state. By applying the above linear projection of the hourly production at each time step, the ACCUMULATOR is incremented. A rough plot of the transient response of the system can be developed. This plot is shown in Figure 10.10.

The transient plot of the single-laborer masonry system is also shown in Figure 10.10 As expected, the transient of the single-laborer system lags that of the two laborer process. Bunching of units exiting the "PLACE BRICK" work task leads to sharp discontinuities at several points on the two plots. As the systems reach steady state, the flow units along the production cycle will distribute themselves equidistantly, leading to a relatively flat production curve. That is, $d\text{PROD}/dt$ approaches 0 as the system reaches steady state.

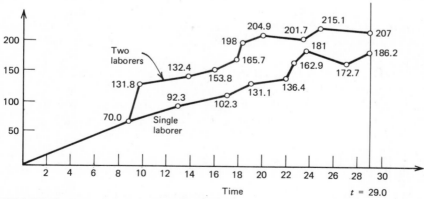

FIGURE 10.10   Masonry system transients.

Again, a slight extension of the hand simulation algorithm allows for the maintenance of the ACCUMULATOR register. Since production is marked following the completion of an activity, a check inserted in the ADVANCE phase following release of flow units from terminated work tasks allows for incrementing of the ACCUMULATOR tally. Following completion of the simulation, the tally is multiplied by the QUANTITY parameter (if one is defined). The modified hand simulation algorithm to include the accumulation of the occupancy, summary and distributed QUEUE node statistics, and ACCUMULATOR tally is presented in Figure 10.11.

10.5 USE OF THE INTRINSIC MONITORING FUNCTIONS

In order to determine the types of information that can be developed using intrinsic statistics, consider a system that is an extension of the 20T-15T hauler system discussed in Section 4.10. The occupancy and time-integrated average delay statistics on the QUEUE nodes and the accumulator productivity statistics on the system provide insights into the performance of the system. The process to be modeled consists of hauling units with nonidentical loader rates and bucket sizes. There are four production units involved in the system.

1. Twenty-ton hauler with top speed of 20 miles per hour.

2. Fifteen-ton hauler with top speed of 30 miles per hour.

3. Three-ton loader with mean cycle time of 30 seconds.

4. Five-ton loader with mean cycle time of 50 seconds.

A model designed to handle the problem of making the loaders compatible to each hauler type is shown in Figure 10.12. This model illustrates the use of several constructs introduced in Chapters Eight and Nine. Parameter sets defining the model transit times* are given in Table 10.4. The table indicates that Station A and C both have associated cycle times with mean values of 0.5 minutes, while Stations B and D have 0.835-minute mean cycle times. This indicates that the smaller 3-ton loader operating at Stations A and C cycles slightly faster than the larger 5-ton loader. The travel times associated with activities 8 and 20 reflect a difference in speed between the 20-ton haulers and the smaller 15-ton machines. The mean transit (travel-dump-return) time associated with the 20-ton hauler at 8 is 4.0 minutes, while the mean transit time for the 15-ton units at 20 is 3.6 minutes. The transit delays associated with the routing activities (3, 10, 16, and 21) represent the maneuver time required to move to the load position. For both the 15- and 20-ton units, the maneuver time is assumed, for simplicity, to be constant, the larger 20-ton

*Time values in the parameter sets are given in minutes.

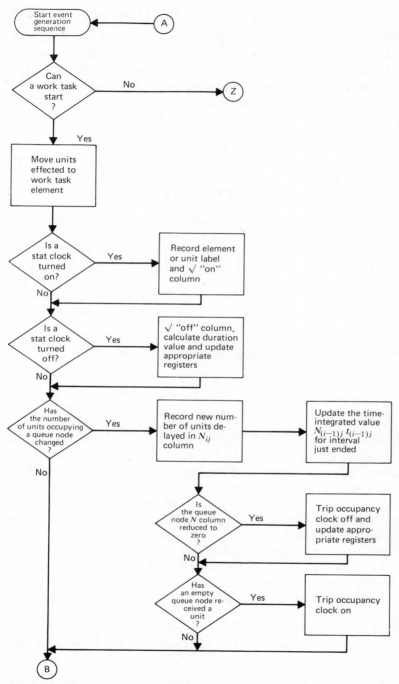

FIGURE 10.11 Revised hand simulation flow graph. (k = constant value of delay.)

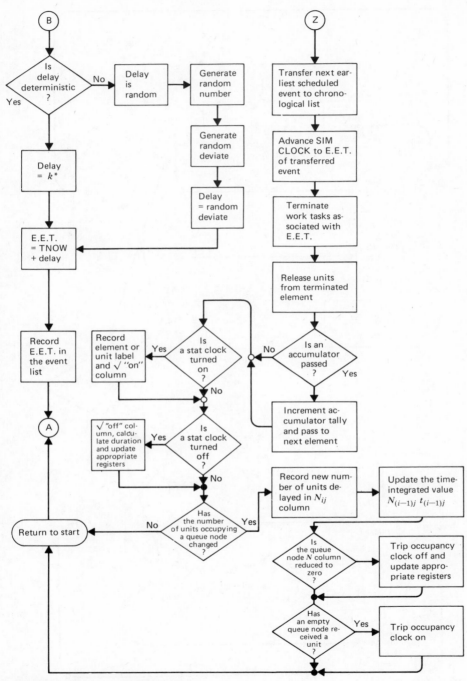

FIGURE 10.11 (*Continued*).

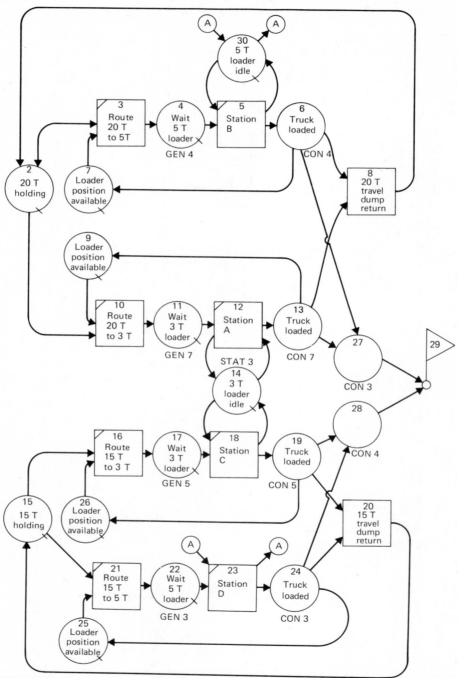

FIGURE 10.12 Nonidentical loader and transit unit network model.

322

Table 10.4 WORK TASK DISTRIBUTION TYPES AND
PARAMETER SETS.

Work Task Number	Distribution Type	Set Number	Mean	Lower Limit	Upper Limit	Standard Deviation
3	Constant	2	0.200	0	0	0
5	Lognormal	3	0.835	0.40	1.9	0.25
8	Lognormal	5	4.00	2.00	10.0	1.00
10	Constant	2	0.200	0	0	0
12	Lognormal	4	0.500	0.25	1.2	0.15
16	Constant	6	0.15	0	0	0
18	Lognormal	4	0.500	0.25	1.2	0.15
20	Lognormal	7	3.600	2.00	10.0	0.15
21	Constant	6	0.150	0	0	0
23	Lognormal	3	0.835	0.40	1.9	0

machines requiring 0.2 minutes (i.e., 12 seconds) and the 15-ton units requiring 0.15 minutes (i.e., 9.5 seconds).

The initial entity locations in the system are given in Table 10.5. All haul units start at their respective holding areas, 2 and 15. Loader permits indicating position availabilities are initialized in the permit mechanisms at 7, 9, 25, and 26. A 5-ton loader is defined at 30 and two 3-ton loaders are initialized at 14. The labeling rule results in both loaders giving priority to the 20-ton haul units.

In this example, the different capacities of the haul units require use of the CONSOLIDATE function to define a common system output quantity that

Table 10.5 FLOW UNIT INITIALIZATION

Category	Description	Number	Initialized at QUEUE Node
A	20-ton haulers	3	2
B	15-ton haulers	5	15
C	5-ton loader	1	30
D	3-ton loader	2	14
E	Loader permit 1	1	7
F	Loader permit 2	1	9
G	Loader permit 3	1	25
H	Loader permit 4	1	26

can be used in conjunction with the accumulator at 29. The common output unit of 60 tons is equivalent to three 20-ton hauler loads and four 15-ton hauler loads. In order to develop 60-ton quantity units for the accumulator, output measures from nodes 6 and 13 are consolidated at 27 in 3 load groups (i.e., three 20-ton loads) and from 19 and 24 in 4 load groups (i.e., for 15-ton loads). The QUANTITY parameter associated with 29 is specified as 60.0 tons, and the system will be allowed to cycle 100 times.

The following delay statistics for the QUEUE nodes representing the truck queues as well as the loader idle queues were generated by simulating the system.

QUEUE node	Title	Occupancy Factor (Percent of Time Units Delayed or Idle)	Time-Integrated Average Number of Units
2	20-Ton hauler holding area	12	0.12
14	3-Ton loader idle	25	0.30
15	15-Ton hauler holding area	72	1.22
30	5-Ton loader idle	1	0.01

Furthermore, the time the loader positions are free can be determined from the status of the permit units at 7, 9, 25, and 26.

Control Queue	Position Controlled	Occupancy Factor (Percent of Time Available)
7	20-Ton hauler at 5-ton loader	18
9	20-Ton hauler at 3-ton loader	46
25	15-Ton hauler at 5-ton loader	2
26	15-Ton hauler at 3-ton loader	1

Tables 10.6 and 10.7 summarize the system production and idle times for both 3-ton and 5-ton loaders, and the percent of the time that trucks are delayed in the holding areas awaiting service at the loader. The systems that are evaluated vary from configurations containing two to four 20-ton haulers and four to six 15-ton haulers. The summary data is not intended to be exhaustive, but it does yield some interesting information regarding the system's characteristics when operating near its maximum output point. A deterministic analysis of the system based on 120 3-ton loader cycles per hour and 72 5-ton loader cycles per hour establishes the system maximum output to be 1080 tons

Table 10.6 SYSTEM PRODUCTION (TONS PER HOUR)

Number of 15-Ton Haulers	Number of 20-Ton Haulers		
	2	3	4
4	811.49 829.91	914.7 920.07	979.67 984.37
5	884.49 897.06	941.63 955.27	986.88 990.62
6	909.40 910.94	951.65 954.43	1003.88 999.80

Run 1
Run 2

Table 10.7   LOADER IDLE TIME (PERCENT)

Number of 15-Ton Haulers	Number of 20-Ton Haulers		
	2	3	4
4	7 / 5 ... 44 / 43	3 / 2 ... 29 / 29	1 / 0 ... 16 / 15
5	3 / 3 ... 36 / 36	1 / 1 ... 25 / 25	0 / 0 ... 15 / 15
6	2 / 3 ... 36 / 36	1 / 1 ... 24 / 24	0 / 1 ... 14 / 15

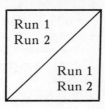

Percent of Time
5-Ton Loader Idle

Run 1
Run 2

Run 1
Run 2

Percent of Time
3-Ton Loader

325

Table 10.8   HAULER DELAYS (PERCENT)

Number of 15-Ton Haulers	Number of 20-Ton Haulers		
	2	3	4
4	0 / 0 / 33 / 34	9 / 8 / 41 / 41	18 / 19 / 61 / 62
5	0 / 0 / 60 / 61	9 / 8 / 80 / 78	18 / 17 / 91 / 92
6	0 / 0 / 90 / 91	8 / 7 / 96 / 95	15 / 14 / 99 / 99

Percent of Time 20-Ton Units Delayed at Loader

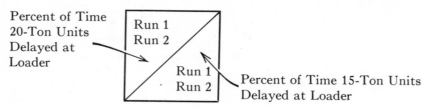

Run 1
Run 2

Run 1
Run 2

Percent of Time 15-Ton Units Delayed at Loader

per hour.* The system using four 20-ton haulers and six 15-ton haulers has a production of approximately 1000 tons per hour. However, this system results in significant delays for the 15- and 20-ton haulers, as noted in Table 10.8. Increase in the number of 20-ton units, in general, causes more delays for both types of units, as can be seen from the rows of Table 10.8. Delays at the 3-ton and 5-ton loaders are given in Tables 10.9 and 10.10. These tables, although not exhaustive, do indicate the large amount of information that is readily avail-

*Maximum output = (one 5-ton loader × 72 cycles per hour × 5 tons per cycle)
+ (two 3-ton loaders × 120 cycles per hour × 3 tons per cycle)
= 360 + 720 = 1080 tons per hour

Table 10.9   DELAYS AT 3-TON LOADER

Number of 15-Ton Haulers	Number of 20 Ton Haulers		
	2	3	4
4	30 32 81 80	52 54 88 86	74 75 94 96
5	34 33 93 95	52 ·53 98 96	74 73 100 100
6	29 32 99 99	53 54 100 99	74 74 100 100

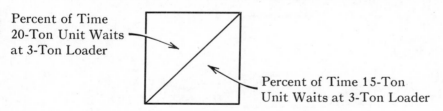

Percent of Time 20-Ton Unit Waits at 3-Ton Loader

Percent of Time 15-Ton Unit Waits at 3-Ton Loader

able on the QUEUE nodes and various delays in the system. The information available is a function of the design of the network model to include the specification of ingredience constrained activities and the use of the statistics function.

10.6 DEFINED MONITORING FUNCTIONS

The defined statistics functions give the modeler a great deal of added flexibility in auditing system performance. These functions are defined at FUNCTION nodes inserted into the network for the purpose of information collection. As with the intrinsic statistics, examination of the collection mech-

Table 10.10   DELAYS AT 5-TON LOADER

Number of 15-Ton Haulers	Number of 20-Ton Haulers		
	2	3	4
4	55 56 / 89 87	74 73 / 93 93	93 90 / 97 98
5	57 57 / 96 96	78 74 / 99 100	92 92 / 100 100
6	58 56 / 100 100	76 75 / 100 100	91 89 / 100 100

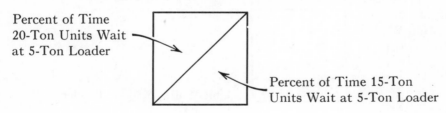

Percent of Time 20-Ton Units Wait at 5-Ton Loader

Percent of Time 15-Ton Units Wait at 5-Ton Loader

anism is helpful in understanding the monitoring information provided by the various defined statistics. Just as with the intrinsic statistics, defined statistical data can be presented in a summary or distributed format. The number of observations is a function of the type of statistic considered. The *first* statistic, for instance, is generated only once per simulation run. Therefore the number of observations associated with a first statistic will be the same as the number of simulation runs. This is in contrast to the "between" and "interval" statistics for which many observatons within a single simulation run will be generated.

Again, referring to the two-laborer masonry system of Figure 10.13, a FUNCTION node can be inserted into the system following the ACCUMULA-

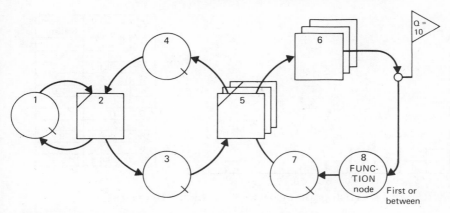

FIGURE 10.13   Masonry model with FUNCTION node.

TOR element and used as a take-off point for defined statistics. If it is assumed that a FIRST statistic is associated with the FUNCTION node, only one observation will occur during the 29.0 minutes of simulated time shown in Table 10.3. Moreover, it makes no difference whether the simulation run is 30 minutes or 3000 minutes, the first statistic for the inserted FUNCTION node remains 8.5 minutes (which coincides with the time of the first AC-CUMULATOR tally). If 10 30-minute simulations are made, 10 FIRST statistics will be accumulated. If only one 300-minute simulation is made, only a single FIRST statistic is accumulated.

Associating a BETWEEN statistic with the FUNCTION node, 10 observa-tions* of the statistic are accumulated in the first 29.0 minutes of simulation. Obviously, if the simulation run length is greater than 29.0 (say 300 minutes in duration), a considerably greater number of observations occur during this single run. Additional runs yield additional sets of between statistics. The distribution of the between statistics for the inserted FUNCTION node is shown in Figure 10.14 in a histogram format for the 29.0-minute simulation.

The histogram shown is plotted by developing the between statistic from the data shown in Table 10.3. The tabular listing shows the time at which each unit arrived at the FUNCTION node. In this situation, the arrival times at the FUNCTION node are the same as the tally times (the ACCUMULATOR element always has zero duration). The between statistic is calculated simply by subtracting the $(N-1)$th arrival time for the Nth arrival time. That is,

$$\text{B Stat} = t_n - t_{n-1}$$

Since only 10 observations are generated in the first 29.0 minutes, the shape of the histogram as shown is still quite amorphous. Figure 10.15 shows the

*This corresponds to the number of ACCUMULATOR tallies registered.

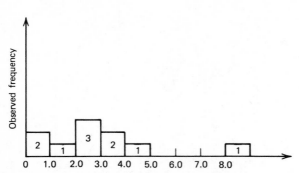

Obs.	t_i	Between Statistic
—	0	—
1	8.5	8.5
2	9.1	0.6
3	13.6	4.5
4	15.6	2.0
5	18.1	2.5
6	18.1	0
7	20.5	2.4
8	23.8	3.3
9	25.1	1.3
10	29.0	3.9

FIGURE 10.14   Histogram of between statistic (10 cycles).

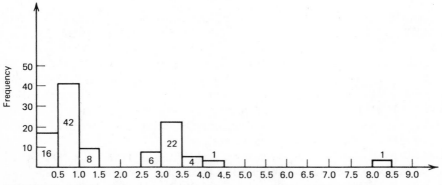

FIGURE 10.15   Histogram of between statistics (100 cycles).

histogram for the between based on two runs of 100 cycles each. This provides a clearer indication of the distributed shape of the observations.

In order to study the action of interval statistics, insert two additional FUNCTION nodes into the masonry problem. One of these will be inserted following COMBI 5 and preceding QUEUE node 4. This FUNCTION node will be used as a MARK point and will define the beginning of the interval to be observed. The second FUNCTION node will be inserted following COMBI 2 and preceding QUEUE node 3. This node will perform the same function as the CLOCK OFF switch in Figure 7.11. It defines the end of the interval to be measured and the point at which transit time from the MARK point is recorded. These two nodes permit the measurement of the time from release of a space position (stack position) until the laborers refill it with another 10-brick packet. This is effectively the time the stack flow unit is

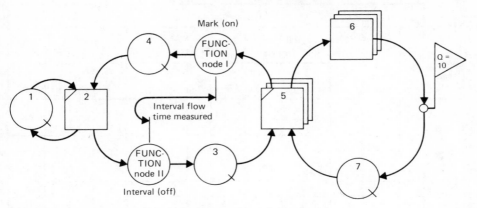

FIGURE 10.16   System with interval statistic.

delayed in QUEUE node 4 plus the processing time in COMBI 2. The system with the two FUNCTION nodes inserted to measure the trace or interval statistic is shown in Figure 10.16.

Referring again to the data generated in Table 10.3, this statistic can be plotted as a histogram. The time associated with the MARK node is the same as the E.E.T. for COMBI 5. The INTERVAL node time is the same as the E.E.T. for COMBI 2.* The interval statistic is a flow unit based statistic and, therefore, the collection procedure is similar to the routine developed in Section 7.6. This is in contrast to the between statistic, which is element based. An extension of Table 10.3 including clock columns for the collection of the stack interval statistics is shown in Table 10.11. In the table, a separate clock for each of the stack units is shown. Because of the speed of laborer one, the third stack position (S3) is filled more quickly than the second (S2). This is apparent in the table, since the first interval measured on the S2 clock terminates at 4.9 minutes, while the S3 clock was tripped off at 4.1. This makes the cycle order S1, S3, S2. Following the first cycle, however, the flow units, S1, S2, and S3, reorder themselves in order S1, S2, S3. This order is maintained throughout the following cycles.

Two factors play a role in the relative positions of the individual units one to another. The first consideration is that units may pass one another. In this particular example, the units initialized at QUEUE node 4 are processed at COMBI 2 in the order S1, S2, S3. Laborer one processed S1 and S3, while laborer 2 was processing S2. Because of the random time generated, laborer one managed to complete two stack cycles in less time than laborer two required to complete one. This caused stack marker three (S3), in effect, to pass marker two leading to the S1, S3, S2 ordering at QUEUE node 3.

*FUNCTION nodes have zero-associated time durations.

Table 10.11 INTERVAL STATISTICS

Ac-cumu-lated Tally	Act	Sim Time	Clock (Stack Unit Int. Statistic) S1				Clock Unit S2				Clock Unit S3			
			Unit	On	Off	Dur	Unit	On	Off	Dur	Unit	On	Off	Dur
0		0.	S1	✓			S2	✓			S3	✓		
	2	3.0			✓	3.0								
	5	4.0	S1	✓										
	2	4.1											✓	4.1
	2	4.9							✓	4.9				
	5	5.1									S3	✓		
	5	5.9					S2	✓						
	2	6.4	S1		✓	2.4								
	2	6.9					S2		✓	1.0				
	2	6.9									S3		✓	1.8
1	6	8.5												
2	6	9.1												
	5	9.6	S1	✓										
	2	10.1	S1		✓	0.5	S2	✓						
	5	10.1												
	2	10.5					S2		✓	0.4				
3	6	13.6												
	5	14.6									S3	✓		
4	6	15.6												
	5	16.6	S1	✓										
	2	17.4									S3		✓	2.8
5	6	18.1												
6	6	18.1												
	5	19.1					S2	✓						
	5	19.1									S3	✓		
	2	19.2	S1		✓	2.6								
7	6	20.5												
	2	20.6					S2		✓	1.5				
	2	21.2									S3		✓	2.1
	5	21.5	S1	✓										
8	6	23.8												
	5	24.8					S2	✓						
9	6	25.1												
	2	26.0			✓	4.5								
	5	26.1									S3	✓		
	2	27.8							✓	3.0				
10	6	29.0												

The second factor relates to the ordering of units in the QUEUE node. The normal ordering is that units are removed from the QUEUE node for processing in the order of their arrival. This is the so-called first-in first-out queue discipline (FIFO). In the second cycle of the stack markers, this means that S3 arriving at QUEUE node 4 at 5.1 remains temporarily ahead of S2, which does not arrive until 5.9. However, the shorter random processing time for S2 in passing COMBI 2 on this cycle leads to scheduled E.E.T. values for both S2 and S3 of 6.9 minutes. That is, both units complete stacking at the same moment. In this instance, S2 is allowed to pass S3, giving the ordering S1, S2, S3. No further passing occurs in the simulated period, so adherence to the FIFO queue discipline maintains this ordering until the simulation is stopped at T = 29.0.

Examination of the histogram shown in Figure 10.17 reveals the total number interval observations for the period as 14. The summary values for the distribution are:

1. Maximum value: 4.9.
2. Minimum value: 0.4.
3. Mean value: 2.47.
4. Standard deviation: S.D. = 1.344.

The next statistic of interest is the delay statistic. This monitoring function records the elapsed time between the arrival of the first unit at a CONSOLI-DATE FUNCTION node and the realization of the node. This statistic can best be illustrated by referring to the haul process model originally shown in Figure 7.17. This model is shown again in Figure 10.18 for convenience. This system consists of one loader initialized at QUEUE node 1 and two trucks initialized at QUEUE node 3. The loading of each truck requires five cycles

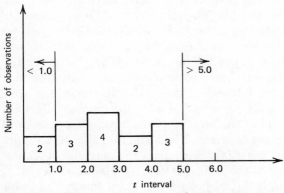

FIGURE 10.17   Interval statistic histogram.

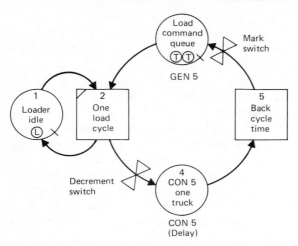

FIGURE 10.18   Simple haul process model with delay statistic.

of the loader. Therefore, each time a truck arrives at QUEUE node, it is "exploded" into five load commands by the action of a GENERATE function. The FUNCTION node at 4 has an associated CONSOLIDATE function, which accumulates five loader cycles (CON5) and then releases a single unit representing the loaded truck. The segment of the CHRONOLOGICAL List for the processing of the first two truck loads is shown in part as Table 10.12.

All statistics are the accumulation of *all* the arrival times of flow units at the FUNCTION node with which the statistic is associated. This statistic indicates the number of units arriving at an element in a given time interval. In order to illustrate this, consider again the FUNCTION node in Figure 10.13 and associate an ALL statistic with it. The ALL values are easily developed from TABLE 10.3, since they are the same as the E.E.T. values for element 6. The occurrence of the ALL statistic observations coincides with the tallies of the ACCUMULATOR in this example. The tabulation of the values and a histogram showing their distribution is given in Figure 10.19. Five-unit class intervals with a lower limit of 5.0 have been used in the distribution shown.

Examination of the histogram indicates the fluctuation of production tallies in each class interval across a period of time. The summary presentation of ALL statistics is not particularly helpful in revealing these time fluctuations. Therefore, this type of statistic is normally presented as a histogram. As the system reaches steady state, the variation of the number of observations per cell will settle down and remain more or less constant. The histogram shown in Figure 10.20 illustrates this effect. This histogram represents the outcome of a computer simulation of 100 cycles (ACCUMULATOR C = 100) with an ALL statistic associated with the FUNCTION node 8 of Figure 10.13. The lower

Table 10.12 CHRONOLOGICAL LIST WITH DELAY STATISTIC COLLECTION

| | | Chronological List | | | | | | | | | | | |
| | | Gen—Con I | | | | Delay Clock (T1) | | | | Delay Clock (T2) | | | |
Act	Sim Time	Mark	Dec	Off	Rem	Unit	On	Off	Dur	Unit	On	Off	Dur
5	0.0	5a											
5	0.0	5b											
2	0.2		1a		4a	T1	✓						
2	0.45		1a		3a								
2	1.35		1a		2a								
2	1.93		1a		1a								
2	2.37		1a	✓	0			✓	2.17				
2	3.00		1b		4b					T2	✓		
2	3.38		1b		3b								
2	3.91		1b		2b								
2	4.39		1b		1b								
2	5.00		1b		0							✓	2.00
5	6.87	5a											
2	7.2		1a		4a								
2	7.87		1a		3a								
2	8.39		1a		2a								

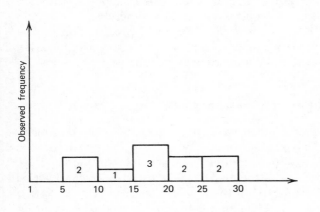

Tally	All statistics
1	8.5
2	9.1
3	13.6
4	15.6
5	18.1
6	18.1
7	20.5
8	23.8
9	25.1
10	29.0

FIGURE 10.19 Distributed ALL statistics.

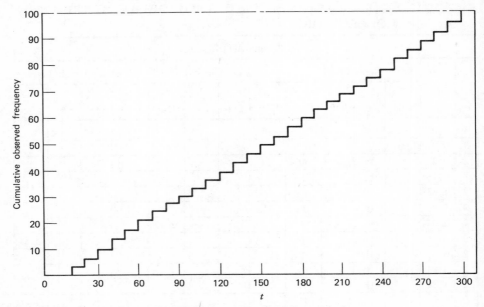

FIGURE 10.20   Cumulative histogram of ALL statistics.

limit was defined as 10.0, and 30 class intervals of 10.0 units width are defined from 10.0 to 310.0 units. Since the FUNCTION node with which the ALL statistic is associated is inserted directly following the ACCUMULATOR, the histogram of the statistic can be read directly as a production curve.

The six defined functions, as given in Table 10.2, provide the modeler with a box of monitoring devices or probes with which a wide range of information regarding system performance can be developed. These functions, in conjunction with the intrinsic QUEUE node and ACCUMULATOR functions, allow the modeler to perform a sensitivity analysis and to arrive at a maximization of productivity and minimization of idle time in a proposed operation by resource balancing.

10.7 SYSTEM SENSITIVITY TO PARAMETER VARIATION

As stated earlier in this chapter, the manager is continually attempting to evaluate the impact of his decisions on the existing state of the system and its productive output. He is always interested in knowing what will happen if he changes the level or setting of one of the management variables over which he has control. He is, in short, interested in the system's sensitivity to variation of the controlled parameters available to management. By experimenting with these controls, he can determine how they effect the system and whether he is working with a "brake" or an "accelerator." Furthermore, he can determine the magnitude of braking or acceleration associated with his management perogatives. The ability to go "inside" the system with the intrinsic and defined statistics gives him a capacity for determining not only that the management variable represents a brake or an accelerator, but also allows the determination of why the braking or acceleration is taking place. This ability may lead to new insights into the system's relationship and may indicate a requirement for redesign.

In order to see how the sensitivity of a system can be investigated to determine the correct levels for management-controlled parameters, consider a problem involving the operation of a precast concrete panel facility. A laboratory experiment on this paper-and-pencil model will be carried out to establish the system's response to parameter variation. This approach is commonly referred to as a "sensitivity" analysis.

The CYCLONE model of the plant's process line operation is presented in Figure 10.21. The system demonstrates the use of several Class I and Class III structures. The flow units defined in the system, together with their initial location and flow paths, are given in Table 10.13. The ingredience-constrained activities and their associated ingredience sets are defined in Table 10.14. The number of ingredients required at the ingredience-constrained activities range from four at activity 2 to two at activities 6 and 19. The unit category types are:

1.   Space—set and cure positions.

2.   Materials—forms and batches.

3.   Equipment—crane and trucks.

4.   Labor—Crews.

Table 10.13   ENTITY PATH TABLE

Category	Name	Number Initialized	Initialized At	Path
A	Set positions	3	23	23–2–3–4
B	Forms	5	24	24–2–3–4–5–6–16–19
C	Batches	20	20	20–21–22–2–3–4–5–6–7–8–9–10–11–12–13–14–26
D	Crews	2	15	15–2/15–6/15–19
E	Cure position	8	18	18–8–9–10–11–12
F	Crane	1	25	25–8/25–12
G	Trucks	4	17	17–12–13–14
H	Batch permit	1	27	21–27

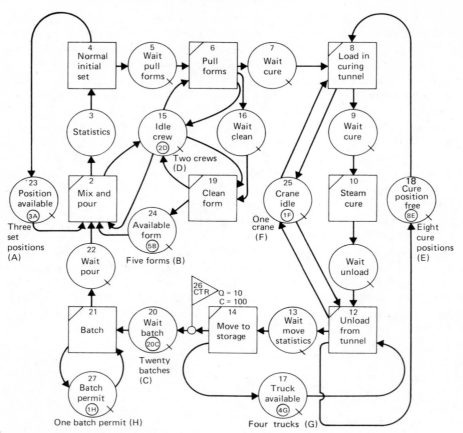

FIGURE 10.21   Precast concrete element plant.

The crews at QUEUE node 15 and the crane at QUEUE node 25 are shared units. The crew works alternately on activities 2, 6, and 19. The crane works between activities 8 and 12. The two-space units (set and cure positions) act as simple permits releasing activities 2 and 8, respectively, as appropriate. Once all set positions are committed, no mix and pour activity can commence until a position becomes free. Similarly, a slab (batch) cannot be lifted into the curing tunnel until a cure position is available at 18. The system productivity is measured by the accumulator element at 26. The QUANTITY parameter is set to 10.0 and indicates that 10 elements are associated with each production flow unit. This can be thought of as 10 elements that are simultaneously batched and poured using a single form. The progress of elements through the system from batch to storage is a function of several constraining

Table 10.14   INGREDIENCE-CONSTRAINED WORK TASKS

			Act		
	2	6	8	12	19
Ingre-dient	Mix and Pour	Pull Forms	Load in Curing Tunnel	Unload from Tunnel	Clean Form
1	Crew at 15	Batch and form at 5	Batch at 7	Batch at 11	Crew at 15
2	Batch at 22	Crew at 15	Cure position at 18	Truck at 17	Form at 16
3	Set position at 23		Crane at 25	Crane at 25	
4	Form at 24				

factors. The cyclic flow of the forms controls the frequency of the pour and mix operation by virtue of its presence in the associated ingredience set. However, the flow of forms in the system is itself a function of the availability and task selection of the crew units stationed at 15. The labeling sequence is such that the crew units give priority to mixing and pouring, pulling forms, and cleaning forms, in that order. The crane unit at 25 controls flow through the curing tunnel activity and gives priority to loading units over unloading. Truck units at 17 and the availability of the crane control the removal of concrete elements from the curing tunnel. Once units have passed the accumulator element, they are reentered into the system at a rate defined by the time parameters associated with batch work task 21. Twenty batch units are used initially. This does not result in system constraint for the smaller two- and three-crew systems. However, for systems using four and five crews and six to eight forms, system production constraints will result because 20 batches are used. That is, the mix and pour work tasks are at times constrained by the requirement to wait for batches. Therefore, the number of batch concrete element units allowable in the system must be increased. Time data for the activities of this process are as shown in Table 10.15.

Data reflecting system response are given in Tables 10.16 to 10.20. This data is based on 100-cycle simulations of systems in which the number of crews and forms available has been varied through a range of values. The number of crews available was varied from two to five, while the number of forms varies from two to eight. These values are considered for the purposes of this illustration to be manager-controlled variables, while other values, such as the number of set positions, were assumed to be fixed and not subject to the

Table 10.15   SYSTEM TIME ATTRIBUTES

Element Number	Distribution	Set Number	Mean	Lower Bound	Upper Bound	Standard Deviation
2	Normal	2	30.0	20.0	40.0	5.0
4	Constant	3	50.0	0	0	0
6	Constant	4	20.0	0	0	0
8	Normal	5	15.0	8.0	22.0	5.0
10	Constant	6	120.0	0	0	0
12	Normal	5	15.0	8.0	22.0	5.0
14	Normal	7	25.0	10.0	40.0	7.5
19	Normal	8	45.0	17.0	83.0	20.0
21	Normal	9	18.0	9.0	27.0	6.0

control of the manager. This tabular presentation gives only a small subset of the data generated by the intrinsic and defined performance indicating functions associated with this system. However, the tables provide sufficient information to evaluate system response and draw certain conclusions about its performance. The system values monitored are:

1. System productivity in units per hour (Table 10-16).
2. Crew availability time (Table 10-17) based on element 15.
3. Form cycle delays (Table 10.18) to include:
   (a) Delays because of form pulling (5).
   (b) Delays for form cleaning (16).
   (c) Delays awaiting other units at the mix and pour work task (24).
4. Curing delays information (Table 10.19) to include:
   (a) Delays awaiting entry to the curing tunnel (7).
   (b) Percent of time cure positions are free (18).
5. Crane idle time (Table 10.20 based on 25).

Table 10.16   SYSTEM OUTPUT (ELEMENTS PER HOUR)

Number of Crews	Number of Forms			
	2	4	6	8
2	7.98	12.01	12.16	12.15
3	7.98	15.07	17.02	17.06
4	7.98	15.50	18.02	18.56
5	7.98	15.50	17.88	18.01

Table 10.17   CREW AVAILABILITY TIME (PERCENT)

Number of Crews	Number of Forms			
	2	4	6	8
2	0.685 / 54%	0.032 / 3%	0.020 / 2%	0.020 / 2%
3	1.680 / 100%	0.549 / 14%	0.118 / 9%	0.077 / 5%
4	2.680 / 100%	1.460 / 100%	0.723 / 51%	0.682 / 48%
5	3.678 / 100%	2.450 / 100%	1.552 / 78%	1.592 / 80%

Time-Integrated Average of Crews

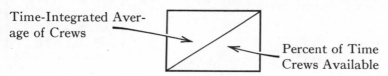

Percent of Time Crews Available

Normal distributions have been associated with all transit time activities with the exception of those having constant durations. Times are developed from field data on observed transit times. The selection of the distribution is also based on examination of the distribution of field transit times.

Examination of the performance data tables indicates certain general information regarding the operation of the system. It will be noted that with the exception of the crew idleness values (Table 10.17), the performance statistics in column one of all tables do not vary as a function of the number of crews. The production of each of these systems is 7.98 units per hour and the form cycle delays are all zero. This happens because the system is totally constrained by the low number of forms being used, and the system is insensitive to the variation of other control parameters. Crew availability, of course, increases as the number of crews increases, since two crews are adequate and the additional crews above this value add no productive output.

Another aspect of general system performance that is not indicated in the tables, but is given in the system statistics, is the delay from the time of system

Table 10.18   FORM CYCLE DELAYS

Number of Crews	Number of Forms			
	2	4	6	8
2	0%	36%	52%	69%
	0%	34%	97%	97%
	0%	24%	36%	39%
3	0%	5%	40%	62%
	0%	4%	43%	45%
	0%	11%	50%	86%
4	0%	0%	2%	10%
	0%	0%	4%	13%
	0%	9%	72%	100%
5	0%	0%	0%	1%
	0%	0%	0%	2%
	0%*	0%	75%	100%

*No delay of forms prior to batching.

Pull Delay (5)
Clean Delay (16)
Delay at Mix
and Pour (24)

"start-up" until the first production concrete unit arrives at the accumulator element. This is normally referred to as the "pipe-line" delay. For all systems examined, the "pipe-line" delays ranged from 270 to 290 minutes. This means that over 4 hours are required for plant start-up.

System production increases as a nonlinear function of both the number of crews and forms. Figure 10.22 gives a graphical representation of the information present in Table 10.16. The production for a system using two crews becomes constrained at values of four or more forms, since at this value and above the crews are totally occupied (i.e., saturated). This is verified by the values in the crew availability timetable that shows availability time as 2% from four forms and above. The maximum production of this system peaks at 12 elements per hour. Similarly, the three-crew system becomes constrained by the number of crews when six or more forms are available with a peak output of 17 units per hour. The four- and five-crew systems peak at output values of 18.56 and 18.01, respectively. At this point, the system begins to be constrained by the availability of the crane at the curing tunnel. The high output of the forming cycle begins to saturate the crane with work, passing

Table 10.19 CURING DELAY INFORMATION. PERCENT TIME DELAY AND TIME-INTEGRATED AVERAGE

Number of Crews	Number of Forms			
	2	4	6	8
2	6%/0.063	5%/0.051	8%/0.08	9%/0.09
	100%/5.92	100%/4.85	100%/4.79	99%/4.73
3	6%/0.063	16%/0.162	22%/0.234	26%/0.30
	100%/5.92	100%/3.79	100%/2.96	94%/2.86
4	6%/0.063	19%/0.214	75%/3.14	65%/2.73
	100%/5.92	100%/3.65	21%/0.71	32%/0.87
5	6%/0.063	19%/0.214	83%/4.89	87%/4.67
	100%/5.92	100%/3.65	16%/0.59	11%/0.53

Wait Cure (7)
Cure Position
Free (18)

Table 10.20 CRANE IDLE TIME

Number of Crews	Number of Forms			
	2	4	6	8
2	62%	38%	37%	37%
3	62%	22%	11%	15%
4	62%	21%	5%	6%
5	62%	21%	5%	5%

production control to the curing segment of the system. The waiting times of the curing facility (element 7) go up significantly from 22% for the three-crew system to 75% for the four-crew system. The average number of curing positions drops from 2.96 to 0.71.

The crew idle time data indicates that the two- and three-crew systems become saturated at form availability values of six and eight, respectively. Additionally, with only two forms available, the crews in the two-crew system are fully committed only 46% of the time. The time-integrated average of

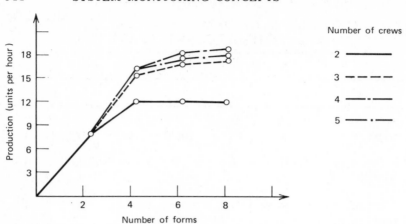

FIGURE 10.22   Production versus number of forms and number of crews.

units idle indicate that 0.68 and 1.680 units on the average are idle in the two- and three-crew systems. Figure 10.23 gives a graphical presentation of the percent time available figures contained in Table 10.17. Although the three-crew system drops from 9 to 5% availability between the six and eight-form values, the associated increase in production is negligible.

The values given in Table 10.18 indicate the percent of the time concrete units are delayed at the activities in the form cycle (i.e., form pulling, cleaning, awaiting mix, and pour). The delays for pulling and cleaning are directly related to the number of forms and inversely related to the number of crews. This is indicated in the two plots shown in Figure 10.24. Figure 10.24a shows the increase in delay of units at the pull form operation (work task 6) for the three-crew system. This delay increases as the number of forms defined in the system increases. Similarly, Figure 10.24b shows that for a constant number of forms in the system, the delay at the clean form operation (work task 19) decreases as the number of crews is increased. In the two- and three-crew systems, delays are large, since the eight forms tend to distribute themselves at these lower-priority activities. Since the crews give priority to the mix and pour activity (based on the labeling rule), these "back" cycle activities must wait longer for a crew assignment. As the number of crews increases, the crew-to form ratio improves, causing a corresponding reduction in form cycle delays.

Form cycle delays at the mix and pour activity (because of other units) increase as the number of forms defined in the system increase. This is due to the effect of the fixed number of three initial set positions that, in this illustration, cannot be varied by the manager. This constraint places an upper limit on the rate at which elements can be generated from the pouring-forming

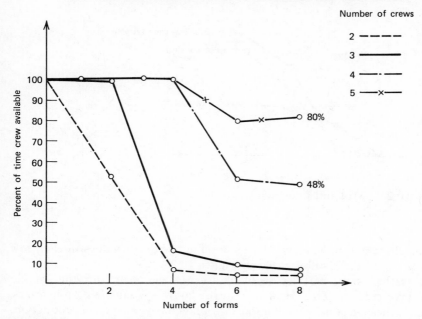

FIGURE 10.23   Crew availability (percent).

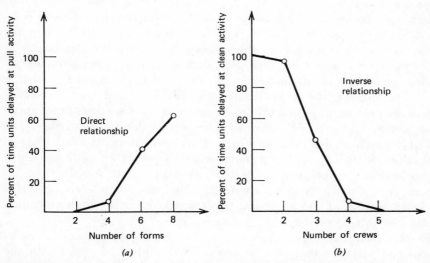

FIGURE 10.24   Form cycle delay relationships. (*a*) Constant number of crews (3). (*b*) Constant number of forms (8).

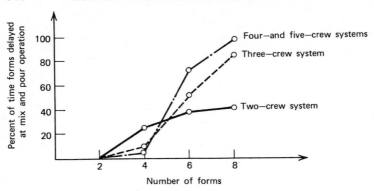

FIGURE 10.25   Mix and pour delays.

cycle. This effect can be observed by examining the delay values in each row from left to right. This is illustrated in Figure 10.25.

Delays at the curing activity segment are not significant for the smaller two- and three-crew systems. However, in the larger systems (e.g., five crews, eight forms), these delays become important and constrain system production. The delay is a function of both availability of cure locations in the curing tunnel and the availability of the crane to load and unload units. For the five crew, eight form system, Table 10.19 shows that the percent of the time cure positions are available awaiting new elements is only 11%. This indicates that all positions are committed 89% of the time. Furthermore, the crane is committed 95% of the time (5% idle time), as indicated in Table 10.20. As noted in discussing system productivity, the saturation of the crane also begins to constrain the system and causes the peak system output to the governed by the curing network segment instead of by the form cycle, as in the smaller systems. The decrease in curing location availability and the increase in element delay time at the curing tunnel loading operation are most pronounced in the 4–6, 4–8, 5–6, and 5–8 systems.* This is also the case with the crane availability, which is illustrated in Figure 10.26. As the crane availability is reduced, the system approaches a deterministic limit of 20 units per hour. The crane makes an average of four loads/unloads per hour based on the 15-minute mean load and unload duration associated with these activities. This results in an average of two loads and two unloads per hour. Each unloading work task generates 20 elements for storage.

In the section on productivity, it was noted that higher system productivities were achieved with the four-crew systems (18.02 and 18.56) than with the five-crew systems (17.88 and 18.01). This reduction in system productivity

*The first value indicates the number of crews and the second the number of forms.

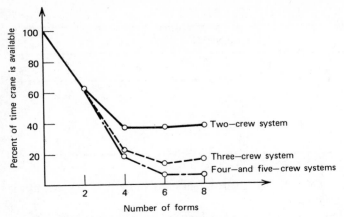

FIGURE 10.26   Crane availability.

can be traced to congestion in the curing tunnel segment of the model. This congestion is increased by the higher arrival rate of elements at the curing tunnel from the forming cycle. Table 10.21 shows the interarrival rates for the nine largest systems investigated. The increase in crews and forms leads to a higher production of elements from the forming cycle and a corresponding smaller interarrival rates at the curing tunnel. Table 10.19 indicates that the percent of the time curing locations are available in the four-crew system is considerably greater than in the five-crew systems. This results in a faster throughput and higher overall productivity. This is indicated by the inter-arrival rates at work task 13 following the curing tunnel (values in Table 10.21 in parentheses). This indicates, for instance, a greater interarrival rate at 13

Table 10.21   INTERARRIVAL RATE AT CURING TUNNEL IN MINUTES (ELEMENT 7)

Number of Crews	Number of Forms		
	4	5	6
3	38.11	33.11	33.36
4	37.15	28.90 (32.95)	28.50 (32.13)
5	37.15	27.90 (33.26)	27.60 (33.12)

Table 10.22   COST PER ELEMENT VARIATION

Number of Crews	Number of Forms			
	2	4	6	8
2	5.02	3.33	3.26	3.26
3	7.54	3.98	3.52	3.52
4	10.06	5.18	4.43	4.31
5	12.58	6.45	5.60	5.53

for the 5–8 system than the 4–8. Obviously, delays in the larger system have occurred in the curing segment.

The analysis conducted in this example has focused only on the plant output in terms of units per hour production. A deeper analysis must also address the cost per unit, which requires additional input features designed to associate cost attributes with productive units (crews, crane, etc.) as appropriate. Using the present system, cost concepts can be included by hand calculation of associated costs. For instance, if it is assumed that the hourly cost associated with each crew is $20, the cost range involved in the experiments conducted above varies from $40 to $100.* Table 10.22 shows a cost per element variation based on these assumed crew costs. The cost analysis indicates that the two-crew system yields the lowest cost per unit ($3.26) with six or more forms. However, if demand is strong, the best combination of production and element cost probably occurs in the three-crew, six-form system, which produces 17 elements per hour at a cost of $3.52 per element.

The object of this presentation has not been to examine the precast plant system exhaustively but, instead, to indicate the insights into system operation that can be developed using simulation and the modeling simplicity afforded by the CYCLONE system format. The interaction of the various flow units and the process activities becomes considerably clearer after an investigative simulation of the system such as the one conducted here on the concrete plant problem. If the manager desires a more detailed analysis, this can be carried out using more sophisticated statistical analysis methods. If, however, the manager simply wants to get the "feel" of the system and the interactive forces and constraints at work, the simple qualitative analysis carried out in this example can be used. The depth of analysis is obviously a function of the detail and statistical reliability of the information desired and the resources available. In most construction systems, the precision of the input data (e.g., transit times, distribution types) is based on the engineering estimates and a sampling of field data. This precision seldom justifies the high

---

*This assumes other costs are fixed.

degree of statistical confidence appropriate in other applications. It does, however, provide sufficient accuracy to carry out an insight-oriented analysis such as the one just developed. In most cases, this type of analysis is most appropriate for construction management purposes.

PROBLEMS

1. Develop occupancy and time-integrated statistics for the masons defined in Problem 7.7.
2. Develop occupancy and time-integrated statistics for the trucks defined in the model shown in Figure 4.10.
3. By inserting COUNTER elements following the lift activities in the three commodity system of Figure 9.5 and Problem 9.6, estimate the number of steel, brick, and duct loads lifted in a typical 8-hour shift.
4. Using a between statistic, estimate the time between arrivals at the welder's station (Activity 18) in the sewerline project model shown in Figure P7.6.
5. Using the following activity times and initializing two trucks at the "WT LOAD" position, estimate the idle time of the asphalt spreader in the model developed in Problem 5.3.

	Mean (Minutes)	Standard Deviation (Minutes)
Load at plant	5.0	0.75
Travel to job	30.0	5.00
Dump into spreader (spread asphalt)	7.0	2.00
Reposition for pass	15.0	3.00
Compact asphalt section	20.0	5.00
Finish Section	30.0	4.00

6. Develop the time-integrated statistics for QUEUE node 11 (WT SPACE IN HOPPER) for the building construction model shown in Figure 7.21. Times for the activities defined in the model are as given in Table 7.9.
7. Investigate the transient response of the haul system shown in Figure 4.10 by positioning all trucks at the loader queue (as shown) at $t = 0.0$ and comparing this to the system response when all trucks are started at the spotter's queue (QUEUE node 4) at $t = 0.0$. Use the LOAD, TRAVEL, DUMP, and RETURN times given in Problem 10.5.

8. Conduct a hand simulation of the precast concrete element plant (Figure 10.21) to determine the interval statistic for units passing from QUEUE node 7 (WAIT CURE) to 13 (WAIT MOVE).

9. Although presented in this chapter in the context of a precast element plant operation, the model developed in Figure 10.21 can be used with appropriate time modification and task redesignation to model the casting operation of a steel plant. Assume that steel is poured in its molten state into a crucible in which it is allowed to cool to form an ingot. After initial cooling, each ingot is removed from the crucible and lifted into a cooling pit until it has reached the proper temperature for rolling. Following the cooling pit, the ingot is removed and rolled into the desired shape. A crew is involved in removing the ingot from the crucible after initial cooling and is also charged with cleaning the ingot for reuse. The parallel between the precast element plant operation and this process is fairly easy to detect. Modify the model of Figure 10.21 to handle the casting-rolling process just described and verify the system's operation using hand simulation. The times for the various operations are:

Operation	Time (in Minutes)
Pour ingot	10.0
Initial set	75.0
Lift into pit	20.0
Lift from pit	20.0
Cool in pit	200.0
Deslag (remove crucible)	15.0
Roll shape	50.0
Smelt	90.0
Clean Crucible	12.0

Determine the idle time for the crew and the crane in the process.

CHAPTER ELEVEN

Computer Processing and List Structures

The computer processing of data structures using algorithms and programmed packages is efficient, fast, and economic compared to manual processing techniques. In fact, the mechanical advantage achieved through computer processing is so large that problems that are computationally infeasible by manual methods become feasible and economic with computer processing.

The computer offers the advantage of speed of simulation for the types of computational processes discussed in Chapters Six and Seven. Since simulation procedures are largely iterative in nature, and iterative procedures are amenable to algorithmic formulation, it is not surprising that computer routines and program packages are relevant, and available, for the simulation analysis of problems.

In order to get useful results from the simulation analysis of a problem, it is necessary to process enough cycles of the system to cover the transient period of system start-up and the period of steady state. The use of hand simulation to generate more than a few cycles of the system is very tedious and time-consuming. The computer, however, can generate random numbers, accomplish the ADVANCE and GENERATE phases of the simulation procedure, and collect statistics, all within a fraction of a second. Therefore the computer offers a tremendous advantage in the simulation analysis of system problems.

This chapter investigates some of the mechanisms and techniques used in automating the simulation procedures presented in Chapter Seven.

11.1 LIST STRUCTURES

Essentially, the simulation procedure reduces to the process of maintaining a list of end event times (E.E.T.'s) and the transfer of these times to the CHRONOLOGICAL list, thus accomplishing the advance of the SIM CLOCK. Other procedures are required for collection and maintenance of statistics, as described in Chapter Ten. Random numbers are used to generate E.E.T. values. The computer production of these numbers was presented in Chapter Six. List structures define the location and format of data arrays.

Since list structures are so important in the simulation of discrete entity flow systems, it is beneficial to discuss them at this point. Two types of lists are of importance.

1. Dynamic lists store information that is pertinent to the movement of flow entities and that is constant only during a period of time. In other words, dynamic lists contain time-variant information. In simulation programs, dynamic lists are used for (a) storing E.E.T.'s; (b) storing information regarding units held in QUEUE nodes so that delay times can be developed; and (c) other information of a temporary nature.

2. Permanent lists are used to store data that is constant throughout the period of simulation. In simulation programs, permanent lists contain data concerning permanent attributes associated with each network element. This includes numerical label, duration parameters and associated distribution type, and associated network logic. Permanent lists can be handled as permanent arrays in the computer program.

In the hand algorithm the selection of the next unit as E.E.T. to be advanced is accomplished by inspecting the E.E.T. list and moving the E.E.T. closest to TNOW from the EVENT list to the CHRONOLOGICAL list. In computer processing efficiency and reduction of storage required demands that only the part of the EVENT list that has not been transferred to the CHRONOLOGICAL list is maintained. In order to select which E.E.T. is the priority candidate for selection as TNOW, the events are continuously ordered and reordered according to an established discipline. In the case of the EVENT list, the E.E.T.'s are ordered or stacked so that the lowest or least E.E.T. remaining to be transferred is always at the top of the list. This insures tnat the E.E.T. at the top of the list following the GENERATE phase is always the next EVENT time to be transferred and therefore establishes the new value of TNOW.

If the hand procedure as described is slightly modified, this operation becomes clear. Assume that instead of recording the E.E.T.'s (as generated) in the EVENT list, the E.E.T.'s are written on individual cards that are held in the hand. At the end of each GENERATE phase the cards are reordered so that the card with the lowest E.E.T. is on the top of the stack, the second card in the stack is the next lowest, and so on. Then the selection of the E.E.T. that establishes the new TNOW is simply a matter of taking the top card from the stack and advancing the TNOW value to the time on the card. Continuing the analogy, at this time (i.e., after advancing the TNOW value to that on the top card), the card is removed from the stack and discarded. The next event time temporarily will be the E.E.T. on the card below the just discarded top card. However, this card may be preempted by an E.E.T. generated during the GENERATE phase. Therefore, it may or may not maintain its top position in the stack.

In the hand simulation algorithm, the E.E.T. selected is not discarded but, instead, is recorded in the CHRONOLOGICAL list. As a practical matter, in computer processing the E.E.T., once it is removed from the EVENT list, is moved to the TNOW location, where it is maintained until the next AD-VANCE phase, at which time it is replaced. Therefore, a CHRONOLOGICAL list is not maintained by the computer. The computer maintains TNOW, which is a single line of the CHRONOLOGICAL list. If a CHRONOLOGICAL list is desired, the value of TNOW can be printed each time it is changed. This results in a listing of the CHRONOLOGICAL trace. However, it is meaningless to store the CHRONOLOGICAL list in the computer and, there-fore, a record of each E.E.T. ceases to exist once it is replaced as the TNOW value. That is, the recorded life of an E.E.T. begins when it is written on a card of events to be processed (or entered into the EVENT list) and ends when it is replaced as the TNOW value.

List structures provide a method of ordering entries in a dynamic or changing list. Associated with each entry is a tag or attribute that contains the address of the entry that is to follow (i.e., which it precedes). The address acts as a link that provides directions as to where the following entry is located. This type of format leads to the term "linked" list. The list, in fact, becomes a chain of entries linked by the addresses associated with them. Normally, each list has a header that indicates the location of the topmost or first entry in the list. The last entry in the list contains a "flag" in its address attribute indicat-ing that the entry is the last in the list. For instance, the last entry may have an associated address of 9999, indicating that there is no following entry and that the entry is the last. A schematic for a dynamic list containing four entries is shown in Figure 11.1.

11.2 LIST STRUCTURES FOR MASONRY EXAMPLE

As shown in Figure 11.1, each entry has five ($N = 5$) attributes (e.g., locations for associated information such as time, duration, and followers) associated, and the $N + 1$, or sixth, attribute contains the address of the

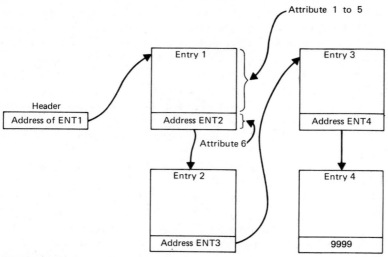

FIGURE 11.1   List structures.

next entry. The $N + 1$ attribute that points to the next entry is sometimes referred to as a pointer.

To illustrate this, consider the first six entries that are generated into the EVENT list for the masonry simulation of Chapter Seven (see Table 7.5). The appearance of the linked list at the event times between 0 and 8.5 is shown in Figure 11.2.

First, it will be noticed that two attributes (i.e., the element label and the E.E.T.) are stored attributes, making the $N + 1$ or third attribute the address of the next entry (i.e., the pointer). The addresses refer to the line number in the list, starting with line 1 just below the header and numbering sequentially to line 6 (the last line in the list segment shown). The first column in the list is reserved for the element number, the second for the E.E.T., and the last for the next entry address or pointer.

At $t = 0$, the list is empty and, therefore, the header contains the end of list designator 9999. Following the end of the GENERATE phase, a single entry is inserted in the list at line 1. Its pointer is set to 9999. The header value is changed to 1, indicating the first entry in the list is stored at line 1. Following the ADVANCE phase, TNOW is set to 3.0 and the entry is removed from the list. The header value is changed to 9999, indicating that the list is again empty.

The GENERATE phase with $t = 3.0$ results in the scheduling of two E.E.T.'s and their insertion into the list. The first E.E.T. (7.9) is inserted in line 1 and the second (4.0) in line 2. Since 4.0 is nearer to 3.0, it is the next scheduled

(a) T = 0.0

Stored attributes N = 1, 2 / Address Attribute

Header

9999		
0	0	0
0	0	0
0	0	0
0	0	0
0	0	0
0	0	0

End of Generate Phase

1		
2	3.0	9999

(b) T = 3.0

Following Advance Phase

9999		
0	0	0
0	0	0
0	0	0
0	0	0
0	0	0
0	0	0

End of Generate Phase

2		
2	7.9	9999
5	4.0	1

(c) T = 4.0

Following Advance Phase

1		
2	7.9	9999

End of Generate Phase

1		
2	7.9	2
6	8.5	9999

(d) T = 7.9

Following Advance Phase

2		
0	0	0
6	8.5	9999

Following Generate Phase

2		
2	9.0	9999
6	8.5	3
5	8.9	1

(e) T = 8.5

Following Advance Phase

3		
2	9.0	9999
0	0	0
5	8.9	1

Following Generate Phase

3		
2	9.0	9999
0	0	0
5	8.9	1

FIGURE 11.2   List dynamics during period $T = 0$ to $T = 8.5$. (a) $T = 0.0$. (b) $T = 3.0$. (c) $T = 4.0$. (d) $T = 7.9$. (e) $T = 8.5$.

event and is the first entry in our linked list. Therefore, the header value is set to 2, indicating that the first entry in the list is located at line 2. The pointer for line 2 is set to 1, indicating that the entry following line 2 is line 1. The development of the remaining list states shown in Figure 11.2c to 11.2e is left as an exercise.

11.3 OPERATION OF LINKED LIST STRUCTURES

Empty lines are always filled from lowest to highest number. Therefore, when the line 1 entry is vacated following the ADVANCE phase at TNOW = 7.9, the next generated event is inserted in line 1 and the pointers are recalculated accordingly.

Computer routines are normally written for the two operations involved in operating a linked list structure—removing and inserting entries. Whenever an entry is entered or removed from a list, the pointers must be updated. Flow diagrams for the insertion and removal of entries in a linked list structure are given in Figure 11.3.

Linked lists are also very useful in controlling the flow of units delayed at QUEUE nodes. The units delayed are released from the QUEUE node in an order or sequence as processors (i.e., following COMBI elements) become available. The sequence of release is a function of the QUEUE discipline associated with the QUEUE node. In CYCLONE, this is assumed to be a first-in-first-out (FIFO) discipline. In other words, priority is given to the unit that has been delayed longest. Other disciplines, such as last-in-first-out (LIFO), are possible but, in construction processes, the FIFO procedure is most common. The order of release is particularly important when statistics regarding the length of time that units are delayed are maintained. In order to calculate the length of delay associated with a given unit, a record must be made of its arrival time in the QUEUE node. Maintenance of these times results in a list in which the units are ordered such that units arriving earlier are stacked above later arriving units. The time of the latest arriving unit is the last entry in the list when FIFO discipline is in effect.

In order to understand this better, consider QUEUE node 4 of the masonry system simulation. At the start of the simulation, four stack markers are initialized in the QUEUE node representing four stack positions. As they enter the QUEUE node, they must be marked or ordered to insure that the FIFO discipline is maintained. In this case the associated list of entry times is ordered so that the first entry stays at the top of the stack or list and the last unit arriving is inserted at the bottom of the stack. The linked list structure associated with a FIFO QUEUE node is illustrated schematically in Figure 11.4. Entries enter the bottom of the stack, move from bottom to top, and then exit.

The states of the linked list for QUEUE node 4 at times between TNOW = 0 and TNOW = 9.0 are shown in Figure 11.5. The times used in the example are taken from the simulation shown in Table 7.5. The stored attribute in this

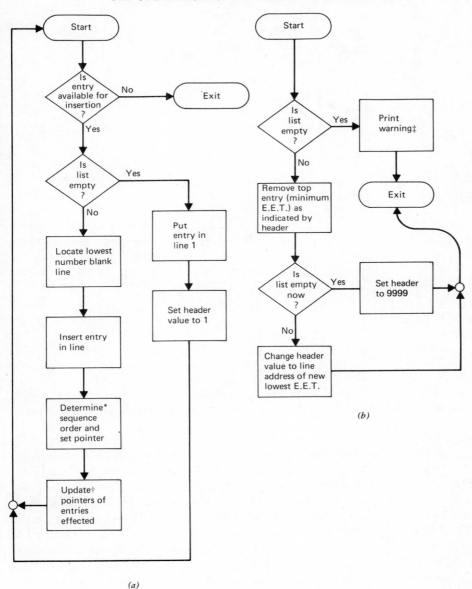

(a)

(b)

*In case of E.E.T. ,list is simply the ranking of the new E.E.T. in a lowest to highest ordering.
†Pointers associated with E.E.T. values directly above and below the new inserted value in the list must
‡If the EVENT LIST is empty at the end of a GENERATE phase the simulation is stopped, since there are no units to ADVANCE.

FIGURE 11.3   Flow diagrams for entry insertion and removal. (a) Insertion routine. (b) Removal routine.

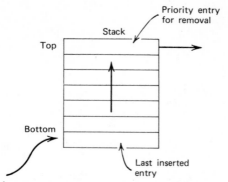

FIGURE 11.4   FIFO list structure.

Header	
1	
0.0	2
0.0	3
0.0	4
0.0	9999

Header	
1	
0.0	2
0.0	3
0.0	9999

Header	
1	
0.0	9999

(a)   (b)   (c)

FIGURE 11.5   QUEUE node Four list states. (a) $t = 0$ (start). (b) $t = 0$ (following GENERATE phase). (c) $t = 7.9$ (following GENERATE phase).

case is the unit arrival time at the QUEUE node that will be used later to calculate the unit delay time (i.e., DELAY = EXIT TIME − ARRIVAL TIME). The $N + 1$ or second entry in the table is the pointer indicating the unit's position in the list. Figure 11.5a indicates that all four brick stack markers are initialized in the QUEUE node list at TNOW = 0. Following the GENERATE phase at TNOW = 0, one unit is moved to the "RESUPPLY STACK" work task (COMBI 2) and its delay time is calculated [DELAY = TNOW (EXIT) − TNOW (ENTER) = 0 − 0 = 0]. At this point the list contains only three entries. At TNOW = 3.0 and 7.9, units are removed from the list in a similar fashion and moved to COMBI 2. The delay times are simply 3.0 and 7.9, since their enter times were TNOW = 0.0. The appearance of the QUEUE node list at TNOW = 7.9 is shown in Figure 11.5c. The last QUEUE node list entry is removed at TNOW = 9.0, and the header is set to 9999. The four delay times accumulated for statistical purposes are 0, 3.0, 7.9, and 9.0.

Examination of the units entering and exiting QUEUE node 4 at and following TNOW = 18.6 illustrates a more typical operation of the QUEUE node list. By consulting Table 7.5 it can be established that a unit exited COMBI work task 5 and entered QUEUE node 4 at TNOW = 18.6. In general, the list for QUEUE four is affected when units exit COMBI work task 5 and units commence or complete COMBI work task 2. Completion of element 5 causes a unit to enter QUEUE node 4 and thus requires insertion of an entry into the associated list. Commencement of element 2 resulted in the movement of a unit from QUEUE node 4 and the removal of an entry from the list. Since completion of COMBI 2 may result in a removal from QUEUE node 4, E.E.T.'s associated with 2 are also monitored. By observing the E.E.T.'s for elements 2 and 5, the list state diagram for QUEUE node 4 can be developed. This is shown in Figure 11.6 for the period TNOW = 18.6 to 28.8.

At TNOW = 18.6, a unit arrives from COMBI 5 and is immediately processed on to COMBI 2 with a delay of zero, since COMBI work task 2 is available. At TNOW = 19.3, another unit arrives from COMBI 5 and is entered in the list. A delay results since the COMBI work task is still busy processing the previous unit from QUEUE node 4 [E.E.T.(2) = 21.4]. At 21.4, the processing at COMBI 2 is completed and the unit in the list is removed and moved to the COMBI. Its delay in the list is calculated as 2.1. Verification of the other list states (Figure 11.6d to 11.6j) is left as an exercise for the reader. A flowchart for the insertion and removal of entries in this FIFO type of list is given at Figure 11.7. The removal routine is the same as shown in Figure 11.3.

11.4 QUEUE NODE LISTS

Maintenance of delay statistics at each QUEUE node in the system results in the requirement to establish lists for each of the delay elements. For this masonry system the requirement is to maintain four lists, one for each of QUEUE nodes in the system. Obviously, it is most efficient to maintain one master list in which the four sublists required are accessed by headers that indicate where the first entry in each sublist is located. This allows consolidation of the many sublists that develop in larger systems. The use of such a consolidation will be considered in the context of the masonry system.

First, a directory consisting of the location of the first (top of stack) entries in each of the sublists is required. This directory will be called HEADER and will consist of four locations corresponding to the four sublists required for the QUEUE nodes in the system. Following initialization of the system at TNOW = 0.0, the directory and list will appear as shown in Figure 11.8. Location (1) in the directory indicates the first entry in the QUEUE node 1 sublist. Location (2) indicating the first entry in the sublist for QUEUE node 3 is fitted with a 9999, since the sublist is empty at TNOW = 0.0. Locations

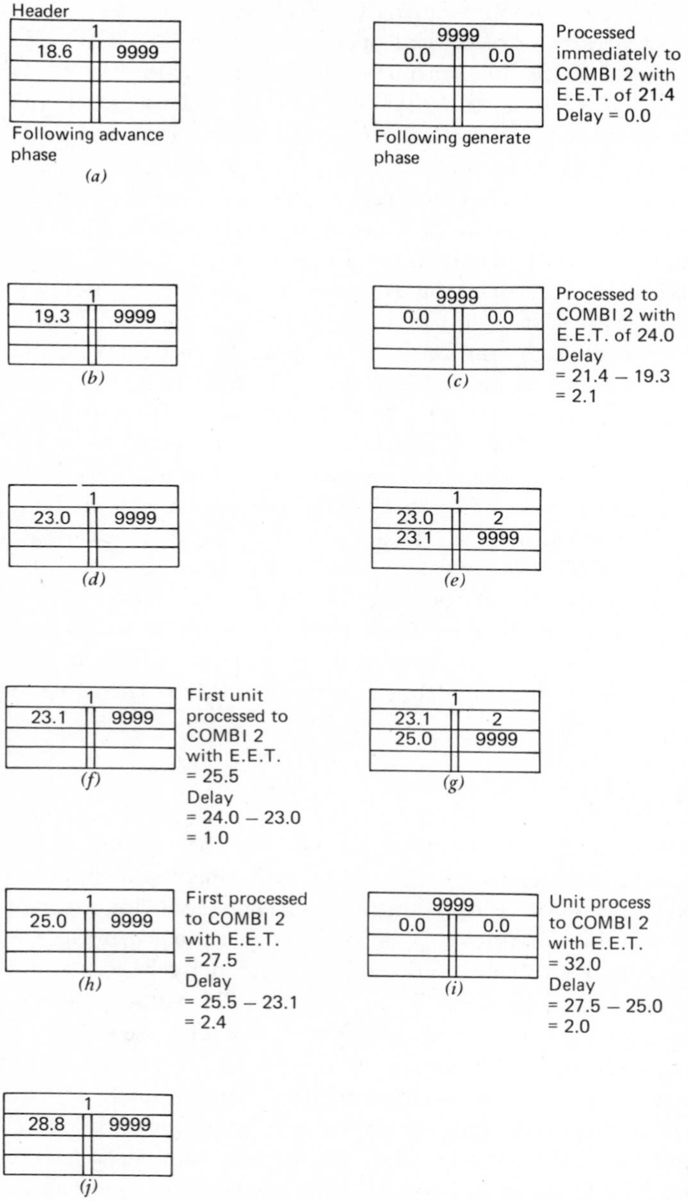

Header

1	
18.6 ‖ 9999	

Following advance
phase

(a)

9999	
0.0 ‖ 0.0	

Processed
immediately to
COMBI 2 with
E.E.T. of 21.4
Delay = 0.0

Following generate
phase

1	
19.3 ‖ 9999	

(b)

9999	
0.0 ‖ 0.0	

(c)

Processed to
COMBI 2 with
E.E.T. of 24.0
Delay
= 21.4 − 19.3
= 2.1

1	
23.0 ‖ 9999	

(d)

1	
23.0	2
23.1	9999

(e)

1	
23.1 ‖ 9999	

(f)

First unit
processed to
COMBI 2
with E.E.T.
= 25.5
Delay
= 24.0 − 23.0
= 1.0

1	
23.1	2
25.0	9999

(g)

1	
25.0 ‖ 9999	

(h)

First processed
to COMBI 2
with E.E.T.
= 27.5
Delay
= 25.5 − 23.1
= 2.4

9999	
0.0 ‖ 0.0	

(i)

Unit process
to COMBI 2
with E.E.T.
= 32.0
Delay
= 27.5 − 25.0
= 2.0

1	
28.8 ‖ 9999	

(j)

FIGURE 11.6 List states commencing at TNOW = 18.6. (*a*) TNOW = 18.6. (*b*) TNOW = 19.3. (*c*) TNOW = 21.4. (*d*) TNOW = 23.0. (*e*) TNOW = 23.1. (*f*) TNOW = 24.0. (*g*) TNOW = 25.0. (*h*) TNOW = 25.5. (*i*) TNOW = 27.5. (*j*) TNOW = 28.8.

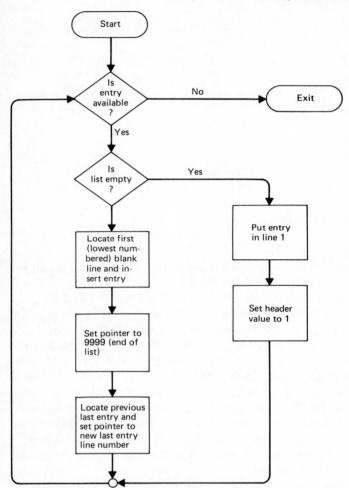

FIGURE 11.7   INSERT routine flow diagram for FIFO discipline.

(3) and (4) in HEADER contain the addresses of the first entries in the QUEUE node 4 and 7 sublists, respectively. All entry times are set to 0.0, since all units are initialized at TNOW = 0.0. The maximum size of list is eight lines, since the maximum number of units that can be in the QUEUE nodes at any time is constrained to eight (i.e., one laborer, four brick stack positions, and three laborers).

It is informative to look at the development of the master list over the first few time steps to see how it changes its configuration. The directory and list

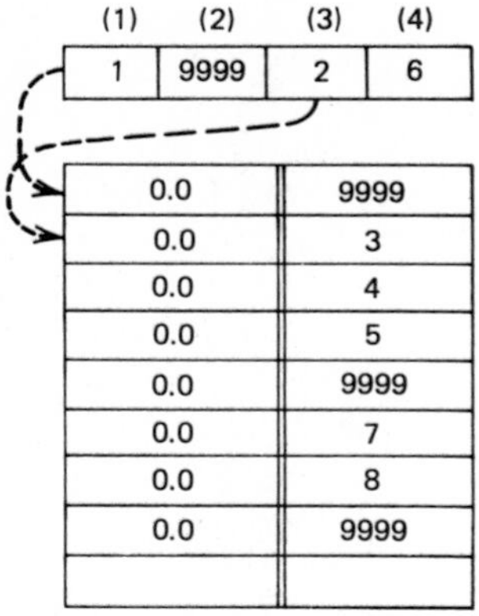

FIGURE 11.8   Master list and directory.

contents for the period TNOW = 0.0 to 9.0 are shown in Figure 11.9. All list states shown in the figure follow the GENERATE phase.

The first operation that can occur after the GENERATE phase at TNOW = 0.0 is the commencement of COMBI work task 2. This results in the removal from the master list of entry 1, representing the helper in QUEUE node 1 and the first entry in the sublist representing QUEUE node 4. Lines 1 and 2 in the master are blanked out. The header locations associated with QUEUE nodes 1 and 4 are also changed to reflect the fact that the sublist for QUEUE node 1 is empty and the first entry in the QUEUE node 4 sublist is located at line 3. At TNOW = 3.0, another unit is removed from the QUEUE node 4 sublist. The unit completing COMBI work task 2 at this time moves directly to COMBI work task 5, resulting in the removal of one of the QUEUE node 7 entries (representing a mason). At this point four of the original eight entries in the master list have been removed.

At TNOW = 4.0, the first brick stack marker completes COMBI work task 5. This unit is released back to QUEUE node 4 and is inserted into the master list. Since the first entry in the list is blank, it is inserted at line 1. It is now the last entry in the QUEUE node 4 sublist, so the pointer on the line 5 entry is changed from 9999 to 1 to reflect this. Again, the remaining modifications of the master table are left as an exercise for the reader.

QUEUE node delay statistics developed in conjunction with the master list states shown in Figure 11.9 are presented in Table 11.1. Delays are calculated by finding the difference between the enter QUEUE node times as recorded in the sublists and the TNOW at the time the unit exists the QUEUE node (i.e., the corresponding entry is removed from the sublist). The importance of ordering the units as they transit the QUEUE node in order to calculate delay statistics should now be apparent.

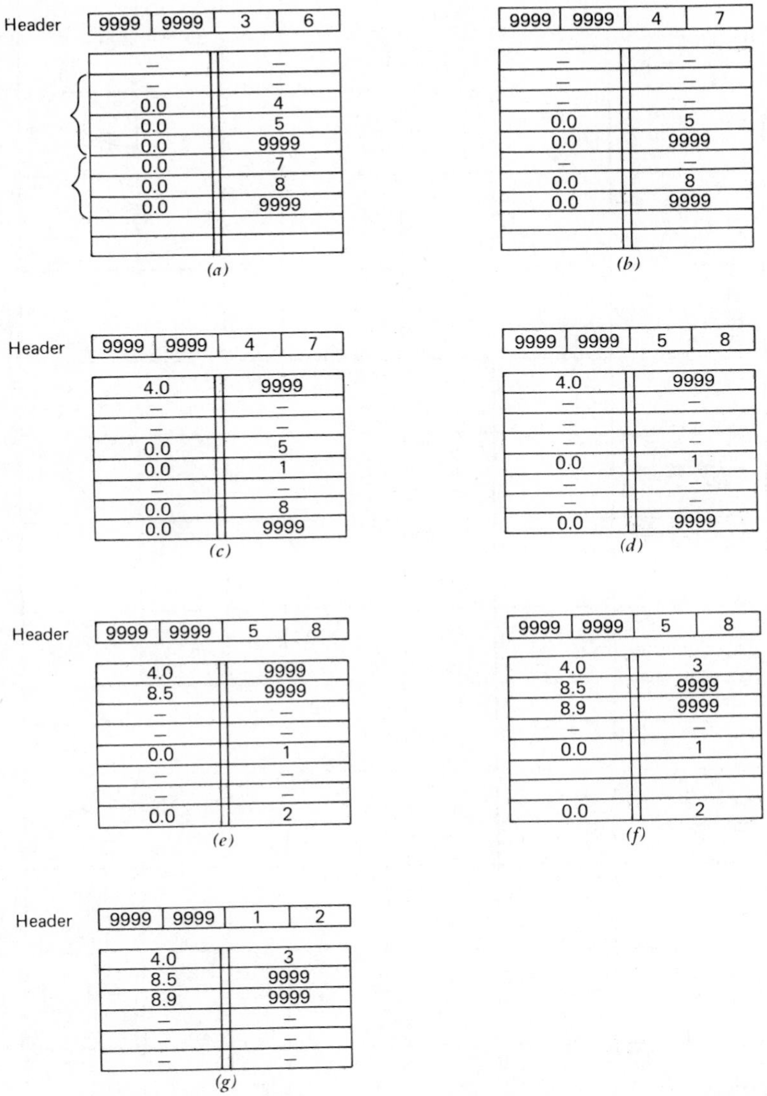

Header (a):

9999	9999	3	6

—	—
—	—
0.0	4
0.0	5
0.0	9999
0.0	7
0.0	8
0.0	9999

(a)

(b):

9999	9999	4	7

—	—
—	—
—	—
0.0	5
0.0	9999
—	—
0.0	8
0.0	9999

(b)

Header (c):

9999	9999	4	7

4.0	9999
—	—
—	—
0.0	5
0.0	1
—	—
0.0	8
0.0	9999

(c)

(d):

9999	9999	5	8

4.0	9999
—	—
—	—
—	—
0.0	1
—	—
—	—
0.0	9999

(d)

Header (e):

9999	9999	5	8

4.0	9999
8.5	9999
—	—
—	—
0.0	1
—	—
—	—
0.0	2

(e)

(f):

9999	9999	5	8

4.0	3
8.5	9999
8.9	9999
—	—
0.0	1
0.0	2

(f)

Header (g):

9999	9999	1	2

4.0	3
8.5	9999
8.9	9999
—	—
—	—
—	—

(g)

FIGURE 11.9 Directory and master list states. (a) TNOW = 0.0. (b) TNOW = 3.0. (c) TNOW = 4.0. (d) TNOW = 7.9. (e) TNOW = 8.5. (f) TNOW = 8.9. (g) TNOW = 9.0.

Table 11.1 DELAY STATISTICS

TNOW	QUEUE Node 1			QUEUE Node 3			QUEUE Node 4			QUEUE Node 7			Action
	Enter	Exit	Delay	Enter	Exit	Delay	Enter	Exit	Delay	Enter	Exit	Delay	
0.0	L1	L1	0.0				S1 S2 S3 S4	S1	0.0	M1 M2 M3			Commence 2
3.0	L1	L1	0.0	S1	S1	0.0		S2	3.0		M1	3.0	Complete 2 Commence 5 Commence 2
4.0							S1						Complete 5 Commence 6
7.9	L1	L1	0.0	S2	S2	0.0		S3	7.9		M2	7.9	Complete 2 Commence 5 Commence 2
8.5										M1			Complete 6
8.9							S2						Complete 5 Commence 6
9.0	L1	L1	0.0	S3	S3	0.0		S4	9.0		M3	9.0	Complete 2 Commence 5 Commence 2

This example has been presented primarily to show the dynamics of a linked list structure and the important role played by pointers. In the computer, it would be very inefficient to move all entries to maintain the order of sublists given originally established in Figure 11.8. Moreover, because of the rule of filling the blanks at the top of the master list, entries of one sublist become interwoven between those of another sublist. At TNOW = 8.5, for instance, the entry (mason) in the QUEUE node 7 sublist is located visually at line 2 between the second and last entries on the QUEUE node 4 sublist. The top entry in the sublist for QUEUE node 4 is located at line 5 below the other two entries. The computer, of course, "sees" these entries only in terms of their assigned pointers and, therefore, the physical stacking of entries in the list is of no consequence to the processing order.

In FIFO list structures it is common to have two headers associated with each list or sublist. One indicates the top entry in the list (or sublist), and the other indicates the last entry in the list (or sublist). Since the latest entry is always inserted at the bottom of the list, it is advantageous to know the address of the previous last entry. Otherwise, the location of the previous last entry would have to be found by "marching" through the list from top to bottom. In some applications, it is desirable to enter the list from either top or bottom and "march" the list from either end. In such instances, two pointers are associated with each entry. One pointer indicates the following entry, and the other indicates the preceding entry. Such lists with two way access are called double-threaded or two-way lists.

11.5 COMPUTER IMPLEMENTATION

It is impossible within the scope of this text to present the details of the CYCLONE system computer program. However, to demonstrate the implementation of the hand algorithm for solution by computer, a small program will be developed incorporating the features of the CYCLONE graphical notation. Four major functions must be performed by the program to include:

1. Network definition.
2. GENERATE phase implementation.
3. ADVANCE phase implementation.
4. Statistics gathering.

Inherent in these functions are the basic operations of random number and variate generation, list and sublist maintenance, read-in of network defining data, and output of relevant reports. The demonstration program will be developed with a structure general enough to handle any problem defined in the CYCLONE graphical format. Specifically, the program will be used to simulate the two laborer program presented in Chapter Ten and to output the EVENT and CHRONOLOGICAL lists as shown in Table 10.3.

A data structure is required to establish the logical relationships among the COMBI, NORMAL, and QUEUE nodes. The data structure utilizes the concept of pointers to link the various input items required for processing. The organization of the input to be compatible with the data structure is handled by an input routine. The input routine can be designed to utilize a problem-oriented language such as the one presented in Chapter Twelve. The data structure used in this example is given in Figure 11.10. The structure consists of four linked lists that contain permanent data and two dynamic

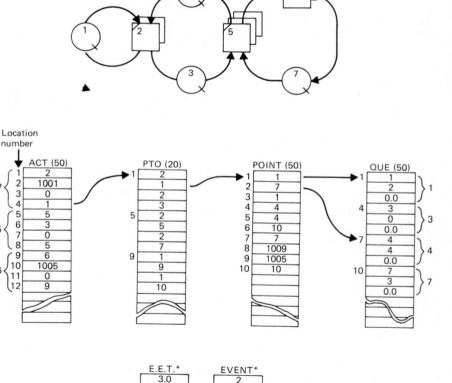

FIGURE 11.10   Data structure at TNOW = 0.0.

lists that maintain data regarding the EVENT list. The five permanent lists are defined as arrays with structure and characteristics as follows:

1. Work Task Descriptor—Array ACT.

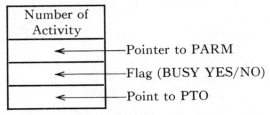

The work task descriptor is a subcomponent of Array ACT and consists of four elements. The first element stores the numerical label of the work task. The second element stores the pointer or address of the parameters which are to be used generating delay times at the work task. These parameters are stored in Array PARM. The third element contains a flag indicating whether the work task, if a COMBI, is busy or available to commence. If the COMBI is available, the flag value is zero. If any non-zero value appears in this element, the COMBI is busy and blocked for the processing of further units. The fourth element contains the point to information in the PTO work task in the network.

2. QUEUE Node Descriptor—ARRAY QUE.

←	—Number of QUEUE node
←	—Number of Units in QUEUE node at TNOW
←	—Time-integrated average statistics for the QUEUE node

The first element in the QUEUE node descriptor is the numerical label of the QUEUE node. The second element indicates the number of units delayed in the QUEUE node, and the third location is the current value of the time-integrated average statistic. In this program, no QUEUE node delay sublists will be maintained.

3. *Directory Descriptor—ARRAY PTO.*

←	—Number of elements preceding work task
←	—Header for list of elements preceding work task
←	—Number of elements following work task
←	—Header for list of elements following work task

The PTO array stores information regarding the number of elements preceding and following each work task. It also stores the headers for the sublists of element addresses stored in POINT.

4. *Address Descriptor—ARRAY POINT.*

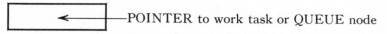

—POINTER to work task or QUEUE node

The addresses of the first elements of the work task* and QUEUE node descriptors in the ACT and QUE arrays, respectively, are contained in this ARRAY. If the address descriptor is $\geq$ 1000, it indicates an address in ACT. The address in ACT is given as (POINT $-$ 1000). For instance, if the POINT entry is 1009, the address of the first element of the work task descriptor is at line 9 in ARRAY ACT. If the entry is <1000, the address is that of the first element of a QUE node descriptor in ARRAY QUE.

5. *Parameter Descriptor—PARM ARRAY.*

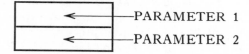

—PARAMETER 1

—PARAMETER 2

If the PARM pointer given in the work task descriptor is <1000, the distribution associated with the work task is uniform and PARM 1 is the lower limit and PARM 2 is the upper limit of the range from which durations will be selected. If the PARM pointer is $\geq$ 1000, the associated distribution is NORMAL with μ = PARM 1 and σ (standard deviation) = PARM 2.

The values in each list at TNOW = 0.0 for the two-laborer masonry system are as given in Figure 11.10. Since there are three work tasks in the system, ARRAY ACT contains 12 entries defining the three work tasks. The label of the first work task (2) is contained in line 1. Line 2 indicates that the distribution associated with work task 2 is uniform ($\geq$1000) and the PARM 1 address is located in the (1001 $-$ 1000) first line of array PARM. The third line is set to zero, reflecting that work task 2 is not busy at TNOW = 0.0. The last entry for work task 2 (line 4) points to the address of the information regarding the logical position of the work task contained in ARRAY PTO. The address of the PTO descriptor describing work task 2 is line 1 of the PTO ARRAY.

The first entry in the PTO descriptor indicates that two elements precede work task 2 and the second entry points to the location in POINT where the first preceding element address is found. Similarly, entries three and four give the number of following elements and the address in POINT of the first

---

*Both COMBI and NORMAL elements are considered to be work tasks in this version.

following element address. The addresses found in POINT refer back to either line addresses in arrays ACT or QUE. If the address given in POINT is ≥ 1000, the reference is to the ACT array (list). If the address is < 1000, the address refers to the QUE array. By entering the POINT array at the address given in line 2 of PTO, the address of the first element preceding work task 2 is found to be 1 (pointing to line 1 in array QUE). PTO indicated there are two elements preceding work task 2, so the next line (2) in POINT gives the address of the second element preceding work task 2. This entry is 7. Therefore, the descriptor for other element preceding work task 2 is contained in the set beginning at line 7 of the QUE array.

Examining the indicated descriptors in array QUE, we find that the first element preceding work task 2 is QUEUE node 1, which contains two units at TNOW = 0.0 and has a time-integrated average statistic of 0.0. Similarly, the second element preceding work task 2 is QUEUE node 4, with an initial complement of four units at TNOW = 0.0. Its time-integrated average statistic is also 0.0. Data regarding the other two work tasks in the masonry system is stored in a similar fashion in the five arrays of the data structure. As an exercise, the reader should verify the storage of information regarding work tasks 5 and 6 in the data structure as was done above for work task 2. Following the generate phase at TNOW = 0.0, the two dynamic lists E.E.T. and EVENT each contain one entry indicating that work task 2 is scheduled to complete at TNOW = 3.0.

11.6 GENERATION PHASE

Once the initialization of the data structure is complete, simulation proceeds using the following basic algorithm.

1. March the ACT array, work task by work task, starting with the first one (i.e., ACT 2).

2. For each COMBI work task that is not busy, locate pointers to the preceding QUEUE nodes by going through PTO to POINT.

3. CHECK the QUE descriptors for the preceding QUEUE nodes to see if units are delayed and available for COMBINATION.

4. If units are available in all of the preceding QUEUE nodes, the COMBI work task can commence. Decrement a unit from each of the preceding QUEUE node delay tallies. Set the COMBI work task descriptor element to "busy."

5. GENERATE an E.E.T. for the work task and insert the work task label in the EVENT list.

A FORTRAN computer routine to implement this algorithm is shown in Figure 11.11.

GENERATION PHASE

```
  1 DO 10 I=1,NUMACT
        PT = (I-1)*4+1
        FLAG = 0
        DO 11 J=1,PTO(ACT(PT+3))
            QN = POINT(PTO(ACT(PT+3)+1)+J-1)
            IF ( QN .GE. 1000 ) GO TO 11
            IF ( QUE(QN+1) .EQ. 0) GO TO 10
            FLAG = 1
 11     CONTINUE
        IF ( FLAG .EQ. 0) GO TO 10

        IF ( ACT(PT+2) .NE. 0 ) GO TO 10
        DO 23 J=1,PTO(ACT(PT+3))
            QN = POINT(PTO(ACT(PT+3)+1)+J-1)
            IF ( QN .GE. 1000 ) GO TO 23
            QUE(QN+1) = QUE(QN+1) - 1
 23         CONTINUE
        ACT(PT+2) = 1
        ACTNUM = ACT(PT)
        NUMEV = NUMEV + 1
        EET(NUMEV) = TNOW + DIST(ACT(PT+1))
        EVENT(NUMEV) = ACTNUM
        CALL LINE( EVENT(NUMEV),EET(NUMEV),TNOW,1)
 10 CONTINUE
```

FIGURE 11.11   GENERATION phase routine.

The address of the first work task address is calculated as PT $= (I - 1)*$ $4 + 1$. The range of I values considered is from 1 to NUMACT, where NUMACT is the number of work tasks in the network (e.g., NUMACT $= 3$ for the masonry system). Therefore, the first value calculated for PT is 1. The work task is assumed initially not to meet ingredience requirements, and the FLAG value is set to zero. The action of the DO loop (11) is to check all QUEUE nodes preceding the work task and determine whether work task ingredience requirements are satisfied (i.e., units available in all preceding QUEUE nodes). DO loop 23 decrements the preceding QUEUE nodes by one unit, and the appropriate work task descriptor element (PT $+ 2$) is set to 1, indicating that the work task is "busy." Following this, an E.E.T. for the work task is generated using the function DIST. The E.E.T. and work task label are stored in the appropriate lists.

11.7 ADVANCE PHASE

The algorithm for the ADVANCE phase was given in Table 7.1. Computer code to implement this algorithm is given in Figure 11.12. The nested DO loops,

```
ADVANCE PHASE

DO 13 K=1,NUMEV
    IP1=K+1
    DO 14 L=IP1,NUMEV
        IF ( EET(K) .GE. EET(L) ) GO TO 14
        TEMP = EET(K)
        EET(K) = EET(L)
        EET(L) = TEMP
        TEMP = EVENT(K)
        EVENT(K) = EVENT(L)
        EVENT(L) = TEMP
14      CONTINUE
13  CONTINUE
30  CALL LINE ( EVENT(NUMEV),EET(NUMEV),0,2)
    DO 28 K=1,NUMACT*4,4
        IF (EVENT(NUMEV) .NE. ACT(K) ) GO TO 28
        PT = K
    GO TO 40
28  CONTINUE
40  ACT(PT+2) = 0
    OLDEET = EET(NUMEV)
    DO 20 K=1,PTO(ACT(PT+3)+2)
        QN = POINT(PTO(ACT(PT+3)+3)+K-1)
        IF ( QN .GT. 1000 ) GO TO 21
        QUE(QN+1) = QUE(QN+1) +1
        GO TO 20
21      IF ( ACT(QN-1000+2) .EQ. 1) GO TO 20
        EVENT(NUMEV) = ACT(QN-1000)
        EET(NUMEV) = DIST(ACT(QN-1000+1)) + TNOW
        ACT(QN-1000+2) = 1
        CALL LINE(EVENT(NUMEV),EET(NUMEV),TNOW,1)
        NUMEV = NUMEV+1
20  CONTINUE
    NUMEV = NUMEV-1
    IF ( NUMEV .EQ. 0 ) GO TO 29
    IF ( EET(NUMEV) .EQ. OLDEET ) GO TO 30
29  CALL STAT (TNOW,OLDEET)
    TNOW=OLDEET
    IF ( TNOW .LE. TUPTO) GO TO 1
    WRITE (6,111)
111 FORMAT(1H1,20X,'THE TIME INTEGRATED STATS FOR THE QNODES ARE'//)
    DO 2 I=1,NUMQN
        QN = (I-1)*3+1
        FQ = QUE(QN+2) / 100.0
        WRITE(6,110) QUE(QN),FQ
110     FORMAT(1H0,25X,'Q(',I3,' )',F10.3)
    2 CONTINUE
```

FIGURE 11.12   ADVANCE phase routine.

13 and 14, select the next event (in order) that implements advance of the
SIM clock. The selected event is then printed under the CHRONOLOGICAL
LIST heading. The ACT array entry with the selected E.E.T. is determined,
and the "busy" flag, ACT (PT + 2), is reset to zero, indicating that the work
task is no longer blocked by a processing unit. DO loop 20 implements the
release of units to following elements. If the follower element is a NORMAL

work task, it is immediately generated and inserted in the EVENT list (array E.E.T.) by the segment between statements 21 and 20. The counter, NUMEV, indicating the number of E.E.T. values in the EVENT list is decremented to indicate the removal of the selected event and TNOW is advanced to the time of the selected E.E.T. Statistics are collected by calling subroutine STAT.

```
      THIS SUBROUTINE LINES UP THE OUTPUT

      SUBROUTINE LINE (P1,P2,P3,L)
      COMMON/SWITCH/GRAPH
      COMMON/PROG/PARM(10),QUE(50),NUMQN,QN1,QN2,TNOW
      DIMENSION G(5)
      LOGICAL GRAPH
      DIMENSION Q1(20),Q2(20),R1(20),R2(20),R3(20)
      INTEGER QUE,QN1,QN2,QNP1,QNP2
      DATA G/'*       ','**      ','***     ','****    ','***** '/
      DATA LAST1,LAST2/0,0/
      INTEGER Q1,Q2
      INTEGER P1
      IF ( L .EQ. 2) GO TO 10
      LAST1 = LAST1+1
      Q1(LAST1) = P1
      R1(LAST1) = P2
      R2(LAST1) = P3
      GO TO 20
   10 LAST2 = LAST2+1
      Q2(LAST2) = P1
      R3(LAST2) = P2
   20 IF ( LAST1 .EQ. 0 .OR. LAST2 .EQ. 0) RETURN
      IF ( .NOT. GRAPH ) GO TO 30
      GRAPH1 = 6H*
      GRAPH2 = 6H*
      QNP1 = (QN1-1)*3+1
      QNP2 = (QN2-1)*3+1
         DO 50 I=1,5
         IF (QUE(QNP1+1) .EQ. I-1 ) GRAPH1 = G(I)
         IF (QUE(QNP2+1) .EQ. I-1) GRAPH2 = G(I)
   50 CONTINUE
   30 WRITE(6,100) Q1(1),R1(1),R2(1),Q2(1),R3(1),
     *TNOW,GRAPH1,GRAPH2
  100 FORMAT(1H ,20X,I5,2(5X,F5.2),5X,I5,11X,F5.1,3X,F5.1,6X,A6,9X,A6)
      K = MAX0(LAST1,LAST2)
      DO 40 J=1,K
         Q1(J) = Q1(J+1)
         Q2(J) = Q2(J+1)
         R1(J) = R1(J+1)
         R2(J) = R2(J+1)
         R3(J) = R3(J+1)
   40 CONTINUE
      LAST1 = LAST1-1
      LAST2 = LAST2-1
      GO TO 20
      END
```

FIGURE 11.13 Subroutine LINE.

11.8 STATISTICS GATHERING

In this program statistics collection has been limited to maintaining the time-integrated statistics for the four QUEUE nodes in the system. In addition, a graphical display of the number of units delayed in QUEUE nodes 1 and 3 is also implemented by the subroutine LINE (Figure 11.13).

Subroutine LINE is used to align the output from the program. It insures that entries are printed in the event table in the order in which they are generated. Entries in the CHRONOLOGICAL list are printed in the proper sequence required for advancing the SIM CLOCK. If the routine is called from the GENERATE PHASE, the newly generated E.E.T. is stored to be printed at a later time in conjunction with a CHRONOLOGICAL list entry. If the routine is called from the ADVANCE PHASE, the entry at the top of the E.E.T. stack (FIFO discipline) is printed as well as the appropriate EVENT and TNOW times for the clock advance.

11.9 GENERATION OF RANDOM VARIATES

In this program implementation, for simplicity, only uniform and normal variates have been considered. The FUNCTION DIST listed in Figure 11.14 determines whether a uniform or a normal variate is required. If the PARM value in the activity descriptor is equal to or greater than 1000, a uniform deviate is required, and control shifts to statement 10. The uniform distribution parameters are located using the index, IPT. DIST is calculated using the random number Z and the range values PARM(IPT) and PARM(IPT+1). If the PARM value in the ACT designator is less than 1000, a normal variate is required, and FUNCTION NORMAL is called.

The normal variate is generated using the DIRECT METHOD as described in (Naylor et al., 1966), using two random number calls. The argument values

```
COMPUTE THE APPROPRIATE DISTRIBUTION

FUNCTION DIST ( POINTR)
COMMON/PROG/PARM(10),QUE(50),NUMQN,QN1,QN2,TNOW
INTEGER QUE,QN1,QN2
INTEGER POINTR
REAL NORMAL
IF ( POINTR .GE. 1000 ) GO TO 10
DIST = NORMAL(PARM(POINTR),PARM(POINTR+1))
RETURN
10 IPT = POINTR - 1000
CALL RANDOM (Z)
DIST = (PARM(IPT+1)-PARM(IPT)) * Z + PARM(IPT)
RETURN
END
```

FIGURE 11.14   Routine for delay generation (DIST).

```
GIVEN THE MEAN AND STD. DEV. COMPUTE A NORMAL DEVIATE
REAL FUNCTION NORMAL(A,B)
CALL RANDOM(Z)


RA = Z
CALL RANDOM(Z)
RB = Z
V = (-2.0*ALOG(RA))**.5*COS(6.283*RB)
NORMAL = V*B + A
RETURN
END
```

FIGURE 11.15   Function NORMAL routine.

A and B are the mean and standard deviation values, respectively (see Figure 11.15). The routine for random number generation is given in Figure 11.16.

11.10 SYSTEM OUTPUT

Two types of system output are defined in this introductory program. First, the two lists generated during hand simulation are printed by the computer. The EVENT LIST to include each end event time (E.E.T.) generated, the associated event and the time value (TNOW) is printed (see Figure 11.18). The entries are printed in their order of generation. Similarly, the CHRONOLOGICAL list to include the time at the beginning of the time step (TNOW) and the new event time (SIM TIME) are printed with the associated event. The number of units in the indicated QUEUE node

```
SUBROUTINE FOR GENERATING RANDOM NUMBERS HAVING A UNIFORM
DISTRIBUTION, BY THE MIXED MULTIPLICATIVE CONGRUENTIAL METHOD.

  SUBROUTINE RANDOM ( Z )
  DATA I/1/
  INTEGER A,X
  IF ( I .EQ. 0 ) GO TO 1
      I = 0
      M = 2 ** 20
      FM = M
      X = 566387
      A = 2**10+3
1 X = MOD(A*X,M)
  FX = X
  Z = FX/FM
  RETURN
  END
```

FIGURE 11.16   Random Number Generation Routine.

[e.g., Q(1) and Q(3)] are also aligned with the time step during which they are delayed. The single asterisk indicates an empty queue. Two asterisks indicate an occupancy level of one unit for the corresponding time period. Therefore, in time step 40.5–41.3, there is one unit in Q(1), and two units are delayed in Q(3).

In addition to the running list of E.E.T.'s and clock advances, following completion of the simulation, the program prints the integrated time statistics for each QUEUE node in the system. This is shown in Figure 11.17a. QUEUE node 3, with a value of 107.76, has the highest average integrated time statistic. This indicates that stack markers spend a considerable amount of time at the "POSITION AVAILABLE" QUEUE node and that the system is constrained by the laborer resupplying brick packers to the scaffold. The comparable statistics for a two-laborer system are shown in Figure 11.17b.

11.11 CONCLUSION

The computer routines used in this introductory program implement a computerization of the hand simulation algorithm. However, these routines are typical of those required for a general simulation system computer program. The basic functions of random time GENERATION, SIM CLOCK ADVANCE, statistics gathering and output, as well as list structure manipulation and utilization of "pointers," are typical of those required in more complex simulation systems. Essentially, these functions are required whether or not a single masonry problem or a complex tunneling simulation is to be undertaken. General purpose simulation "languages" such as those referred to in Chapter Two are available for certain types of simulation applications. In general, these languages require a specialist for the design of a systems model and are too complex for use in job site situations. However, the CYCLONE computer system retains the simplicity of the circle and square notation developed throughout this text. It allows the modeler to define the CYCLONE model to the computer in a simple problem-oriented language that retains the spatial and interactive clarity of the idle and active state process model.

THE TIME INTEGRATED STATS FOR THE QNODES ARE:

Q (1)	43.340	Q (1)	94.610
Q (3)	107.760	Q (3)	161.650
Q (4)	35.250	Q (4)	22.650
Q (7)	35.080	Q (7)	31.580
(a)		*(b)*	

FIGURE 11.17   Time-integrated average statistics. (a) Single-laborer system. (b) Two-laborer system.

EVENT LIST			CHRONOLOGICAL LIST			OCCUPANCY GRAPHS	
EVENT	EET	TNOW	EVENT	SIM TIME	TNOW	Q(1) 012345	Q(3) 012345
2	2.73	.00	2	2.7	.0	*	*
2	5.27	2.73	5	3.9	2.7	*	*
5	3.88	2.73	2	5.3	3.9	*	*
6	6.18	2.73	6	6.2	5.3	*	*
6	7.66	5.27	5	6.3	6.2	*	*
2	6.29	5.27	2	7.7	6.3	*	*
5	14.92	6.18	5	8.8	7.7	*	*
6	10.11	6.18	2	10.1	8.8	*	*
2	8.82	7.66	5	11.5	10.1	*	*
5	12.66	7.66	2	12.7	11.5	*	**
5	11.48	10.11	6	14.9	12.7	*	*
6	15.41	10.11	2	15.4	14.9	*	*
2	15.57	12.66	5	15.6	15.4	*	**
5	18.37	14.92	2	18.4	15.6	*	**
2	20.79	15.41	6	20.5	18.4	**	***
6	20.47	15.41	5	20.8	20.5	**	***
2	21.58	18.37	2	21.6	20.8	**	****
5	26.66	20.79	6	24.1	21.6	**	***
6	24.14	20.79	2	26.7	24.1	**	***
2	27.60	21.58	6	27.6	26.7	**	****
6	32.63	26.66	2	30.2	27.6	**	****
2	30.21	26.66	5	32.6	30.2	**	****
6	33.68	27.60	2	33.7	32.6	**	***
5	40.46	32.63	6	36.5	33.7	**	****
2	36.45	32.63	5	40.5	36.5	**	*****
5	41.29	33.68	2	41.3	40.5	**	****
6	44.14	40.46	6	43.3	41.3	**	****
2	43.32	40.46	2	44.1	43.3	**	***
6	44.87	41.29	5	44.9	44.1	**	****
2	49.67	44.14	6	47.4	44.9	**	****
6	47.39	44.14	6	49.7	47.4	**	****
2	50.59	44.87	5	50.6	49.7	**	***
5		49.67					

FIGURE 11.18 System output.

PROBLEMS

1. Set up the directory and master list at $t = 0.0$ for the tunnel model of Figure 5.19, assuming that 10 units are defined at QUEUE node 6 and a single unit is initialized at 13, 15, 16, 17, and 19.

2. Develop the initial data structure for the haul system (Figure 4.10) in conformance with the arrays ACT, PTO, POINT, and QUE as defined in Figure 11.10.

3. Develop the directory master lists for the first five steps of the building construction simulation shown in Figure 7.10.

CHAPTER TWELVE

Cyclone Computer System

In order to access the power of computer simulation in analyzing construction processes, the manager requires a method of reducing the CYCLONE model to a data representation that can be processed quickly and efficiently. Such a method must retain the simplicity of the CYCLONE modeling system itself, allowing the encoding of the model directly into a data representation without intermediary steps. The drawbacks associated with existing simulation packages (see Section 2.8) have to do mainly with the complexity of the method of defining the process model to the computer. GPSS and similar simulation languages emphasize various programming functions that must be performed during simulation. This results in a translation of the initial conceptual model into a program functional model understandable by the simulation package being used. Since designing and programming these functional models require a systems analyst, this translation is not normally performed by the manager. The systems analyst, being unaware of the management considerations, often develops a model that is efficient from the programming standpoint but deemphasizes and, in some cases, totally preempts the practical aspects of the process that are of most interest to the manager. Moreover, the computer model developed to interface with the simulation program bears little or no resemblance to the physical process being considered. Since the information developed is referenced to the computer model, this information must be retranslated so that it is accessible to the manager. Usually this retranslation is done by the analyst while setting up this computer model.

To illustrate this point, consider the two-link system presented in Section 2.8. In this system, three loaders load a variable number of trucks. The CYCLONE system model for this model is shown in Figure 12.1a, and the GPSS model is shown again in Figure 12.1b. The conceptual model in CYCLONE system format consists of two idle states and two work tasks. A total of five conceptual elements are required to define the system. The GPSS representation consists of 15 elements representing a wide range of functional operations that must be defined in order to simulate the system. The variety of symbols and functions used in GPSS leads to large models for relatively simple construction processes. Such models require a manager who is knowledgeable in simulation techniques a considerable amount of time to develop and interpret. In addition, they bear little resemblance to the conceptual model. Clearly, it would be advantageous to be able to define the conceptual model directly to the computer.

The CYCLONE computer language is designed to retain the features of the conceptual model and use many of the input procedures common to existing time scheduling network programs. In defining CYCLONE system networks, the modeler utilizes a problem-oriented language (POL) that allows direct specification of the model developed in CYCLONE system format without translation into a functional model. The CYCLONE system POL uses a word set that specifies each of the CYCLONE elements in terms understandable both to the modeler and the computer. A description of the number of work tasks involved and their attributes and interrelationships defines information that is sufficient to organize the data for simulation. The definition of logical relationships between CYCLONE model elements is the same as that used in critical path and PERT scheduling programs employing precedence (i.e., circle) notation. In general, the design of the network specification language is such as to minimize the number of new concepts that must be learned. The CYCLONE system POL relies on a problem specification structure similar to that already familiar to managers using time scheduling networks.

12.1 CYCLONE SYSTEM SPECIFICATION

The total problem specification of a CYCLONE system model requires five types of information. This data is submitted to the computer in free format. The input categories are indicated schematically as a segmented card deck in Figure 12.2a and define the following model components.

1. General systems information.
2. Network input segment.
3. Flow unit input segment.
4. Histogram input segment.
5. Time parameter input.

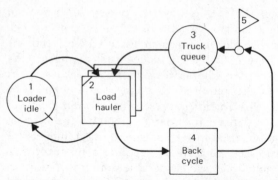

FIGURE 12.1   CYCLONE and GPSS Models. (*a*) CYCLONE system conceptual model. (*b*) GPSS Model.

Each input category is distinguished by an initial unique header or procedural statement. These headers are NAME, NETWORK INPUT, EQUIPMENT INPUT, REPORT HISTOGRAMS, and DURATIONS, respectively. The ENDDATA header statement is used to signal end of CYCLONE system definition.

Figure 12.2*b* shows the CYCLONE system input required for the haul-load model of Figure 12.1*a*. The categories of input are indicated. The hierarchy of CYCLONE system words available for system definition is given in Figure 12.3.

Consider first the NETWORK INPUT segment of the data shown in Figure 12.2*b*. Each statement of the NETWORK INPUT segment specifies an element and its attributes and logical relationship to other elements in the network. Element 1 is a QUEUE node preceding the COMBI processor 2. The defining statement for this element is

ACT 1 Q "LOADER IDLE" FOLLOW OPS 2.

The input as shown indicates that element 1 is a QUEUE node and is followed by element 2. Notice that the topological relationship of the QUEUE node to the system elements is captured by the specification of following nodes only. The ARC modeling elements are implied.

The general form of the CYCLONE system statement defining a QUEUE node is*

(ACT) {numerical label} {Q|QUEUE} (STATISTICS) ("Title")

(FOLLOW OPS) {numerical label(s) for following element(s)}

---

*A symbolic notation will be used in describing the input data associated with each element. The symbols are:

   ( ) This symbol indicates that an item is optional,

   {   } This symbol indicates that one or a list of items is required.

   | This vertical line indicates "or."

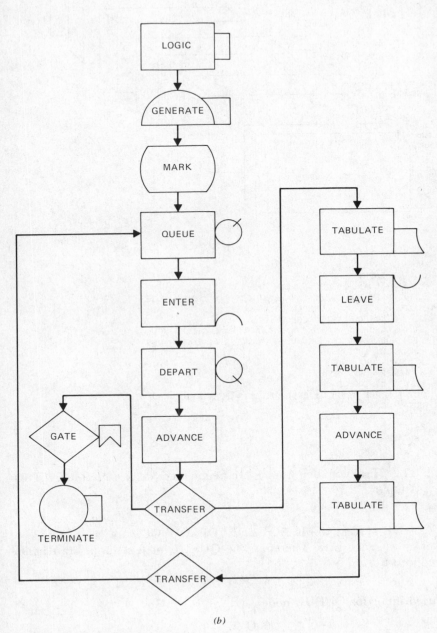

(b)

FIGURE 12.1   (Continued).

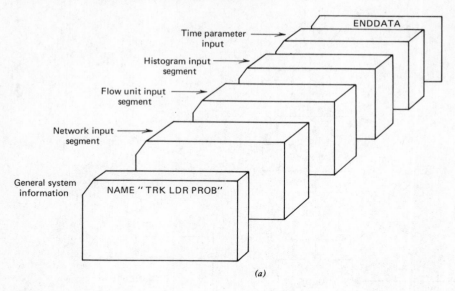

(a)

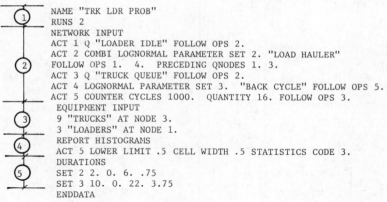

```
        NAME "TRK LDR PROB"
  ①     RUNS  2
        NETWORK INPUT
        ACT 1 Q "LOADER IDLE" FOLLOW OPS 2.
        ACT 2 COMBI LOGNORMAL PARAMETER SET 2. "LOAD HAULER"
  ②     FOLLOW OPS 1.  4.   PRECEDING QNODES 1. 3.
        ACT 3 Q "TRUCK QUEUE" FOLLOW OPS 2.
        ACT 4 LOGNORMAL PARAMETER SET 3.  "BACK CYCLE" FOLLOW OPS 5.
        ACT 5 COUNTER CYCLES 1000.  QUANTITY 16. FOLLOW OPS 3.
         EQUIPMENT INPUT
  ③     9 "TRUCKS" AT NODE 3.
        3 "LOADERS" AT NODE 1.
        REPORT HISTOGRAMS
  ④     ACT 5 LOWER LIMIT .5 CELL WIDTH .5 STATISTICS CODE 3.
        DURATIONS
  ⑤     SET 2 2. 0. 6. .75
        SET 3 10. 0. 22. 3.75
        ENDDATA
```

(b)

FIGURE 12.2   Problem definition. (a) Segmented deck. (b) TRK LDR
PROBLEM input.

The CYCLONE system words ACT and FOLLOW OPS are optional, as is
the use of an element title. Therefore the QUEUE node 1 input statement
can be reduced to simply:

1 Q 2.

The reduced input for QUEUE node 3 is:

3 Q 2.

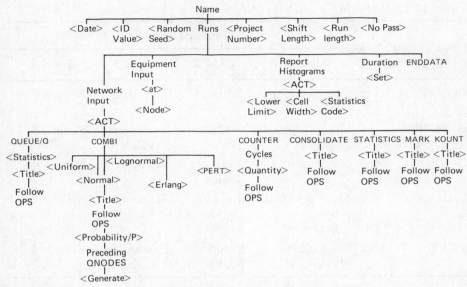

FIGURE 12.3   CYCLONE POL hierarchy. (⟨ ⟩ = optional.)

If statistics other than intrinsic occupancy and time-integrated averages are to be maintained on the QUEUE node, the procedural word STATISTICS is used, identifying the QUEUE node as a collection point.

Definition of the COMBI element requires information regarding the associated entity transit time as well as the QUEUE nodes preceding the element. The attributes associated with a COMBI element are:

1. Element numerical label (2).*
2. Element type (COMBI).
3. Associated distribution type (lognormal).
4. Associated parameter set (set 2).
5. Work task title ("LOAD HAULER").
6. Following operations (1 and 4).
7. Preceding QUEUE nodes (1 and 3).

As shown above and in Figure 12.2b, the input statement indicates that element 2 is a COMBI preceded by QUEUE nodes 1 and 3 and followed by elements 1 and 4. The associated entity transit delay distribution is LOGNORMAL and the distribution parameters are defined in parameter set 2. Six types of

*The items given in parentheses pertain to the system in Figure 12.2b.

Table 12.1   PARAMETER SET VALUES

Distribution Type	Parameter Defined In			
	Field 1	Field 2	Field 3	Field 4
CONSTANT	Constant time value	Not used	Not used	Not used
NORMAL LOGNORMAL	Mean value	Minimum value	Maximum value	Standard deviation
UNIFORM	Not used	Minimum value	Maximum value	Not used
ERLANG	Mean divided by value in field 4	Minimum value	Maximum value	Number of exponential deviates
PERT	Most likely value, M	Optimistic value, A	Pessimistic value, B	Not used

entity transit delay distributions can be associated with a COMBI element (see Table 12.1).

1. **NORMAL.**
2. **UNIFORM.**
3. **ERLANG.**
4. **LOGNORMAL.**
5. **PERT.**
6. **CONSTANT.**

In addition, probabilities can be associated with the ARCs exiting the element. These probabilities are assigned to each following element using the procedural words P or PROBABILITY. If the defined element had followers, as shown in Figure 12.4a, the CYCLONE system input would be:

FOLLOWING OPS 5.   P 0.2   6.   P 0.2   7.   P 0.6

If a GENERATE function is defined in conjunction with any of the QUEUE nodes preceding the COMBI element, the N value specifying the number of units generated is established in the COMBI element input statement. The input statement for the QUEUE nodes preceding COMBI element 9 in Figure 12.4b would appear as:

PRECEDING QNODES 4.   GENERATE 10.   6.   7.

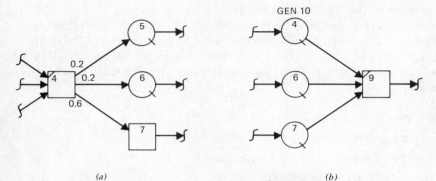

(a)   (b)

FIGURE 12.4   Follower probabilities and GENERATE function example.
(a) Follower probabilities. (b) Generate function.

This indicates that the COMBI is preceded by QUEUE nodes 4, 6, and 7, and 10 units are generated at 4 for each arriving entity.

The general form of the CYCLONE system input statement for a COMBI element is:

(ACT) {numerical label} {COMBI} $\left(\begin{array}{l}\textbf{CONSTANT}\\ \textbf{NORMAL}\\ \textbf{UNIFORM}\\ \textbf{ERLANG}\\ \textbf{LOGNORMAL}\\ \textbf{PERT}\end{array}\right)$

(PARAMETER SET set number) ("title")

(FOLLOW OPS) $\Big\{$ label(s) of following element(s)

$\left(\left\{\begin{array}{l}\textbf{PROBABILITY}\\ \textbf{|P}\end{array}\right\}\left(\begin{array}{l}\textbf{decimal value}\\ \textbf{from 0.0 to 1.0}\end{array}\right)\right)\Big\}$

{PRECEDING QNODES ⟨label(s) of queue nodes⟩ (GENERATE

⟨value of number generated⟩)}

The default specification if no entity transit distribution type is defined is CONSTANT. If no parameter set is specified, the default value sets the entity transit time to 0.0 (zero). The input statement for the COMBI element in Figure 12.2b can be shortened to:

2 COMBI LOGNORMAL PARAMETER SET 2.

FOLLOW OPS 1.   4.

PRECEDING QNODES 1.   3.

Specification of the NORMAL element 4 is similar to that of the COMBI, except that no preceding QUEUE nodes must be considered. The input statement of Figure 12.2*b* indicates that element 4 has an associated LOGNORMAL delay distribution, is not a COMBI node, and therefore must be a NORMAL work task element. The attributes required to specify a NORMAL work task element are:

1. Element numerical label (4).
2. Following operations (5).
3. Associated distribution type (lognormal).
4. Associated parameter set (3).
5. Work task title ("BACK CYCLE").

Elements without element type or function attributes are assumed to be NORMAL elements. The parameter values of the distribution are given in SET 3. Default values for the parameter set and distribution type are the same as for the COMBI processor. In its reduced input format, the specification of NORMAL work task 4 is:

4 LOGNORMAL PARAMETER SET 3. FOLLOW OPS 5.

Probabilistic exit arcs are defined as for the COMBI element.
Definition of the ACCUMULATOR element involves five attributes.

1. Element numerical label (5).
2. Element type (COUNTER).
3. Following operations (3).
4. Number of cycles (1000).
5. QUANTITY parameter (16).

The element type is established by use of the procedural word COUNTER. Definition of both the CYCLES and QUANTITY parameters is optional. The QUANTITY value has a default value of 1.0, and the number of cycles can be defined using a parameter in the GENERAL INFORMATION segment.

The header words EQUIPMENT INPUT are used to introduce the segment initializing units into the system. As indicated, nine trucks are initialized at element 3 and three loaders at QUEUE node 1. In its simplest form, this input can be reduced by deleting all except the numerical input items (i.e., 9 3. and 3 1.)

Statistical data other than the intrinsic statistics associated with the QUEUE nodes can be displayed in histogramatic format. This is implemented using the REPORT HISTOGRAMS header statement. In this example, a statistics type 3 (between) has been associated with the ACCUMULATOR element. The statistics specified by the numerical code following the words STATISTICS CODE are as given in Table 12.2. The CYCLONE system program provides for presentation of a histogram consisting of 34 cells. One additional cell is provided for underflow and one for overflow (i.e., total cells = 36). The LOWER LIMIT defines the base value. All observations below this value are accumulated in the underflow cell. The CELL WIDTH establishes the class interval widths in the histogram and the range considered (RANGE = 34 × CELL WIDTH). In this example the range is defined from 0.5 to 17.5. All observations above 17.5 are accumulated in the overflow cell.

The parameter sets referenced in defining COMBI element 2 and NOR-MAL element 4 are specified following the header word DURATION. Four fields are associated with each set. The values specified in each field for the distributions available are given in Table 12.1. In the "TRK LDR PROB," the four parameters given pertain to the LOGNORMAL distribution and indicate the mean value, minimum value, maximum value, and standard deviations (e.g., mean = 2.0, minimum value = 0, maximum = 6.0, and standard deviation = 0.75).

Table 12.2 DEFINED STATISTICS

Name	Description	Numerical Code
First	This statistic records the time at which the first entity arrives at the associated FUNCTION node.	1
ALL	This statistic records the time at which each entity arrives at the associated FUNC-TION node. An average value is printed.	2
Between	This statistic records the time between ar-rival of entities at the associated FUNC-TION node.	3
Interval	This statistic records the time between pas-sage of a preceding MARK node and ar-rival at the associated function node.	4
Delay	This statistic is used in conjunction with the CONSOLIDATE function. It records the time between the first entity of the group to be consolidated and the last entity of the group.	5

A variety of GENERAL INFORMATION can be defined by the modeler in this segment. Details regarding the GENERAL INFORMATION segment and the CYCLONE system computer program are presented in Appendix C. In this example, the minimum amount of information required has been specified. A NAME is required for identification purposes on the computer output. The RUNS parameter establishes the number of times the simulation will be repeated (i.e., the number of simulations of "N" CYCLES as specified by the ACCUMULATOR element).

12.2 CYCLONE SYSTEM DEFINITION WITH GEN-CON FUNCTIONS

To illustrate further problem definition using the CYCLONE system POL, consider the 20T–15T hauler system used to demonstrate the intrinsic monitoring functions in Section 10.5. This system required several GENERATE and CONSOLIDATION combinations to balance production between loaders and haulers as well as CONSOLIDATION functions to provide a common measure for system productions. The graphical model for this system is shown in Figure 12.5. Information regarding the work task entity delay distributions and the flow units initialized in the system was given in Tables 10.4 and 10.5. The NETWORK INPUT segment of the CYCLONE system definition of this model is shown in Figure 12.6. GENERATE–CONSOLIDATE (GEN–CON) combinations are used at COMBI elements 5, 12, 18, and 23. For this reason, it is worthwhile to look at the input defining these COMBI elements. The input statement defining COMBI element 5 is:

ACT 5 COMBI LOGN PARAMETER SET 3.   "STATION B"
FOLLOW OPS 6.   30.

PRECEDING QNODES 4.   GENERATE 4.   30.

This results in the generation of four load units for each single unit arriving at QUEUE node 4. This is compatible with the fact that it takes four cycles of the 5-ton bucket to load a 20-ton truck. Following four cycles of the loader, the truck departs "station B" as a single unit. Therefore a CONSOLIDATE function is defined at FUNCTION node 6 following the COMBI processor 5. The CONSOLIDATE FUNCTION element 6 is defined as:

ACT 6 STATISTICS CONSOLIDATE 4. "TRK LOADED"
FOLLOWING OPS 7.   8.   27.

This input statement indicates that two functions are associated with the element. The procedural word STATISTICS specifies that the node has a statistics collection function. To determine what type of statistics are to be collected, the REPORT HISTOGRAM segment of the input must be ex-

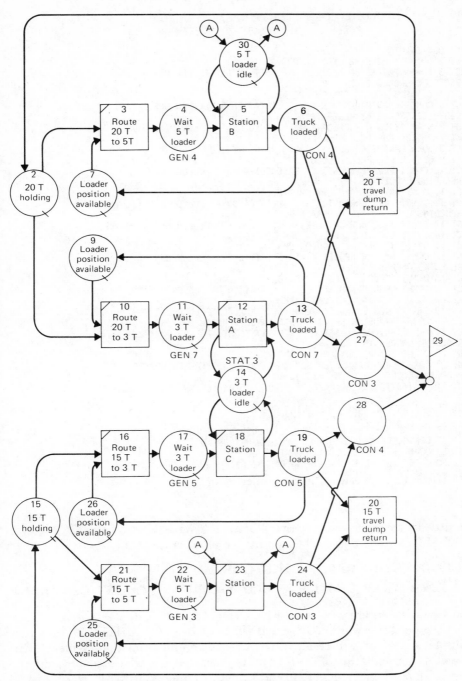

FIGURE 12.5   Twenty-ton 15-ton hauler model.

```
NETWORK INPUT
ACT 2 Q '20 T HOLDING' FOLLOW OPS 3. 10.
ACT 3 COMBI PARAMETER SET 2. 'RTE 20 T TO 5T' FOLLOW OPS 4.
PRECEDING QNODES 2. 7.
ACT 4 Q 'WT 5 T LDR' FOLLOW OPS 5.
ACT 5 COMBI LOGN PARAMETER SET 3. 'STATION B' FOLLOW OPS 6. 30.
PRECEDING QNODES 4. GENERATE 4. 30.
ACT 6 STATISTICS CONSOLIDATE 4. 'TRK LOADED' FOLLOWING OPS 7. 8. 27.
ACT 7 Q 'LDR POS FREE' FOLLOW OPS 3.
ACT 8 LOGN PARAMETER SET 5. '20T TRAVEL' FOLLOW OPS 2.
ACT 9 Q 'LDR POS FREE' FOLLOW OPS 10.
ACT 10 COMBI PARAMETER SET 2. 'RTE 20T TO 3T' FOLLOW OPS 11.
PRECEDING QNODES 2. 9.
ACT 11 Q 'WT 3T LDR' FOLLOW OPS 12.
ACT 12 COMBI LOGN PARAMETER SET 4. 'STATION A' FOLLOW OPS 13. 14.
PRECEDING Q NODES 11. GENERATE 7. 14.
ACT 13 STATISTICS CONSOLIDATE 7. 'TRK LOADED' FOLLOW OPS 8. 9. 27.
ACT 14 Q '3T LDR IDLE' FOLLOW OPS 12. 18.
ACT 15 Q '15 T HOLDING' FOLLOW OPS 16. 21.
ACT 16 COMBI PARAMETER SET 6. 'RTE 15T TO 3T' FOLLOW OPS 17.
PRECEDING QNODES 15. 26.
ACT 17 Q 'WT 3T LDR' FOLLOW OPS 18.
ACT 18 COMBI LOGN PARAMETER SET 4. 'STATION C' FOLLOW OPS 14. 19.
PRECEDING QNODES 14. 17. GENERATE 5.
ACT 19 STATISTICS CONSOLIDATE 5. 'TRK LOADED' FOLLOW OPS 20. 26. 28.
ACT 20 LOGN PARAMETER SET 7. '15T TRAVEL' FOLLOW OPS 15.
ACT 21 COMBI PARAMETER SET 6. 'RTE 15T TO 5T' FOLLOW OPS 22.
PRECEDING QNODES 15. 25.
ACT 22 Q 'WT 5 TON LDR' FOLLOW OPS 23.
ACT 23 COMBI LOGN PARAMETER SET 3. 'STATION D' FOLLOW OPS 24. 30.
PRECEDING QNODES 22. GENERATE 3. 30.
ACT 24 STATISTIC CONSOLIDATE 3. 'TRK LOADED' FOLLOW OPS 20. 25.
28.
ACT 25 Q 'LDR POS FREE' FOLLOWING OPS 21.
ACT 26 Q 'LDR POS FREE' FOLLOWING OPS 16.
ACT 27 CONSOLIDATE 3. ' 60 TON LD' FOLLOWING OPS 29.
ACT 28 CONSOLIDATE 4. '60 T LD' FOLLOWING OPS 29.
ACT 29 COUNTER CYCLES 500.   QUANTITY 60.
ACT 30 Q '5T LDR IDLE' FOLLOWING OPS 5. 23.
```

FIGURE 12.6   Network input hauler model.

amined. In this case the fact that "between" statistics are being collected at FUNCTION node 6 is established by the input statement:

PRODUCTION REPORT HISTOGRAM
ACT 6 LOWER LIMIT 0.2 CELL WIDTH 0.2 STATISTICS CODE 3.

Similar functions are associated with elements 13, 19, and 24.

The procedural word CONSOLIDATE followed by 4 indicates that four units arriving at 6 will result in release of one unit from that element. This leads to a consolidation of the four units generated at QUEUE node 4, and the release of a single truck unit for transit to the NORMAL work task 8 (TRAVEL–DUMP–RETURN). The specifying input for COMBI element 5 and FUNCTION node 6 is typical of that required by GEN–CON Combinations.

Output from the upper portion of the network in 20-ton increments is routed to FUNCTION node 27 from 6 and 13. Similarly, 15-ton production increments are routed to FUNCTION node 28 and from 19 and 24. To establish a common basis for production measurement, the units are consolidated at 27 and 28 into 60-ton increments. This is achieved by using CONSOLIDATE functions at these two nodes. The defining input is:

ACT 27 CONSOLIDATE 3.   "60 TON LD" FOLLOW OPS 29.

ACT 28 CONSOLIDATE 4.   "60 TON LD" FOLLOW OPS 29.

Three 20-ton units and four 15-ton units are consolidated to create 60-ton loads that exit to the ACCUMULATOR 29 for measurement. The ACCUMULATOR is defined as:

ACT 29 COUNTER CYCLES 500.   QUANTITY 60.

12.3 PRECAST PLANT INPUT

One further example of the CYCLONE system problem-oriented input will suffice to illustrate its use. Consider the precast plant operation described in Chapter Ten. The network diagram for this system is presented again for reference as Figure 12.7. The CYCLONE system POL model to include all five segments of defining input is shown in Figure 12.8.

This input indicates the flexibility of spacing made possible by the use of the free format feature of the CYCLONE system POL. In defining the elements following COMBI 12, for instance, the numerical labels begin on the line following the procedural words FOLLOW OPS. This offers a considerable advantage over the fixed format input systems, which not only prohibit breaking lines in this fashion but also specify exact column placement of input items on the card image (e.g., between columns 11 and 20).

It will be also noticed that slight departures from the procedural words are understood by the system. For instance, the use of the words FOLLOWING OPS instead of FOLLOW OPS causes no problem. This is possible since the program keys on only the first N letters of any procedural word. N is defined by the number of characters that can be stored in a location in the computer system. For IBM 360 systems, N is 4, so that only the first four characters must match the procedural word, as shown in Figure 12.3. Therefore, the use of the words STAT CODE in defining the type of statistic at element 3 is acceptable.

Since the CYCLONE system program is written in FORTRAN IV, it can be implemented (with slight modifications) on any computer system that has a FORTRAN compiler. The modifications required have to do primarily

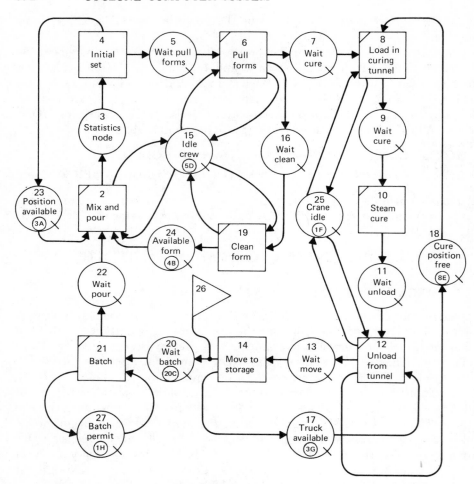

FIGURE 12.7   Precast concrete element plant.

with adaptation of the random number generation routines and changes required in some input routines because of differences in word size.

As noted, the input can be reduced by removing certain of the procedural words (e.g., ACT) that are included to make the reprinted input more readable. These words, however, are not required for program execution. The header for the report histogram definition, for instance, can be reduced simply to the single word HISTOGRAMS. On the other hand, the system understands the terms PRODUCTION REPORT HISTOGRAMS and will accept all of these words in processing data.

Use of the default characteristics of the system is also illustrated by the precast plant model input. For instance, no distribution type is specified for element 10. Since no FUNCTION or element defining words are associated with it, it must define a NORMAL work task element. Also, since no distribution type is specified, the transit delay is CONSTANT. By consulting the parameter set data following the header DURATIONS, the delay duration is established as a constant value of 120 time units (i.e., minutes).

12.4 SYSTEM OUTPUT

The output formats in the CYCLONE system are designed to aid the user in quickly checking the correctness of the defined network model and indicate delay and idle times associated with COMBI work task processors and their preceding QUEUE nodes. Additionally, as noted, histogram output of statistics defined within the system is available. The system production in terms of the units associated with the ACCUMULATOR element is also defined as an output item.

In order to aid in checking the input for errors, the CYCLONE program reprints the data in the format given in Figure 12.9. This data summary presents the network in a precedence network format that can easily be checked to see if the model has been properly specified. The data summary shown in Figure 12.9 is for the precast plant model. The format is similar to that used in summary representations of scheduling networks that use precedence notation (i.e., activity on node).

The intrinsic statistics associated with each QUEUE node are printed following each simulation run. These statistics indicate the time-integrated average number of units at each preceding QUEUE node and the percent of the time that the QUEUE nodes are occupied by waiting or delayed units. Typical performance data for the 20T–15T hauler system as printed following a simulation run are shown in Figure 12.10. From this data the following information can be developed.

Queue Number	Title	Percent of Time Units Are Delayed	Time-Integrated Average Number of Units Delayed
2	20-Ton hauler Hauling area	12	0.117
14	3-Ton loader Idle	25	0.30
15	15-Ton hauler Holding area	72	1.22
30	5-Ton loader Idle	1	0.01

```
NAME "PRECAST PLANT PROB"
DATE 2 15 1973
RUNS 1
PROJECT NUMBER  1
NETWORK INPUT
ACT 2 COMBI NORMAL PARAMETER SET 2. "POUR ELEMENT"
FOLLOW OPS 3. 15. PRECEDING QNODES 15. 22. 23. 24.
ACT 3 STATISTICS "STATISTIC NODE" FOLLOW OPS 4.
ACT 4 PARAMETER SET 3. "INITIAL SET" FOLLOW OPS 5. 23.
ACT 5 Q "WT PULL FORM" FOLLOW OPS 6.
ACT 6 COMBI PARAMETER SET 4. "PULL FORMS" FOLLOWING OPS 7. 15.
16. PRECEDING QNODES 5. 15.
ACT 7 Q STATISTICS "WT CURE" FOLLOWING OPS 8.
ACT 8 COMBI NORMAL PARAMETER SET 5. "LOAD CURING TUNNEL"
FOLLOWING OPS 9. 25. PRECEDING QNODES 7. 18. 25.
ACT 9 Q "WAIT CURE" FOLLOWING OPS 10.
ACT 10 PARAMETER SET 6. "STEAM CURE" FOLLOWING OPS 11.
ACT 11 Q "WAIT UNLOAD" FOLLOW OPS 12.
ACT 12 COMBI NORMAL PARAMETER SET 5. "UNLOAD" FOLLOWING OPS
13. 18. 25. PRECEDING QNODES 11. 17. 25.
ACT 13 Q STATISTICS "WAIT MOVE" FOLLOWING OPS 14.
ACT 14 NORMAL PARAMETER SET 7. "MOVE TO STORAGE" FOLLOWING OPS
17. 26.
ACT 15 Q "CREW IDLE" FOLLOW OPS 2. 6. 19.
ACT 16 Q "WT CLEAN" FOLLOW OPS 19.
ACT 17 Q "TRK AVAIL" FOLLOW OPS 12.
ACT 18 Q "CURE POSITION FREE" FOLLOW OPS 8.
ACT 19 COMBI NORMAL PARAMETER SET 8. "CLEAN FORM"
FOLLOW OPS 15. 24. PRECEDING QNODES 15. 16.
ACT 20 Q "WT BATCH" FOLLOWING OPS 21.
ACT 21 COMBI NORMAL PARAMETER SET 9. "BATCH" FOLLOW OPS 22. 27.
PRECEDING QNODES 20. 27.
ACT 22 Q "WAIT POUR" FOLLOWING OPS 2.
ACT 23 Q "AVAILABLE POSITION" FOLLOW OPS 2.
ACT 24 Q STATISTICS "AVAILABLE FORM" FOLLOW OPS 2.
ACT 25 Q "CRANE IDLE" FOLLOWING OPS 8. 12.
ACT 26 COUNTER CYCLES 100. QUANTITY 10. FOLLOW OPS 20.
ACT 27 Q "BATCH PERMIT" FOLLOW OPS 21.
EQUIPMENT INPUT
3 "SET POSITIONS" AT 23.
8 "FORMS" AT 24.
20 "BATCHES" AT 20.
5 "CREWS" at 15.
8 "CURE POSITIONS" AT 18.
1 "CRANE" AT 25.
3 "TRUCKS" AT 17.
1 "BATCH PERMIT" AT 27.
HISTOGRAMS
ACT 3 LOWER LIMIT 2. CELL WIDTH 2. STATIS CODE 3.
```

FIGURE 12.8   Problem specification model.

```
ACT 7 LOWER LIMIT 0. CELL WIDTH 0.5 STATISTICS CODE 3.
ACT 10 LOWER LIMIT 0. CELL WIDTH 5. STATISTICS CODE 3.
ACT 13 LOWER LIMIT 5. CELL WIDTH 5. STATISTICS CODE 3.
ACT 24 LOWER LIMIT 0. CELL WIDTH 0.5 STATISTICS CODE 3.
ACT 26 LOWER LIMIT 0. CELL WIDTH 5. STATISTICS CODE 3.
DURATIONS
SET 2 30. 20. 40. 5.
SET 3 50. 0. 0. 0.
SET 4 20. 0. 0. 0.
SET 5 15. 8. 22. 5.
SET 6  120. 0. 0. 0.
SET 7 25. 10. 40. 7.5
SET 8 45. 17. 83. 20.
SET 9 18. 9. 27. 6.
ENDDATA
```

FIGURE 12.8 *(Continued)*.

Obviously, the 20-ton haulers experience considerably less delay in loading than the 15-ton units. Units are stacked up in the 15-ton hauler holding area 72% of the time versus occupancy of only 12% for the 20-ton holding area. Similarly, the 5-ton loader is almost totally committed (i.e., idle only 1% of the time), while the 3-ton hauler is idle 25% of the time.

For elements that have defined statistics collection functions, output of the simulation results is presented in both summary and histogramatic format. The input for the precast plant problem specified defined functions in conjunction with QUEUE nodes 7, 13, 24. In addition, FUNCTION node 3 has been inserted following the "MIX & POUR" COMBI element in order to determine the rate at which units exit this work task. A defined statistic is also specified in conjunction with the ACCUMULATOR element 26. In each case, a code 3 (between) statistic has been specified. The summary presentation of these statistics as printed in the CYCLONE is shown in Figure 12.11. The distributed output in histogram format is also shown in this figure.

12.5 EXAMPLE PROBLEM

The procedure for defining a model to the CYCLONE program will be illustrated within the context of the tunneling system developed in Chapter Six. A flow graph indicating the steps required for model definition and checkout is shown in Figure 12.12. The graphical model and the data representation are shown in Figure 12.13. Operations 2, 4, 6, and 8 are defined as COMBI work task elements. Elements 10 and 16 are NORMAL work task elements. For preliminary checkout of system integrity, the associated delay times are defined as constants, and only two cycles of the system are simulated. The

INITIAL INFORMATION FOR PROJECT   PRECAST PLANT PROB

SUMMARY OF DATA FOR SIMULATION

NO.	OPERATION		FOLLOWING OPS			
2	POUR ELEMENT	3	15	0	0	0
3	STATISTIC NODE	4	0	0	0	0
4	INITIAL SET	5	23	0	0	0
5	WT PULL FORM	6	0	0	0	0
6	PULL FORMS	7	15	16	0	0
7	WT CURE	8	0	0	0	0
8	LOAD CURING TUNNEL	9	25	0	0	0
9	WAIT CURE	10	0	0	0	0
10	STEAM CURE	11	0	0	0	0
11	WAIT UNLOAD	12	0	0	0	0
12	UNLOAD	13	18	25	0	0
13	WAIT MOVE	14	0	0	0	0
14	MOVE TO STORAGE	17	26	0	0	0
15	CREW IDLE	2	6	19	0	0
16	WT CLEAN	19	0	0	0	0
17	TRK AVAIL	12	0	0	0	0
18	CURE POSTION FREE	8	0	0	0	0
19	CLEAN FORM	15	24	0	0	0
20	WT BATCH	21	0	0	0	0
21	BATCH	22	27	0	0	0
22	WAIT POUR	2	0	0	0	0
23	AVAILABLE POSTION	2	0	0	0	0
24	AVAILABLE FORM	2	0	0	0	0
25	CRANE IDLE	8	12	0	0	0
26	COUNTER	20	0	0	0	0
27	BATCH PERMIT	21	0	0	0	0

FIGURE 12.9   Data summary for precast plant problem.

data summary of the system and the trace of unit flow from time 0.0 to time 170 is shown in Figure 12.14.

The trace of flow unit movement indicates the initialization of the system and TNOW = 0.0, and initial flows. Ten "pipe sections" are initially defined at QUEUE node 15 as "PIPE STOCK PILE" as well as the appropriate units at QUEUE nodes 13, 11, 9, 17, and 12. The COMBI work task 2 "LOWER PIPE SECTION" is also commenced, since its ingredience conditions are realized. COMBI 2 is completed and units are released to QUEUE nodes 3, 11, and 13. Element 4 is also commenced. Study of this trace will indicate whether the desired unit flows are being realized within the defined

FOR QNODE -2 PRECEDING ACT 3 THE AVERAGE NO. OF UNITS HELD PENDING OTHER REQUIRED ARRIVALS IS 0.117
THE PERCENT OF THE TIME ENTITIES WERE WAITING WAS 0.12

FOR QNODE -7 PRECEDING ACT 3 THE AVERAGE NO. OF UNITS HELD PENDING OTHER REQUIRED ARRIVALS IS 0.184
THE PRECENT OF THE TIME ENTITIES WERE WAITING WAS 0.18

FOR QNODE -4 PRECEDING ACT 5 THE AVERAGE NO. OF UNITS HELD PENDING OTHER REQUIRED ARRIVALS IS 1.353
THE PRECENT OF THE TIME ENTITIES WERE WAITING WAS 0.59

FOR QNODE -30 PRECEDING ACT 5 THE AVERAGE NO. OF UNITS HELD PENDING OTHER REQUIRED ARRIVALS IS 0.010
THE PERCENT OF THE TIME ENTITIES WERE WAITING WAS 0.01

FOR QNODE -2 PRECEDING ACT 10 THE AVERAGE NO. OF UNITS HELD PENDING OTHER REQUIRED ARRIVALS IS 0.117
THE PERCENT OF THE TIME ENTITIES WERE WAITING WAS 0.12

FOR QNODE -9 PRECEDING ACT 10 THE AVERAGE NO. OF UNITS HELD PENDING OTHER REQUIRED ARRIVALS IS 0.461
THE PERCENT OF THE TIME ENTITIES WERE WAITING WAS 0.46

FOR QNODE -11 PRECEDING ACT 12 THE AVERAGE NO. OF UNITS HELD PENDING OTHER REQUIRED ARRIVALS IS 1.452
THE PERCENT OF THE TIME ENTITIES WERE WAITING WAS 0.38

FOR QNODE -14 PRECEDING ACT 12 THE AVERAGE NO. OF UNITS HELD PENDING OTHER REQUIRED ARRIVALS IS 0.300
THE PERCENT OF THE TIME ENTITIES WERE WAITING WAS 0.25

FOR QNODE -15 PRECEDING ACT 16 THE AVERAGE NO. OF UNITS HELD PENDING OTHER REQUIRED ARRIVALS IS 1.220
THE PERCENT OF THE TIME ENTITIES WERE WAITING WAS 0.72

FOR QNODE -26 PRECEDING ACT 16 THE AVERAGE NO. OF UNITS HELD PENDING OTHER REQUIRED ARRIVLAS IS 0.021
THE PERCENT OF THE TIME ENTITIES WERE WAITING WAS 0.02

FOR QNODE -14 PRECEDING ACT 18 THE AVERAGE NO. OF UNITS HELD PENDING OTHER REQUIRED ARRIVALS IS 0.300
THE PERCENT OF THE TIME ENTITIES WERE WAITING WAS 0.25

FIGURE 12.10 QUEUE node performance data.

FINAL RESULTS FOR 1 SIMULATIONS

NODE	PROB./COUNT	MEAN	STD.DEV.	# OF OBS.	MIN.	MAX.	NODE TYPE
26	1.0000	32.2224	27.5741	200.	8.7813	258.3335	B
24	1.0000	30.7513	26.7121	208.	0.0	144.4389	B
13	1.0000	32.1568	27.0441	200.	8.7813	245.2169	B
7	1.0000	31.5605	23.6352	204.	8.3223	103.2277	B
3	1.0000	30.9068	25.2850	208.	0.1334	62.2113	B

HISTOGRAMS

FREQUENCIES

NODE	LOWER LIMIT	CELL WIDTH
26	0.0	5.00
24	0.0	0.50
13	5.00	5.00
7	0.0	0.50
3	2.00	2.00

FIGURE 12.11 Summary and histogram output.

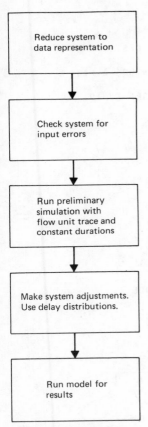

FIGURE 12.12   Computer model definition.

system. Following initial checkout of the model, a run of 50 cycles is conducted to determine occupancy and time-integrated statistics for the system. The results of this run are given in Table 12.3.

QUEUE nodes 3, 5, 7, and 12 have 0.0 values, indicating that flow units experienced no delays at these locations. For QUEUE nodes 9, 11, 13, and 17, the occupancy and time-integrated average statistics are the same. This occurs since only one unit can be delayed at a time in these locations.

The delay at QUEUE node 9 "COLLAR REMVD" indicates that the collar must wait pending lowering of the pipe section into place. At QUEUE node 11, "CREW IDLE," the crew is idle 62.5% of the time, indicating that it might be assigned other tasks without reduction in system output. The crane is also idle 90% of the time. This is expected, since it is slaved to COMBI work task element 2 "LOWER PIPE." Finally, the tunneling machine

```
NAME "TUNNEL PROBLEM"
PROJECT -1
RUNS 1
NETWORK INPUT
ACT 2 COMBI PARAMETER SET 2. "LOWER PIPE" FOLLOW OPS
3. 11. 13. PRECEDING QNODES 11. 12. 13. 15.
ACT 3 Q "PIPE IN POSITION" FOLLOW OPS 4.
ACT 4 COMBI PARAMETER SET 3. "REPL JACKS " FOLLOW OPS 5. 11.
PRECEDING QNODES 3. 9. 11.
ACT 5 Q "ALL IN POS" FOLLOW OPS 6.
ACT 6 COMBI PARAMETER SET 4. "JACK 6 FT" FOLLOW OPS 7. 10.
14. 16. PRECEDING QNODES 5. 14.
ACT 7 Q "COLLAR RDY" FOLLOW OPS 8.
ACT 8  COMBI PARAMETER SET 5.  "REMOVE COLLAR"
FOLLOW OPS 9. 11. PRECEDING QNODES 7. 11.
ACT 9 Q "COLLAR REMVD" FOLLOW OPS 4.
ACT 10 PARAMETER SET 6. "RESET JACK" FOLLOW OPS
12.
ACT 11 Q "CREW IDLE" FOLLOW OPS 2. 4. 8.
ACT 12 Q "SEC REQD" FOLLOW OPS 2.
ACT 13 Q "CRANE AVAIL" FOLLOW OPS 2.
ACT 14 COUNTER CYCLES 99. QUANITY 1. FOLLOW OPS 15.
ACT 15 Q "PIPE STOCK PILE" FOLLOW OPS 2.
ACT 16 PARAMETER SET 7. "MOLE MAINTENANCE"
FOLLOW OPS 17.
ACT 17 Q "MOLE RDY" FOLLOW OPS 6.
EQUIPMENT INPUT
10 "PIPE" AT 15.
1 "CRANE" AT 13.
1 "CREW" AT 11.
1 "COLLAR" AT 9.
1 "MOLE" AT 17.
1 "JACK SET" AT 12.
REPORT HISTOGRAMS
ACT 14 LOWER LIMIT 5. CELL WIDTH 5. STATISTICS CODE 3.
DURATIONS
SET 2 20.
SET 3 30.
SET 4 120.
SET 5 25.
SET 6 30.
SET 7 45.
ENDDATA
```

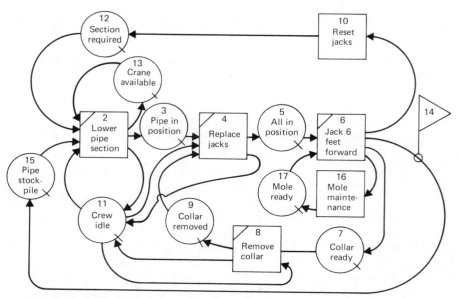

FIGURE 12.13   Tunnel model.

INITIAL INFORMATION FOR PROJECT TUNNEL PROBLEM

SUMMARY OF DATA FOR SIMULATION

NO.	OPERATION	FOLLOWING OPS				
2	LOWER PIPE	3	11	13	0	0
3	PIPE IN POSITION	4	0	0	0	0
4	REPL JACKS	5	11	0	0	0
5	ALL IN POS	6	0	0	0	0
6	JACK 6 FT	7	10	14	16	0
7	COLLAR RDY	8	0	0	0	0
8	REMOVE COLLAR	9	11	0	0	0
9	COLLAR REMVD	4	0	0	0	0
10	RESET JACK	12	0	0	0	0
11	CREW IDLE	2	4	8	0	0
12	SEC REQD	2	0	0	0	0
13	CRANE AVAIL	2	0	0	0	0
14	COUNTER	15	0	0	0	0
15	PIPE STOCK PILE	2	0	0	0	0
16	MOLE MAINTENANCE	17	0	0	0	0
17	MOLE RDY	6	0	0	0	0

TRACE OF UNIT MOVEMENT

AT TIME	ELEMENT REALIZED
0.00	15
0.00	13
0.00	11
0.00	9
0.00	17
0.00	12
0.00	15
0.00	2
0.00	15
0.00	15
0.00	15
0.00	15
0.00	15
0.00	15
0.00	15
0.00	15
20.00	3
20.00	11
20.00	13
20.00	4
50.00	5
50.00	11
50.00	6
170.00	7
170.00	14
170.00	16

Initialization of Pipe Sections at 15

FIGURE 12.14 System checkout of tunnel problem model.

Table 12.3   INTRINSIC STATISTICS (TUNNEL MODEL)

Hourly production = 0.30 Units

Shift production (4.5 hours) = 1.35

QUEUE Node	Occupancy	Time-Integrated Average
3	0.0	0.0
5	0.0	0.0
7	0.0	0.0
9	0.125	0.125
11	0.625	0.625
12	0.0	0.0
13	0.9	0.9
15	1.00	9.145
17	0.18	0.18

("mole") is idle 18% of the time, awaiting the commencement of the jacking work task. The pipe stockpile as defined is never empty and maintains a time integrated level of 9.145 units. The system production figures indicate an hourly output of 0.3 units per hour. Because of the positioning of the ACCU-MULATOR, this indicates a penetration rate of $1.8 \simeq 2.0$ feet per hour.

A between statistic has been associated with the ACCUMULATOR to measure the interval required for 6 feet of production. Following the initial pipe section (which, it can be seen from 12.14, required 170.0 minutes) the system requires 200 minutes per cycle. The cycle on a deterministic basis results from the sequence of work tasks 2, 4, 6, and 10, the durations of which sum to 200 minutes. The between statistics calculated by the program to include the first cycle are:

Element	Mean	Standard Deviation	Number of Observations	Minimum	Maximum
14	199.42	4.10	50	170.0	200.0

Following successful checkout of the system using constant time delays, random distributions from field data are used.

12.6 MODELING BENEFITS

This brief introduction to the characteristics of the CYCLONE computer program illustrates the insight-building potential of computer simulation in better understanding job site processes. It also points out the advantages that can accrue to the field manager equipped with a simple process modeling tool. Development of CYCLONE system models makes realization of certain benefits possible.

1. The CYCLONE elements provide a vehicle for problem information communication and documentation.

2. The system modeling simplicity and accessibility aid in increasing the manager's awareness of simulation as a problem-solving technique.

3. The system provides a framework for an adaptive and experimental approach to systems analysis. This approach allows the manager to experiment with varying strategies and system configurations and thereby gain insight into the system operation by means of the modeling process itself.

The CYCLONE formats fulfill the need for a notation that can be used to document processes. This documentation allows information transfer and provides a common basis for discussing the process of interest. Expertise gained about the operation of systems can be accumulated and passed on in a common language. The only notation of this type that has been available until now relates to time and motion study analysis, which defines processes in a static fashion. The CYCLONE notation provides a modeling format with dynamic properties (e.g., logical routing, permit constraints). The lack of such a simple manager-oriented graphical modeling format or notation has led to limited awareness at the management level of the insights that can be gained by using simulation as an investigative technique. The CYCLONE modeling set provides the graphical format for documenting systems and accumulating information about their performance characteristics.

Because of its lack of accessibility to practicing management personnel, the development of simulation as a problem-solving aid has been oriented more toward obtaining statistically rigorous solutions that are known to be correct within a certain degree of confidence. The development of simulation as an insight-giving tool leading to a better understanding of system characteristics has received less attention. It is, however, just this insight-developing application that constitutes the greatest aid that simulation can offer managers. This approach recognizes that preliminary or pilot models may have to rely on approximate data or engineering estimates as input until the model proves that it is beneficial enough to justify obtaining more refined data. This approach allows an application in which preliminary systems can be refined or scrapped, depending on their utility. This adaptive approach causes the model building and refinement to become a learning experience into which the manager or engineer can enter. The CYCLONE modeling set that allows the line manager to participate in the structuring of the model directly provides a modeling clay that can be formed and reformed until the results are acceptable.

The direct interface between man and machine (e.g., computer) that is possible using the CYCLONE system provides several modes in which the system can be implemented in the field. On some large construction sites (e.g., nuclear power plants, large negotiated contracts), computer access is available through a batch mode card reader at the site headquarters. Such computer terminal facilities are available for problem solving as well as for administrative functions such as payroll and report preparation. On such sites, construction engineers access the CYCLONE system program mounted on secondary storage at the central computer facility through card input, as shown in Figure 12.2. Smaller sites or companies have teletype access to a computer facility. In such cases, the problem specifying input stream can be typed directly on-line and sent to the computer for processing in a time-sharing mode.

With the advent of computer graphics and greater use of cathode ray tube (CRT) input devices, the direct definition and submittal of CYCLONE models in the graphical format will be possible using light pens and other graphical display input devices. By this means, the reduction of the graphical model to a data representation proceeds automatically as the system elements are defined on the CRT and linked by logical arcs. Using such a specification procedure, an engineer at the job site will be able, by punching a button on the CRT console, to define any of the five CYCLONE elements and then, using the light pen, to link the defined element to other elements already specified on the display screen. Using such interfacing, the engineer can quickly define and parametrically evaluate several process configurations in a short session on-line with the computer.

PROBLEMS

1. Write the POL input required to define the sewer line job system of Figure P7.6 to the CYCLONE program.

2. Develop the CYCLONE input for the material hoist model shown in Figure P7.7a.

CHAPTER THIRTEEN

Designing and Analyzing Construction Operations

In the planning phase of any construction operation, certain decisions and projections must be made about the intended development of the works. Usually the decisions and projections are made by considering a variety of situations or scenarios that express what is known about the operation and what can be done with resources in the work face environment. In this way, the construction operation manager thinks through the work task sequences associated with a given construction technology, establishes a feasible work plan, and assesses the adequacy of a resource allocation to the operation. The resulting formulation of construction technology, work sequences resource requirements, and management policies establishes the design details of the construction operation.

13.1 THE DESIGN PROCESS

Traditionally, the process of developing a plan that reduces a concept to a practical statement and format for implementation is referred to as the design process. The design process for construction operations is characterized by a procedure consisting of four major activities.

1. The development of a feasible plan.
2. The equipment and labor selection process.

3. The development of management policies.

4. The monitoring and evaluation of the construction operation performance relative to the efficient use of resources and management goals.

The development of a feasible plan for a construction operation requires the selection of a suitable construction technology and the definition of the work tasks and processes that must be performed according to the technological logic. In CYCLONE modeling, the construction operation plan corresponds to the enumeration, modeling, and labeling of the work tasks, sequences, and logic using the NORMAL, COMBI, and QUEUE nodes together with the ARC modeling element. The plan and construction technology is exposed in the basic static structure of the CYCLONE model.

The equipment and labor selection process establishes the set of resources that are to be made available for the working of the construction operation. The nature of the required resources is defined by the needs of the construction technology, while the number of resources used is a function of the magnitude of the work involved and the desired operational productivity. In many cases the mix of resource types (i.e., equipment fleets and crew compositions) to be used is based on past experience and may only represent acceptable equipment use and efficiency. In some cases field conditions may permit the gradual development of efficient resource mixes by the selective addition and manipulation of individual resources. This method is often used in tunneling, earthwork, and masonry-type operations. It is good practice to develop a complete equipment and tool list for a construction operation as a means of thoroughly understanding the operational requirements and for establishing the field requirements of the operation.

In CYCLONE modeling, the unique specification of input resources to be modeled highlights the number of independent cycles and paths that must exist in the static structure model. This requirement follows, since resource entities can only be initialized at QUEUE nodes and each independent cycle or path must have entities initialized in at least one resource idle QUEUE node state. The essential focus in CYCLONE modeling is on the identification and state modeling of the major resource flows and the labor work assignments that define the construction operation plan and constrain output productivity. Thus, in CYCLONE models, the level of modeling detail may not necessarily require the fine portrayal of the need and use of small consumable equipment items.

Management policies help field agents to cope with situations that may develop when working on a construction site wherein the assessment and selection of a course of action among those alternatives available is greatly simplified. A management policy can be defined as an ordered set of conditions that are predefined to operate in response to a set of anticipated conditions.

Thus field agents meeting these situations need only identify the relevant conditions and initiate the specific course of action based on the relevant preestablished management policy. Management policies are often designed for critical operations such as hoisting priorities on congested sites and are rarely considered for most operations. In most situations, management policies exist by default in that individual field agents establish unique priorities and actions to suit their assessment of situations or as a result of years of experience.

The CYCLONE modeling of management policies is developed by incorporating initializing conditions into the definition of the construction operation model and by establishing routing priorities in the static structural model. This effect is achieved by the use of control structures, mechanisms, and judicious numbering of critical COMBI nodes. In this way it is possible to establish work task priorities for equipment and labor resources, and permissive conditions that must pertain before certain work sequences can be initiated. Management policies play an important role in the way actual construction proceeds and must be fully recognized and considered if relevant and reliable models are to be developed for construction operations.

The proper field management of a construction operation involves the continuous monitoring of productivity and job progress and the manipulation of resources to meet the vagaries of the construction site environment. Thus management should ensure that excessive work loads, or idleness, of labor and equipment resources do not occur, since these factors effect long-term productivities and economies and form the basis for comparison and selection among competing construction technologies of the construction method to be followed in the operation.

In CYCLONE models system response, productivity, and the effective use of resources are monitored at QUEUE, FUNCTION, STATISTICS, and ACCUMULATOR COUNTER nodes. By changing initial conditions and resource specifications, different system responses result, so that the manager can select the mix of resources, work sequences, and technologies that best achieves the objectives. In this way the manager can experiment with the design of the construction operation and evaluate the economies and productivities of competing construction methods.

In many cases where the construction operation is familiar to the contractor, little time is devoted to the design phase unless unusual features exist such as difficult working conditions, magnitude of job, criticality because of time or cost, and so on. Often, if a construction method is relevant and past experience has proved that the methods work, then the design phase reduces to recapitulating how it was done last time, with what resources, and what must be changed this time. The contractor normally asks "where does it pay to look at design? where is the leverage? and where are big savings possible?" Thus design effort is justified by cost and time savings.

13.2 SELECTION OF MATERIALS HANDLING METHOD

A major consideration of any construction project is the extent and nature of materials handling required over the life of the project. In certain types of projects, such as large earth or concrete dams, materials handling constitutes the major construction activity. On congested sites such as those that exist for tunnel projects and most high-rise city building projects, materials handling problems are aggravated by the lineal nature of the facility itself. Often the selection and design of the materials handling system is the major preplan decision area to be considered. The productivity, duration, and economics of a project are affected by the capacity of the adopted materials handling systems.

Many factors influence the selection, design, and capacity of a materials handling system for a project. These include the mass-time profile over the life of the project of the materials to be handled, the ratio of the peak demand of materials handling requirements to the capacity of the materials handling system, and whether general or special purpose systems are involved in different stages of the project.

Consider, for example, problems associated with the placement of concrete on a city building site. Factors influencing the selection and design of the concrete placement system arise out of the nature of the material itself, the quantities involved, the interaction of site delivery, site handling, and required placement rates, and the variety of placement equipment and methods available to the contractor. Once concrete is mixed, a limited placement time exists which, together with its required structural strength and mix design, may exclude concrete pumping techniques or decide between the off-site or on-site manufacture of the concrete. The concrete quantities involved and the requirements of the work face placement rates in correlation with variations in the delivery rate to the site may require the installation of hopper storages and special delivery dock facilities incorporated into the structure of the building itself. In this way hopper capacity may dampen out delivery rate variations at the expense of more stringent requirements on placement productivity because of the limited set time period available for hopper concrete. The accessibility and congestion of the site, the height to which concrete is to be carried, and the availability of cranes, hoists, skips, and pumps affect the selection of the site concrete placement method to be used.

A number of placement methods are available to the contractor, each having unique properties that affect maximum placement rate, equipment capitalization, labor work force size, and the like. Typical concrete placement systems are illustrated in CYCLONE modeling format in Figures 13.1 to 13.4. These delivery systems can be classified according to the technology used at the site in transporting the concrete horizontally and vertically from the delivery point on the site to the work face.

Figure 13.1, for example, incorporates wheelbarrows or concrete buggies for the horizontal delivery system with a material hoist for the vertical delivery

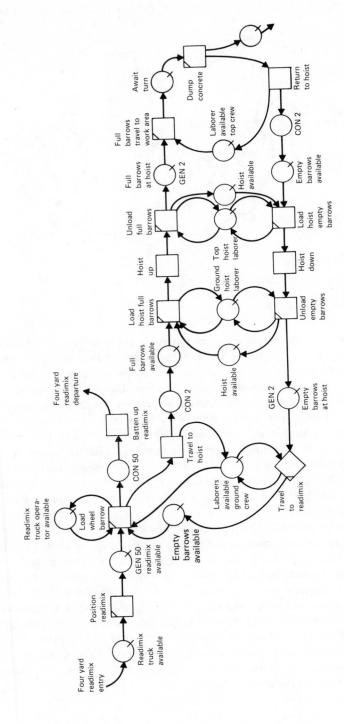

FIGURE 13.1 Concrete delivery (1).

409

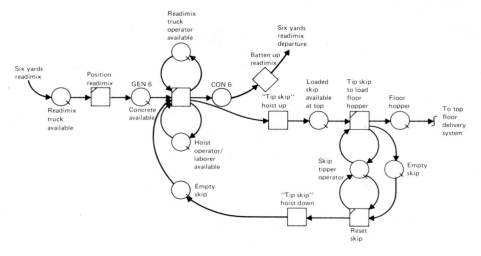

FIGURE 13.2   Concrete delivery (2).

system. Assuming a 4-cubic yard transit mix truck delivery and 2-cubic foot per load wheelbarrows, the CYCLONE model indicates the various work tasks and static structure of the concrete delivery and placement system for a materials hoist of two-wheelbarrow capacity. Design features that require attention are the number of wheelbarrows required, both ground and top labor crew sizes, and production rate. Figure 13.1 assumes that the materials hoist requires both a ground and top hoist laborer-operator, which therefore effectively separates the ground labor crew (cycling wheelbarrows between the transit mix truck delivery point and the hoist) from the top placement delivery labor crew. This placement method is very flexible, can handle a variety of materials, requires little equipment investment, but has limited capacity and placement rate.

Production rates can be increased, as shown in Figure 13.2, when the transit mix trucks can deliver concrete directly to skips in the material hoist system. In this situation larger-capacity transit mix trucks can be used and the ground labor crew can be eliminated. Usually a two-man hoist crew is involved unless an automatic tip-skip hoist is used. The CYCLONE model of Figure 13.2 assumes a laborer-operated top skipping work task and a 1-cubic yard skip. The top floor horizontal delivery system using buggies is not shown. As the vertical hoist delivery system is now specifically oriented to concrete placement and requires more careful erection, and other special consideration, the hoist delivery rates and skip bucket size become design problems that directly affect concrete placement rates.

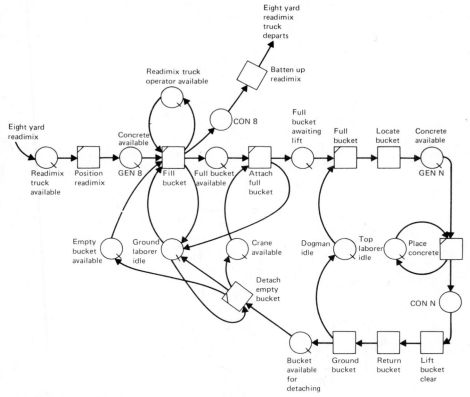

FIGURE 13.3   Concrete delivery (3).

Figure 13.3 illustrates a crane and bucket delivery system. The need for specific horizontal and vertical delivery system is eliminated, since the crane operation is three dimensional. The CYCLONE model of Figure 13.3 assumes the use of 8-cubic yard transit mix trucks and a 1-yard bucket. Design problems associated with this delivery system include the location of the crane itself and the required boom length to command the work face areas and, therefore, the determination of the capacity of the crane. Production rates are therefore related to the hoist speed, bucket size, crane operation, and location and congestion of the work face.

Figure 13.4 illustrates a concrete pump delivery system and includes modeling segments for setting up the concrete pump and the clearing of blockages. Design problems associated with this delivery system include the determination of pump capacity, concrete mix design relative to pipe delivery size, economic volumes and lift heights, and the design of floor pattern pours.

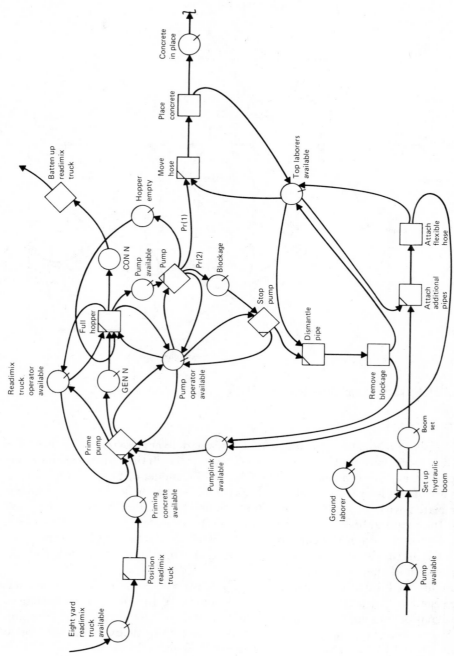

FIGURE 13.4 Concrete delivery (4).

Each of the above placement methods has specific and relative advantages and disadvantages, so that each is preferable if suitable site conditions exist.

13.3 A BUILDING CONSTRUCTION PROJECT

To illustrate the applicability of the CYCLONE modeling format to complex real-world construction processes, consider the development and analysis of the concrete placement sequence on a high-rise hotel construction. The subject of the analysis is the Peachtree Plaza Hotel in Atlanta, Georgia. After completion in late 1975, this building will be the tallest concrete hotel in the world, rising 70 stories and 700 feet (213.36 meters) above the street level. The structure, as it will appear on completion, is shown in Figure 13.5. It consists of a 63-story reflective glass-clad cylindrical tower rising from a 7-story base structure. In the center of the base structure, there is a 7-story central court or atrium, open from the lobby to a massive skylight at the base of the tower.

The construction of the tower section of the building provides a very interesting linearly repetitive operation for study and analysis. The process of concrete placement for each of the 63 tower floors is essentially the same. The tower is slender, with a diameter of 116 feet (35.36 meters). In many respects, the construction of the tower resembles a tunneling project except that in this case the tunnel is constructed skyward along a vertical instead of a horizontal axis.

The model to be developed will focus on the concreting operation. The total job site is extremely congested, covering only 1.3 acres. Lifting of materials on the tower is handled by two tower cranes. The location of these cranes is shown in Figure 13.6. The cranes are "jumped" about once a week to keep up with the changing height of the structure. The jumping of each crane occurs during the weekend or nights when there is little or no work scheduled. The right crane on the tower is the smallest on site, with a radius of 80 feet (24.38 meters). The left side or south crane has a 150-foot (45.72 meter) boom length, allowing it to reach out into street for direct off-loading of vehicles. A concrete hoist is used to move concrete from transit trucks at street level to the level of placement. A man hoist is used for movement of personnel.

The layout of the job site in plan view is shown in Figure 13.7. A lane of the bordering street has been allocated for the queueing of concrete transit mix trucks (A). A bending yard for reinforcing steel is located at B on the roof of a parking garage next to the job site.

Disregarding special features, the sequence of concrete work for each of the 63 tower floors is repetitive. Each floor consists of a round slab with 10 shear walls radiating from lines that intersect at the center and a center core with walls at its boundaries. There is also a hall through the middle of the

FIGURE 13.5   Peachtree Center Plaza Hotel. (Architect: John Portman & Associates, Atlanta.)

FIGURE 13.6   The overall site.

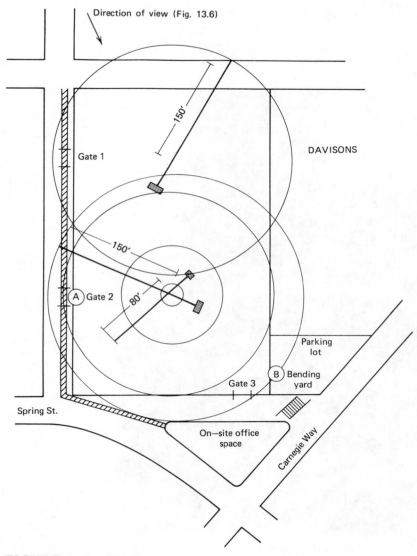

FIGURE 13.7 Site layout.

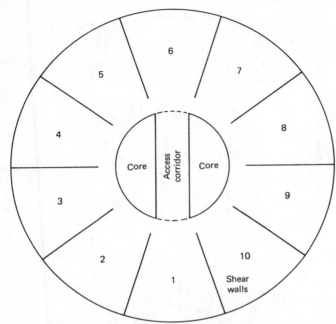

FIGURE 13.8   Typical floor section—plan view.

core that provides access to the elevators. A typical floor section is shown in Figure 13.8.

13.4 THE CONSTRUCTION PROCEDURE

The construction of each new floor begins with the telescoping or "extension" of the interior core form from the previous floor, as shown schematically in Figure 13.9a. The complete floor construction process consists of six steps that are repeated algorithmically for each floor. After extension of the interior core forms, they are cleaned and prepared for concrete. At this time, also, the exterior core forms and the shear wall forms are pulled and lowered to the ground for cleaning.

Following this, wedge-shaped floor slab forms are lifted or "flown" from two floors below to be inserted between the shear walls of the just completed floor. There are two sets of floor slab forms (i.e., 20 sections). This allows the set of forms on the floor immediately below to remain in place, providing temporary shoring. After removal of the slab forms from two stories below, temporary timber shores are inserted on that floor. After insertion and setting of the 10 sections of wedge-shaped flying forms, the reinforcing steel and embeds for the floor slab are placed. In addition, the reinforcing steel cages

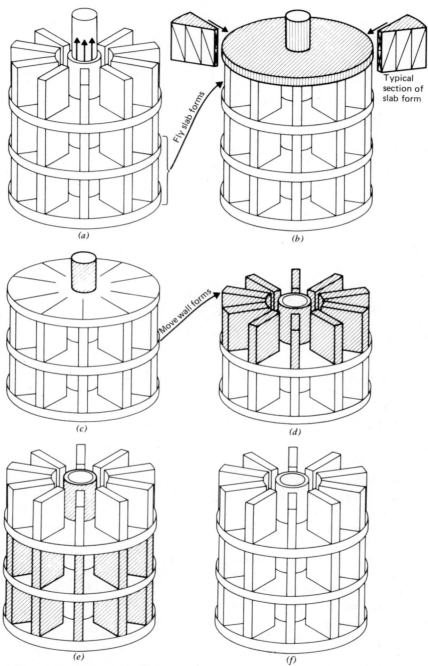

FIGURE 13.9   (See legend on opposite page.)

for the core are lifted from the bending yard and placed in the core. This is indicated schematically in Figure 13.9b. After placement of steel, the floor slab is poured.

The shear wall forms are moved from ground level to the working floor once initial set has been achieved on the floor slab pour. These wall forms are set, and reinforcing steel is placed. Following this, each of the 10 shear walls is poured and cured. Finally, the exterior core forms, which have been cleaned and oiled, are lifted from ground level and set in place. The floor construction is completed with the pouring of the core and the overhead for the elevator access hall. This sequence of tasks is indicated in Figures 13.9c to 13.9f. The placement of floor slab steel and various electrical embeds is shown in Figure 13.10. Figure 13.11 shows the interior core form in its telescoped position preparatory to the installation of the reinforcing steel cages. The concrete hoist frame and skip are visible at the back of this picture. Concrete is lifted from the hoist with a concrete bucket mounted on the smaller (north side) crane. Placement of concrete using the bucket is shown in Figure 13.12. The logical sequence described above is not rigid, and variations occur. For instance, it can be seen that the reinforcing steel for one of the shear walls is already in place as the floor slab pour is being made. However, for preliminary analysis of the process, it will be assumed that work tasks are sequential.

13.5 DEVELOPMENT OF THE CYCLONE MODEL

The design of the process model of the construction procedure described above requires the identification of a "main-line" unit of production as well as specification and allocation of the resources to be considered. The unit of production in this process is the completion of a floor of the hotel tower. Resources of interest include the two cranes used for materials handling on the tower and crews assigned to the placement of steel, the setting of forms, and the pouring of concrete. A model of the process is shown in Figure 13.13. The model focuses on the activities developed in Section 13.4. The system boundary defined is such that the concrete hoist is not included. The lifting and placement of concrete is considered at work tasks 13, 22, and 27 for the floor slab, core, and shear walls, respectively. The lifting segment considered, however, is concerned only with movement of concrete from the hoist skip (see Figure 13.11) to the placement location. Therefore, in this model, the rate of the hoist is not assumed to be a constraining factor.

FIGURE 13.9 Concrete pour sequence on tower. (a) Telescope (move up) interior core form and clean. (b) Fly. Move slab forms from two stories below. Install forms; rebar slab; rebar core; shore under slab forms two floors below. (c) Pour slab concrete. (d) Move shear wall forms from the floor below. Install and rebar. (e) Pour shear walls. (f) Set external core forms; pour core.

FIGURE 13.10   Placement of slab steel and embeds.

Certain assumptions have been made about the division of work tasks between the two cranes. The smaller crane is involved in the placement of the slab and core concrete, lowering of the core forms for cleaning, telescoping of the interior core forms, and lifting of floor slab steel. The larger crane is charged with the lifting of the flying forms and the placement of shear wall cages. It is also used in the placement of shear wall concrete and the lowering of shear wall forms to ground level for cleaning. Both exterior core forms and shear wall forms are lowered to ground level for cleaning because of the restricted working area on the tower itself. The work tasks are divided between the two cranes as follows.

Large Crane		Small Crane	
Act	Work Task	Act	Work Task
6	Fly slab forms	4	Telescope interior core form
11	Lift wall cages	10	Lift slab steel
27	Lift wall forms and pour concrete	17	Pull core forms
36	Pull wall forms	22	Place slab concrete

FIGURE 13.11   View of interior core form.

The lift priorities are established by the numbering sequence. In the lifting of steel, parallel operation is achieved by placing one crew-crane combination on each parallel track (e.g., 9-10-15-16 and 8-11-12).

Two types of crews are described in the model. Crew 1 is assigned work tasks related to slab steel placement, core form cleaning, and concrete placement. Crew 2 is required for lifting and setting shear wall cages and shear

FIGURE 13.12   Pouring floor slab.

wall removal of core and shear wall forms. The division of responsibility between the two crew types is as follows.

Steel Crew(s)		Pour Crews	
Act	Work Task	Act	Work Task
4-7	Fly form sequence	11-14	Lift wall cages and set
10-15-16	Lift floor steel	17	Remove and lower exterior core forms
13	Lift extension core forms and pour core	22	Place slab concrete
29	Clean core forms	27	Lift wall forms and place concrete
39	Clean wall forms	36	Pull wall forms and lower

Other units of interest in the model are the wall forms and exterior core forms that describe Class I constraining cycles between 36-27 and 34-13, respectively.

It is informative to trace the flow of the mainline unit commencing at element 3. The main-line unit subdivides and reconsolidates at various points in its cycle, leading to parallel tracks within the model. A unit is initialized

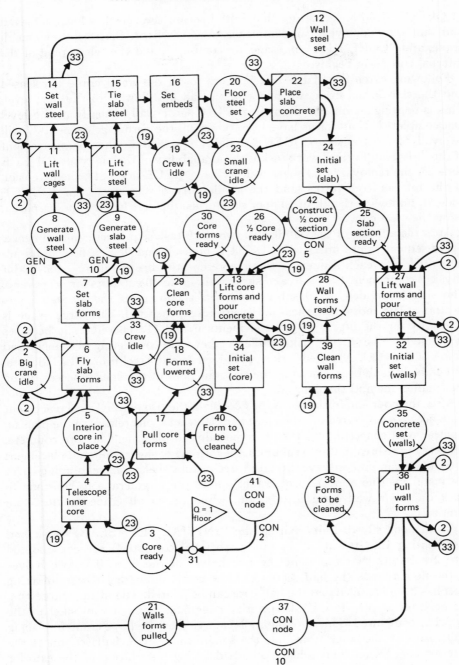

FIGURE 13.13   CYCLONE process model.

at QUEUE node 3, indicating that initial set on the core has been achieved and that the floor cycle can commence. At COMBI 4, the steel crew, small crane, and "CORE RDY" message are combined, and the telescoping of the internal core form begins.

Following extension of the interior core form, the small crane is released and the slab forms from two floors below can be "flown" to the new level. This is contingent on the removal of all 10 shear wall forms from the level just completed. If all forms have been removed, the permit at element 37 will be available to transit to QUEUE node 21. Implicit in the commencement of the "Fly Slab Forms" work task is the availability of the big crane (QUEUE node 2), the removal of all shear wall forms (QUEUE node 21), the extension of the interior core forms, and the availability of the steel crew (QUEUE node 5). Following the lift of all floor slab forms, they are leveled and prepared, using the steel crew.

After installation of the slab forms, two work task sequences can commence. These sequences are initiated at QUEUE nodes 8 and 9. Ten flow units are generated into each sequence; they represent the break out of each major physical component (i.e., slab and shear walls) into 10 subunits for processing. The slab is divided in 10 subsections corresponding to the 10 slab pours (sectors). The individual shear walls are also generated. The large crane is used only for the lifting tasks at elements 6 and 11. Therefore, it becomes available for other lifting tasks, while the crews are completing placement and tie-off of the steel. As noted, priority is given to slab steel lifts. When all steel and embeds have been placed in the slab, the pour of the floor proceeds, using the small crane and the pour crew.

After the pour and initial set of the floor slab, the two parallel steel placement sequences are converted into two new parallel tracks; one relates to the pouring of shear wall concrete, and the other pertains to placement of core concrete. Because of the breakout into subunits, the forming of shear walls and placement of concrete can proceed, even if only three or four shear wall reinforcing cages are installed. The allocation of cranes allows the large crane to continue to work on shear wall form installation, while the small crane handles core form setup.

At element 42, 10 units exiting the "INITIAL SET (SLAB)" task and representing the 10 slab pours (and associated core steel) are consolidated into 2 subunits that represent the two halves of the floor. The two halves of the floor release the half sections of the cores, indicating that reinforcing steel has been installed and the half core can be poured. The lifting and setting of the exterior core forms take priority over placement of wall steel in the model as configured. This allows parallel activity by the large crane in placing wall forms at the same time that the small crane sets a half section of the exterior core forms. This scheme avoids delaying installation of the exterior core forms until all shear wall steel has been placed.

After placement of the exterior core forms, the core is poured in half sections. In order to be compatible with the reduction in flow units from 10 to 2 (implemented by the CON 5 at element 42), two exterior core form units must be initialized at QUEUE node 30. Since the 10 shear walls are handled individually, no reduction is utilized prior to FUNCTION node 37. The shear wall "form and pour" track (27-36) is constrained by the number of prefabricated shear wall forms available. Since the forms must be pulled and lowered to ground level, a number of wall forms less than 10 leads to a constraint. After pour of concrete on the core, the exterior forms are also pulled and lowered to ground level for cleaning because of the extremely tight working conditions on the tower itself.

The floor cycle is completed with the consolidation of the two core half sections at 41 and the 10 shear wall form removals consolidated at 37. These two FUNCTION nodes release permits required to commence the cycle. Production in floors completed is measured by the ACCUMULATOR inserted at 31.

13.6 MODEL INPUT AND RESPONSE

Intrinsic statistics available from the model immediately indicate the percent of the time the large and small cranes are idle and therefore potentially available for other lifting tasks. The response of the system to varying numbers of crews can be easily studied. Additionally, variation of the number of shear wall forms available leads to production variation fluctuations that provide information regarding the dynamic response of the system. Modifications of the model developed can be used to determine the reduction in output to be experienced if only one crane is utilized.

In the model as described, units are initialized as follows.

Number of Units	Unit Type	At Element
1	"INTERIOR CORE FORM"	3
1	"LARGE CRANE"	2
1	"SET OF WALL FORMS"	21
1 (or 2)	"CREW 1"	19
1	"SMALL CRANE"	23
1 (or 2)	"CREW 2"	33
10 (or less)	"WALL FORMS"	28
2	"EXT. CORE FORMS"	30

The CYCLONE network defining input for the graphical model as defined is shown in Figure 13.14a and the data print of input information is given in Figure 13.14b. The response of the system is indicated in Tables 13.1 to 13.7.

NAME 'PEACHTREE PLAZA'
PROJECT -1
RUNS 1
ID 150
NETWORK INPUT
ACT 2 Q 'BIG CRANE IDLE' FOLLOW OPS 6, 11, 27, 36.
ACT 3 Q 'CORE RDY' FOLLOW OPS 4.
ACT 4 COMBI LOGNOR 'TELE INNER CORE' PARAMETER SET 2.
FOLLOW OPS 5, 23. PRECEDING QNODES 3, 19, 23.
ACT 5 Q 'INT CORE IN PL' FOLLOW OPS 6.
ACT 6 COMBI LOGNOR PARAMETER SET 3, 'FLY SLAB FORMS'
FOLLOW OPS 2, 7, PRECEDING QNODES 2, 5, 21.
ACT 7 LOGNOR PARAMETER SET 4, 'SET SLAB FORMS' FOLLOW OPS
8, 9, 19.
ACT 8 Q 'GEN WALL STEEL' FOLLOW OPS 11.
ACT 9 Q 'GEN SLAB STEEL' FOLLOW OPS 10.
ACT 10 COMBI LOGNOR PARAMETER SET 5, 'LIFT FLR STEEL'
FOLLOW OPS 15, 23. PRECEDING QNODES 19, 23, 9, GENERATE 10.
ACT 11 COMBI LOGNOR PARAMETER SET 7, 'LIFT WALL CAGES'
FOLLOW OPS 2, 14. PRECEDING QNODES 2, 33, 8, GENERATE 10.
ACT 12 Q 'WALL STEEL SET' FOLLOW OPS 27.
ACT 13 COMBI LOGNOR PARAMETER SET 6, 'LIFT CORE FMS'
FOLLOW OPS 19, 23, 34. PRECEDINGE QNODES 19, 23, 26, 30.
ACT 14 LOGNOR PARAMETER SET 8, 'SET WALL STL' FOLLOW OPS 12, 33.
ACT 15 LOGN PARAMETER SET 9, 'TIE SLB STEEL' FOLLOW OPS 16.
ACT 16 LOGNOR PARAMETER SET 11, 'SET EMBEDS' FOLLOW OPS 19, 20.
ACT 17 COMBI LOGN PARAMETER SET 19, 'PULL CORE FMS'
FOLLOW OPS 18, 23, 33. PRECEDING QNODES 23, 33, 40.
ACT 18 Q 'FORMS LOWERED' FOLLOW OPS 29.
ACT 19 Q 'STL CREW IDELE' FOLLOW OPS 4, 10, 13, 29, 39.
ACT 20 Q 'FLR STEEL SET' FOLLOW OPS 22.
ACT 21 Q 'WL FMS PULLED' FOLLOW OPS 6.
ACT 22 COMBI LOGN PARAMETER SET 12, 'PLACE SLB CONC'
FOLLOW OPS 23, 24, 33. PRECEDING QNODES 20, 23, 33.
ACT 23 Q 'SML CRANE IDLE' FOLLOW OPS 4, 10, 13, 17, 22.
ACT 24 LOGN PARAMETER 18, 'INIT SET(SLB)'
FOLLOW OPS 25, 42.
ACT 25 Q 'SLB SECT RDY' FOLLOW OPS 27.
ACT 26 Q 'SLAB RDY' FOLLOW OPS 13.
ACT 27 COMBI LOGN PARAMETER SET 13, 'LIFT WALL FMS'
FOLLOW OPS 2, 32, 33. PRECEDING QNODES 2, 25, 28, 33, 12.
ACT 28 Q 'WALL FORM RDY' FOLLOW OPS 27.
ACT 29 COMBI LOGNOR PARAMETER SET 20, 'CL COR FORMS'
FOLLOW OPS 30, 19. PRECEDING QNODES 18, 19.
ACT 30 Q 'CORE FORMS RDY' FOLLOW OPS 13.
ACT 31 COUNTER CYCLES 9, QUANTITY 1, FOLLOW OPS 3.
ACT 32 LOGNOR PARAMETER SET 14, 'INIT SET (WL)'
FOLLOW OPS 35.
ACT 33 Q 'POUR CREW IDLE' FOLLOW OPS 11, 17, 22, 27, 36.
ACT 34 LOGNOR PARAMETER SET 15, 'INIT SET(CORE)'
FOLLOW OPS 40, 41.
ACT 35 Q 'CONC CURED(WL)' FOLLOW OPS 36.
ACT 36 COMBI LOGNOR PARAMETER SET 16, 'PULL WL FMS'
FOLLOW OPS 2, 33, 38, 37.
PRECEDING QNODES 2, 35, 33.
ACT 37 CONSOLIDATE 10, 'CON NODE' FOLLOW OPS 21.
ACT 38 Q 'FM TO BE CL' FOLLOW OPS 39.
ACT 39 COMBI LOGN PARAMETER SET 17, 'CL WL FORMS'
FOLLOW OPS 28, 19. PRECEDING QNODES 19, 38.
ACT 40 Q 'FM TO BE CL' FOLLOW OPS 17.
ACT 41 CONSOLIDATE 2, 'CON NODE' FOLLOW OPS 31.
ACT 42 CONSOLIDATE 5, 'CON NODE' FOLLOW OPS 26.

(a)

INITIAL INFORMATION FOR PROJECT   PEACHTREE PLAZA

SUMMARY OF DATA FOR SIMULATION

NO.	OPERATION		FOLLOWING OPS			
2	BIG CRANE IDLE	6	11	27	36	0
3	CORE RDY	4	0	0	0	0
4	TELF INNER CORE	5	23	0	0	0
5	INT CORE IN PL	6	0	0	0	0
6	FLY SLAB FORMS	2	7	0	0	0
7	SET SLAB FORMS	8	9	19	0	0
8	GEN WALL STEEL	11	0	0	0	0
9	GEN SLAB STEEL	10	0	0	0	0
10	LIFT FLR STEEL	15	23	0	0	0
11	LIFT WALL CAGES	2	14	0	0	0
12	WALL STEEL SET	27	0	0	0	0
13	LIFT CORE FMS	19	23	34	0	0
14	SET WALL STL	12	33	0	0	0
15	TIE SLB STEEL	16	0	0	0	0
16	SET EMBEDS	19	20	0	0	0
17	PULL CORE FMS	18	23	33	0	0
18	FORMS LOWERED	29	0	0	0	0
19	STL CREW IDLE	4	10	13	29	39
20	FLR STEEL SET	22	0	0	0	0
21	WL FMS PULLED	6	0	0	0	0
22	PLACE SLB CONC	23	24	33	0	0
23	SML CRANE IDLE	4	10	13	17	22
24	INIT SET(SLB)	25	42	0	0	0
25	SLB SECT RDY	27	0	0	0	0
26	SLAB RDY	13	0	0	0	0
27	LIFT WALL FMS	2	32	33	0	0
28	WALL FORM RDY	27	0	0	0	0
29	CL COR FORMS	30	19	0	0	0
30	CORE FORMS RDY	13	0	0	0	0
31	COUNTER	3	0	0	0	0
32	INIT SET (WL)	35	0	0	0	0
33	POUR CREW IDLE	11	17	22	27	36
34	INIT SET(CORE)	40	41	0	0	0
35	CONC CURED(WL)	36	0	0	0	0
36	PULL WL FMS	2	33	38	37	0
37	CON NODE	21	0	0	0	0
38	FM TO BE CL	39	0	0	0	0
39	CL WL FORMS	28	19	0	0	0
40	FM TO BE CL	17	0	0	0	0
41	CON NODE	31	0	0	0	0
42	CON NODE	26	0	0	0	0

(b)

FIGURE 13.14   (a) Peachtree Plaza network input. (b) Summary output for Plaza Project.

Table 13.1   FLOOR CYCLE TIMES (in minutes)

Number of CREW 1 \ Number of CREW 2	Ten Shear Wall Forms	Eight Shear Wall Forms	Five Shear Wall Forms	Two Shear Wall Forms
1 \ 1	4471	4471	4578	
2 \ 1	3593	3879	4260	6505
1 \ 2	4448	4448	4534	
2 \ 2	3374	3432	3764	6078

The first item of interest to the construction manager is the question "How long will it require to complete one floor of construction?" The cycle times for floor construction under varying conditions are presented in Table 13.1. This table, as well as the others relating to system response, is organized as a four by four matrix; the rows indicate variations of the crew combinations and the columns indicate the number of shear wall forms available (range 10 to 2 forms). As would be expected, cycle times increase as the number of forms is reduced, causing a constraint in the segment 27-32-35-36. However, when using only one CREW 1-type unit (rows 1 and 3), the reduction to eight forms does not affect cycle time, indicating that no constraint occurs in shear wall concreting segment. This obviously indicates that eight forms are sufficient to handle the input rate from element 24, provided only one CREW 1 unit is active in the slab steel area. When the number of CREW 1-type units is increased to two, the input rate increases, causing the constraint of only eight forms to become operative.

Examination of the 10-form column indicates that addition of an extra crew at 19 (i.e., increase CREW 1 units to two) yields a cycle time improve-

Table 13.2 CRANE IDLE TIMES

Number of CREW 1	Number of CREW 2	Ten Shear Wall Forms (Large)	Ten Shear Wall Forms (Small)	Eight Shear Wall Forms (Large)	Eight Shear Wall Forms (Small)	Five Shear Wall Forms (Large)	Five Shear Wall Forms (Small)	Two Shear Wall Forms (Large)	Two Shear Wall Forms (Small)
1	1	69	77	69	77	70	78		
2	1	62	68	68	71	68	75	78	78
1	2	68	77	68	77	69	77		
2	2	58	68	59	68	64	70	72	73

Percent
Time Idle
(Large Crane)

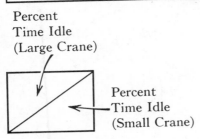

Percent
Time Idle
(Small Crane)

ment of 878 minutes or approximately 15 hours. Assuming that a double shift operation results in 15 hours of productive effort per day, this represents a time saving of approximately 1 day, reducing the cycle time from 5 days to 4 days. By contrast, increase of the number of CREW 2 units does not yield a significant reduction in cycle time (see row 3). This indicates that the system is relatively insensitive to variation of CREW 2 type units. This is further verified by considering the 2/2 configuration of row 4 and comparing it to the 2/1 system of row 2. With 10 shear wall forms, the improvement in cycle time

Table 13.3 CREW IDLE TIMES

Number of CREW 1	Number of CREW 2	Ten Shear Wall Forms	Eight Shear Wall Forms	Five Shear Wall Forms	Two Shear Wall Forms
1	1	12 / 52	12 / 52	14 / 54	
2	1	60 / 38	57 / 44	62 / 48	73 / 62
1	2	12 / 85	12 / 85	13 /	
2	2	65 / 87	60 / 88	59 / 87	65 / 86

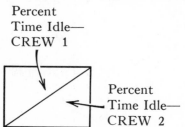

Percent Time Idle— CREW 1

Percent Time Idle— CREW 2

achieved by adding the second CREW 2 unit is (3593 − 3374) minutes or approximately 3-1/2 hours. On a cost effectiveness basis, this addition normally would not be justified.

Utilization of the cranes for the cases presented can be developed from the idle time figures given in Table 13.2. Again, variation of the number of the CREW 2-type units from one to two has negligible effect on system response. The 2/1 system configuration again yields the most economic utilization values for the cranes, resulting in the large crane (at 2) being active on system

Table 13.4 DELAY AT 6

Number of CREW 1	Number of CREW 2	Ten Shear Wall Forms	Eight Shear Wall Forms	Five Shear Wall Forms	Two Shear Wall Forms
1	1	0 / 6	0 / 6	0 / 7	
2	1	3 / 1	9 / 1	16 / 1	31 / 1
1	2	0 / 6	0 / 5	0 / 7	
2	2	0 / 3	2 / 1	2 / 1	29 / 1

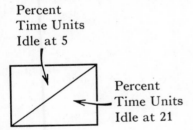

Percent Time Units Idle at 5 (upper-left triangle)
Percent Time Units Idle at 21 (lower-right triangle)

tasks 38% of the time and the smaller crane being active 32% of the time. This activity is confined to tasks defined in the concrete placement model, and the cranes will be available for other tasks 62% and 68% of the time, respectively. Actual observed idle times at the job site are reduced by loading the cranes with lifting tasks not defined in the concreting model (e.g., lifting, glazing and plumbing fixtures, interior finish items). The 2/2 configuration results in only a slight increase in the utilization of the large crane, while the small crane utilization remains the same.

Table 13.5 DELAY AT 4 AND 22

Number of CREW 1	Number of CREW 2	Ten Shear Wall Forms	Eight Shear Wall Forms	Five Shear Wall Forms	Two Shear Wall Forms
1	1	3 / 23	3 / 23	3 / 21	
2	1	1 / 29	1 / 29	1 / 25	1 / 19
1	2	3 / 8	3 / 8	3 / 8	
2	2	2 / 15	1 / 15	1 / 15	1 /

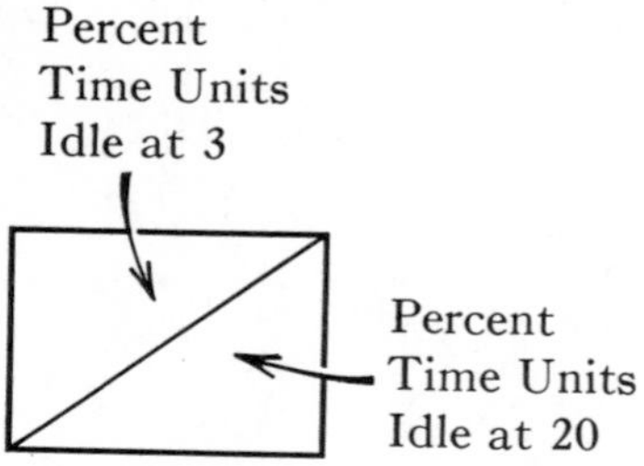

Percent Time Units Idle at 3

Percent Time Units Idle at 20

Crew utilization on systems tasks can be developed from Table 13.3. The occupancy statistics for the 1/1 system configuration indicate that the CREW 1 unit is highly committed being active on system tasks 88% of the time. The CREW 2 unit is busy only 48% of the time in this configuration. Increasing the number of CREW 1 units, with the attendant increased input to the shear wall concreting sequence, causes an increase in the activity of the CREW 2 unit to 62% of the time. The CREW 1 occupancy statistic rises sharply, since there are now two units.

Table 13.6   DELAY AT 27

Number of CREW 1 \ Number of CREW 2		Ten Shear Wall Forms	Eight Shear Wall Forms	Five Shear Wall Forms	Two Shear Wall Forms
1	1	6 / 100	6 / 100	22 / 73	
2	1	17 / 96	17 / 90	25 / 73	46 /
1	2	0 / 99	0 / 99	6 / 76	
2	2	6 / 94	13 / 83	24 / 70	51 / 34

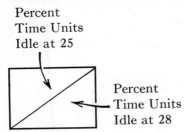

Percent Time Units Idle at 25

Percent Time Units Idle at 28

The other tables presented indicate the delays incurred by units at various critical points in the flow network. The effect of the increased input rate from the slab steel placement sequence (10-15-16) caused by increasing CREW 1-type units can be observed by considering the idle times at QUEUE nodes 28 and 25 as given in Table 13.6. In the 1/1 configuration, units at 25 are delayed only 6% of the time in the 10- and 8-form systems. This is increased to 17% in the 2/1 system, since the addition of an extra CREW 1 unit causes higher productivity along the 10-15-16 track and results in a greater backup

Table 13.7   DELAY AT 13

Number of CREW 1	Number of CREW 2	Ten Shear Wall Forms (Idle 26 / Idle 30)	Eight Shear Wall Forms (Idle 26 / Idle 30)	Five Shear Wall Forms (Idle 26 / Idle 30)	Two Shear Wall Forms (Idle 26 / Idle 30)
1	1	4 / 94	4 / 94	7 / 94	
2	1	0 / 82	0 / 81	0 / 84	0 / 84
1	2	4 / 95	4 / 95	6 / 99	
2	2	0 / 82	0 / 81	0 / 88	0 / 82

Percent Time Units Idle at 26

Percent Time Units Idle at 30

of units at this monitoring point. The 1/2 system records no delays at 25, since the single CREW 1 units now become the constraining unit causing the release units from 12 to be delayed. The constraining track in every case can be determined by examining the preceding QUEUE node delays and evaluating the idle times. In the 2/2 with only eight shear wall forms, the occupancy statistic that previously indicated a very high availability of clean forms at 28 drops sharply, reflecting for the first time potentially significant delays at 27 caused by lack of clean shear wall forms.

The delays at element 6 occur primarily as a result of (1) the requirement to wait until all 10 wall forms have been pulled at 36, or (2) the completion of telescoping of the interior core forms. The delays are indicated by the idle times associated with QUEUE nodes 5 and 21, as given in Table 13.4. In the system with one unit of each crew type, the delay of zero associated with QUEUE nodes established that the permit at 21 is always present when required and that commencement of form movement (flying) is never delayed due to the pulling of shear wall forms. When the number of CREW 1-type units is increased to two, the situation reverses with delays usually occurring because of the requirement to wait until all forms all pulled. That is, the idle time at 5 now usually exceeds that at 21. This develops because the additional crew at 19 expedites flow of the core sections along segment 13-34 as well as at the interior form telescope operation. This effect becomes more pronounced as the number of forms is reduced.

The idle times in Table 13.7 indicate that in the systems utilizing two CREW 1-type units, the core pour is never delayed because of either non-availability of the forms, the crane, or the crew required. The 1/1 and 1/2 system do show some delay at 26 because of the requirement to compete for the single CREW 1 unit with the slab steel placement sequence (10-15-16).

13.7 SITE LEVEL SCHEDULING

The CYCLONE model developed for the concrete placement process used in the Peachtree Plaza project incorporates the cyclic nature of the construction operations and the specification of the number, type, and use of major resources, and it defines operational management policies. To ensure that the cyclic nature of the construction operations as defined by the model is, in fact, realized in the field under the prescribed management policies, it is necessary to communicate this information to the site level work force.

The multiple-activity chart is a field-oriented tool that is directly related to the cyclic structure, work sequence, and system response characteristics of CYCLONE models. As mentioned in Chapter Two, the multiple-activity chart is a modified bar chart consisting of a series of equal length vertical bars drawn to the time scale of the building flow cycle. A vertical bar is used for each crew, hoist, or crane unit involved. The scheduled sequence of work tasks for each resource is identified from the CYCLONE model and suitably scaled according to averaged work task durations and system response predictions of the interacting sequences and idle times. In this way the multiple-activity chart captures and exhibits the cyclic nature of the CYCLONE model and the preplan design of the construction process as influenced by management decisions and policies. The multiple-activity chart then portrays the planned sequence and schedule for all work tasks on the building site relevant to the basic concrete placement cycle. The individual vertical bar portraying the work assignments and schedule for each individual crew and major piece of equipment can be used as a field work order both to measure current status and as a statement of the next work assignment. Finally, the measurement of current status as a multiple-activity chart can be integrated through a line of balance concept to produce a field tool capable of assessing the impact of variations from plan caused by delays of all sorts. In this way the multiple-activity chart provides field managers and agents with a field decision tool for reestablishing the cyclic nature and harmonious interaction of the building operations.

Consider, for example, the development of a multiple-activity chart for the Peachtree Plaza concrete placement sequence given in Figure 13.9. Using the work assignment breakout specified earlier and referring to the CYCLONE process model of Figure 13.13, it is possible to identify 11 structured components of the multiple-activity chart, as shown in Figure 13.15. According to the CYCLONE modeling logic, certain work task sets are repeated. For example, the work task patterns 10-15-16 and 11-14 are repeated 10 times per floor, while the pattern 6-7 is used once per floor.

In order to develop the process multiple-activity chart, it is necessary to scale each of the patterns of Figure 13.15 to their required work durations and to locate them within the floor cycle duration consistent with the technological plan given in the CYCLONE model and the compatible use of resources at any one time. Assuming, for this example, that the various work tasks have deterministic times as follows:

ACT	Minutes
4	90
6	240
7	480
10	30
15	120
16	90
11	24
14	60

Then Figure 13.16 illustrates the start-up development of the process multiple-activity chart based on these deterministic times. This sequence is plotted to a time scale as Figure 13.17, which shows the typical format of the multiple-activity chart. The resource idle times have been identified with the relevent idle QUEUE nodes (Q2 for the big crane, Q23 for the small crane, etc.). The multiple-activity chart can be developed rigorously from the

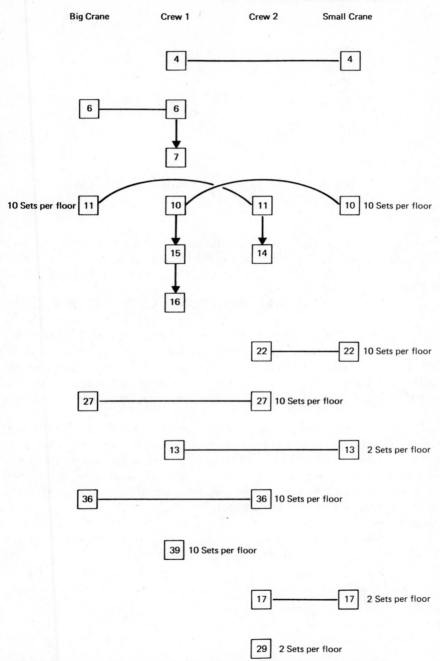

FIGURE 13.15   Structured elements of multiple activity chart.

Big Crane	Crew 1	Crew 2	Small Crane
	S4 0.		S4 0.
	F4 90.		F4 90.
S6 90.	S6 90.		
F6 300.	F6 300.		
	S7 300.		
	F7 480.		
S11 480.	S10 480.	S11 480.	S10 480.
F11 504.	F10 510.	F11 504.	F10 510.
	S15 510.	S14 504.	
	F15 630.	F14 564.	
S11 2nd cycle	S16 630.	S11 2nd cycle	
	F16 720.		
	S10 2nd cycle		S10 2nd cycle

FIGURE 13.16   "Startup" development of multiple activity chart.

CYCLONE model once the work task durations have been specified, since the CYCLONE model has already defined the technological plan and included management policies on the priority of work sequences by the manner of labeling the various COMBI nodes.

Given specific work task durations defined by the CYCLONE DURATION input, it is an easy matter to develop through computer processing on the CYCLONE system a trace printout for node realization times. This printout thus defines on a time scale all the relevant information for the multiple-activity chart. In fact, the information on Figure 13.16 was taken directly from such a trace printout. In this way the multiple-activity chart representation of a CYCLONE model captures the essential characteristic of the system model.

Finally, it is a simple matter to redefine the EQUIPMENT INPUT segment of the CYCLONE model as many times as desired, each time developing through a CYCLONE trace output the consequential multiple-activity chart. In this way decision agents can experiment with the resources for the process and evaluate through the floor cycle durations multiple-activity chart the impact of the various resource strategies.

It should now be obvious to the careful reader that the multiple-activity chart can serve as a field communication tool that can be derived from the CYCLONE system model formulation.

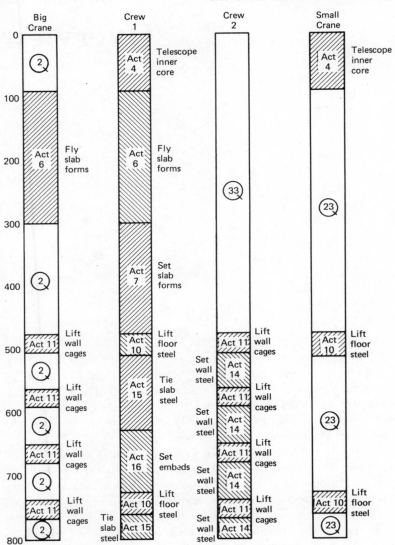

FIGURE 13.17   "Startup" multiple activity chart.

CHAPTER FOURTEEN

Overview

The CYCLONE graphical modeling format together with the CYCLONE problem-oriented language computer processing system establishes a modeling and system analysis methodology for construction operations. The CYCLONE methodology enables construction operations to be described, modeled, analyzed, and designed in whatever level of detail is relevant to the needs of the construction engineer, head office planner, or field agent.

The modeling concepts of the CYCLONE graphical format are simple and versatile and enable the ready portrayal of work sequences, construction technology, and conditional interrelationships among the various work tasks and processes involved in the construction operation. When properly developed and labeled, CYCLONE models are readily understood by field personnel and provide a venue for the identification of crew interaction and individual work assignments for crew members.

The CYCLONE P.O.L. statements enable the construction operation to be described by field agents directly from the CYCLONE graphical model. In this way the CYCLONE statements enable the field agent to approach the digital computer directly, through a CYCLONE processor, with his own interpretation and formulation of the construction operation. The field agent can thus exert in his own terms his professional understanding and viewpoint of the makeup and management of a construction operation. The elimination of the need for mastery of a highly abstract computer language is obviated, as is the deterrent of working through a computer-buff intermediary whenever the field agent is tempted to seek assistance.

The CYCLONE methodology consists of a number of sequential stages in the formulation and development of models. These stages correspond to the various levels of professional effort, decision making, and management of construction operations. The CYCLONE methodological stages of construction operation definition, analysis, monitoring performance, and design parallel head office and field functions and decision processes.

The description of a construction operation using elemental work tasks that are structured together to portray the technology of the construction method and the focus on the active processing of resources provides a comprehensive format for describing the specifics of the construction operation. In this way the formal knowledge of the construction method and the accumulated experience of field agents can be combined to portray the various steps, logic, and procedures of the construction method to be used in the field.

Once the construction method is defined for the construction operation, the next stage requires the definition, enumeration, and assigning by management of the required resources. In the CYCLONE P.O.L. format, this requires the listing of resources in the EQUIPMENT INPUT segment and the statement of the initial posture for commencing the construction operation. In practice this definition is usually done in the preplan phase and confirmed in the field implementation phase. Usually the preplan definition of resources is confined to major plant and equipment items, leaving crew size and individual work assignments to the field implementation phase.

The prediction, monitoring, and assessment of the effects, impact, or productivity over time of the construction operation relative to the assigned input resources and construction method constitutes the third stage of the CYCLONE approach. In the CYCLONE P.O.L. this requires the definition of FUNCTION, STATISTICS, and COUNTER nodes that, together with the intrinsic properties of QUEUE nodes, establishes the reporting format of the CYCLONE system analysis. In field practice this corresponds to field performance observations, and assessments on crew assignments, equipment idle time, work item quantities in place, and the like. These observations form the basis of the field reporting system, which thus enables comparisons to be made with estimate parameters established for the construction operation.

The field design and management decision stage is concerned with the manipulation, adaption, and mix of the assigned resources to best meet field management goals and productivities. This requires the formulation of policy statements on work procedures and work task assignment priorities. In the CYCLONE methodology this stage requires the development and incorporation into the system model of control structures and mechanisms, the embodiment of priority policy statements in COMBI node numbering, and the redefinition of EQUIPMENT (or resource) INPUT statements. In the field this stage takes the form of the monitoring and redefinition over the life of the operation of crew sizes and work assignments together with the establishment by trial and error (i.e., by fiat) of the best sequences of work tasks that eliminate bottlenecks and speeds up production.

The CYCLONE methodology can be utilized at any number of different levels of involvement. Thus the first stage corresponds to a method of describing construction operations that might be useful and relevant enough to stand on its right as a means of instructing field staff in a new operation or as a teaching methodology for construction engineering students. In this way it provides the format for communicating information between different field agents and skilled personnel. The second stage (and, by implication, the last stage) corresponds to the management of in-progress construction operations. In this respect the INPUT EQUIPMENT segment of the CYCLONE P.O.L. could correspond to actual briefing statements by field agents of WHO and WHAT will be available to do WHICH tasks. Finally, the third stage focuses on the specification of a construction operation reporting system that could form the basis of a field reporting document to head office management. This information would then be a means of capturing and documenting field experience. In this way the correlation of a CYCLONE model (and its systems analysis) with field documentation would provide the format for validly portraying field experience and transferring this experience from one situation to another.

The CYCLONE methodology provides the format and opportunity to the user for the repetitive sequencing of the decision phases associated with the selection of technology, design, and the assessment of the implication of a construction operation before actual work committments are made. This capability enables management to check out the impact of a variety of decisions and policies and provides a framework for evaluating alternative competing construction methods and resource allocations and for establishing predictions of field productivities.

CYCLONE models may be usefully developed for the analysis and design of large unique capital intensive undertakings. A typical example is the precast plant layout and operation design described in previous chapters. In this type of situation the capital intensive nature of the project, the long-term impacts that result from the implementation of critical irreversible decisions on material flow layouts, types of curing units, plant productivity and storage facilities, and the like, provide an environment where planners and management are receptive to methodologies that relate to these decision problems.

The development of CYCLONE models alerts management and field agents to features of a construction operation that affect its productivity, cycle duration, and efficiency. Often a close look at an operation is worthwhile if improvements in its design, stability, and implementation are magnified because of the large number of repetitive cycles involved in its work content in the project. Thus CYCLONE models for material handling processes associated with mass concrete dams, earthworks, tunneling, pipeline construction, and extensive pile driving may become feasible. In some cases work volume achieved by a particular construction operation over a number of projects is sufficient to make its analysis worthwhile on a long-term basis although, for each particular project, its impact appears small.

The CYCLONE methodology uses the work task as the elemental building component of the construction operation. While the number of work task elements in an operation may be considerable their definition and focus on the active processing of resources is more receptive to capturing the specifics of a particular work face layout and environment.

The methodology allows for the better analysis of operations using field estimates, since the estimate data input for the procedure is based on small tasks that are relatively clearly defined in terms of resources and their participation in the process and operation. The field agent providing the estimated input data has an intimate knowledge of the work task to a much finer degree than his knowledge of traditionally defined project activities, since the durations involved are in minutes and hours and are more relevant to his thinking and work patterns. It is also possible to gain statistically reasonable data from field measurement and, because the work task definition is based on active and passive states of resources, the arbitrariness of task definition is reduced. The emphasis on resource utilization and idleness provides a sounder, more unique definition than for the activity-oriented breakout of a project.

Field agents do not think in terms of arbitrarily defined project activities related to a physical structure. Instead they think in terms of resources and their maximum utilization to achieve production on a process that may be common to many project activities. The resource emphasis provides insight into the proper balancing of processes.

Activities defined by higher management do serve a useful purpose in organizing time committed to the project and the establishment of goals. But, at the site level, these goals act as a requirement that must be met by deploying resources. The site superintendent is most concerned with the committment of resources, and his planning focus is at this level.

Higher management and site management must work together and find a common framework within which the individual problems can be understood. Higher management provides project goals and time-cost profiles and field management provides expertise in resource and construction operation management. Their interaction establishes the feasibility of the project plan and implementation to achieve company goals. Higher management has available a number of tools and planning and decision methodologies relevant to their time-cost profile problems. Field management, however, has little in terms of a medium for communicating information and analyzing interactions of field resources and processes. Field management often builds ad-hoc reference systems, which are the equivalent of a CYCLONE analysis.

The CYCLONE methodology talks the same language as that used by field agents. It has the right level of abstraction and detail for field management. This fact, plus the simplicity of the CYCLONE modeling format and P.O.L. statements, means that all field agents down to foreman level can participate in the planning and decision processes associated with construction operations.

The CYCLONE system allows higher management to look at work sequences and the details of construction operations through the eyes of field management and to tap their accumulated expertise. Similarly, the CYCLONE system helps in the training of candidates for field management. Thus foremen being groomed for a superintendent position can be introduced to the concepts of the planning, analysis, and design of construction operations as preparation for organizing a job site, work processes, and communicating with management.

The practical and useful implementation of the CYCLONE methodology requires the recognition or perception by field agents that an opportunity for its use exists, a knowledge of the purpose and scope of the CYCLONE technique and the existence of a permissive environment established by management, which encourages field agents to exploit management techniques.

Bibliography

Chapter 1

1. Antill, J. M. and P. W. S. Ryan, *Civil Engineering Construction*, 3rd ed. London, Angus and Robertson, 1967.

2. Havers, J. A. and F. W. Stubbs, Jr. (eds.), *Handbook of Heavy Construction*, 2nd ed. New York, McGraw-Hill, 1971.

3. Peurifoy, R. L., *Construction Planning, Equipment, and Methods*, 2nd ed. New York, McGraw-Hill, 1970.

4. Rapp, William G., *Construction of Structural Steel Building Frames*, New York, Wiley, 1968.

Chapter 2

1. Antill, J. M. and R. W. Woodhead, *Critical Path Methods in Construction Practice*, 2nd ed. New York, Wiley, 1970.

2. Barroso, L. F., T. Nakajima, and R. W. Woodhead, "The Construction Project Daily Labor Allocation Problem," *Proceedings, Internet 72*, Stockholm, Sweden, May 1972.

3. Clark, Wallace, *The Gantt Chart; A Working Tool for Management*, 2nd ed. London, Pitman and Sons, Ltd. 1942.

4. Fondahl, John, Chap. 14, *Handbook of Construction Management and Organization*, Bonny and Frein, (eds.) New York, Van Nostrand Reinhold, 1973.

5. Gordon, G., *System Simulation*, Englewood Cliffs, N. J., Prentice-Hall, Inc., 1969.

6. Halpin, D. W. and R. W. Woodhead, "Flow Modeling Concepts in Construction Management," Preprint 1618, ASCE National Water Resources Engineering Meeting, Atlanta, Georgia, January 24-28, 1972.

7. Meredith, et al., *Design and Planning of Engineering Systems*, Englewood Cliffs, N. J., Prentice-Hall, 1973.

8. Naylor, T. H. et al., *Computer Simulation Techniques*, New York, Wiley, 1966.

9. Gaarslev, Axel, "Stochastic Models to Estimate the Production of Material Handling Systems in the Construction Industry," *Technical Report No. 111*, The Construction Institute, Stanford University, Palo Alto, Calif., August 1969.

10. Sprague, Charles R., "Investigation of Hot-Mix Asphaltic Systems by Means of Computer Simulation," Ph.D. Dissertation, Texas A & M University, College Station, Texas, 1972.

11. Teicholz, P., "A Simulation Approach to the Selection of Construction Equipment," *Technical Report No. 26*, The Construction Institute, Stanford University, June 1963.

Chapter 3

1. Halpin, D. W. and W. W. Happ, "Digital Simulation of Equipment Allocation for Corps of Engineer Construction Planning," *Proceedings of the Seventeenth Conference of Design of Experiments in Army Research*, Washington, D. C., October 1971.

2. Halpin, D. W. and W. W. Happ, "Network Simulation of Construction Operations," *Proceedings of the Third International Congress of Project Planning by Network Techniques*, Stockholm, Sweden, May 1972.

3. "HOCUS—The Management Parlour Game," *Business Management*, April 1969.

4. *HOCUS Manual*, P-E Consulting Group Limited, Surrey, England, 1970.

Chapter 4

1. Halpin, D. W., "An Investigation of the Use of Simulation Networks for Modeling Construction Operations," Ph.D. Dissertation, University of Illinois, Urbana, Ill., 1973.

2. Halpin, D. W. and W. W. Happ, "Network Simulation of Construction Operations," *Proceedings of the Third International Congress on Project Planning by Network Techniques*, Stockholm, Sweden, May 1972.

3. Halpin, D. W. and R. W. Woodhead, "A Network-Based Methodology for the Management Modeling of Complex Projects," *Proceedings of the Third International Congress on Project Planning by Network Techniques*, Stockholm, Sweden, May 1972.

Chapter 5

1. Halpin, D. W., "An Investigation of the Use of Simulation Networks for Modeling Construction Operations," Ph.D. Dissertation, University of Illinois, Urbana, Ill., 1973.
2. *HOCUS Manual*, P-E Consulting Group Limited, Surrey, England, 1970.
3. "Preview of Rapid Excavation and Tunneling Conference," *Civil Engineering*, ASCE, April 1972.

Chapter 6

1. Ang, A. and W. Tang, *Probability Concepts in Engineering Planning and Design*, Vol. I, New York, Wiley, 1975.
2. *Caterpillar Performance Handbook*, 2nd ed. Peoria, Ill., Caterpillar Tractor Company, January 1972.
3. Feller, W., *An Introduction to Probability Theory and Its Applications*, Vol. 1, 2nd ed., New York, Wiley, 1957.
4. *Fundamentals of Earthmoving*, Peoria, Ill., Caterpillar Tractor Co., April 1968.
5. Maynard, Harold B., G. J. Stegemerten, and John L. Schwab, *Methods-Time Measurement*, New York, McGraw-Hill Book Company, 1948.
6. Morgan, W. C. and L. Peterson, "Determining Shovel-Truck Productivity," *Mining Engineering*, December 1968.
7. Naylor, T. H. et al., *Computer Simulation Techniques*, New York, Wiley, 1966.
8. The U.S. Navy Facilities Engineering Command has issued a series of Engineered Performance Standards (EPS) Publications with the following NAVFAC publication numbers and titles: P-700.0, Engineers' Manual, 1963; P-701, General Handbook, 1964; P-701.1, General Formulas, 1962; p-702.0, Carpentry Handbook, 1962; P-702.1 Carpentry Formulas, 1962; P-703.0, Electrical and Electronic Handbook, 1963; p-703.1, Electrical and Electronic Basic Supporting Data, 1957; P-704.0, Heating, Cooling and Ventilating Handbook, 1963; P-704.1 Heating, Cooling and Ventilating Formulas, 1963; P-707.0, Machine Shop, Machine Repair Handbook, 1960; P-707.1, Machine Shop, Machine Repair Formulas, 1959; P-708.0, Masonry Handbook, 1963; P-708.1, Masonry Formulas, 1963; P-709.0, Moving and Rigging Handbook, 1962; P-709.1, Moving and Rigging Formulas, 1962; P-710.0 Paint Handbook 1963; P-710.1, Paint Formulas, 1963; P-711.0 Pipefitting and Plumbing Handbook, 1962; P-711.1, Pipefitting and Plumbing Formulas, 1962; P-712.0, Roads, Grounds, Pest Control Handbook, 1963; P-712.1, Roads, Grounds, Pest Control Formulas, 1963; P-713.0, Sheetmetal, Structural Iron and Welding

Handbook, 1960; P-713.1, Sheetmetal, Structural Iron and Welding Formulas, 1960; P-714.0, Trackage Handbook, 1963; P-714.1, Trackage Formulas, 1963; P-715.0, Wharfbuilding, 1963; P-715.1, Wharfbuilding Formulas, 1963.

9. RAND Corporation. *A Million Random Digits with 100,000 Normal Deviates*, Glencoe, Ill., The Free Press, 1955.

Chapter 7

1. Ackoff, Russell L. and M. W. Sasieni, *Fundamentals of Operations Research*, New York, Wiley, 1968.

2. Gordon, Geoffrey, *System Simulation*, Englewood Cliffs, N. J., Prentice-Hall, Inc., 1969.

3. "HOCUS—The Management Parlour Game," *Business Management*, April 1969.

4. Meredith, D. et al., *Design and Planning of Engineering Systems*, Englewood Cliffs, N. J., Prentice-Hall, Inc., 1973.

Chapter 8

1. Halpin, D. W. and W. W. Happ, "Network Simulation of Construction Operations," *Proceedings of the Third International Congress on Project Planning by Network Techniques*, Stockholm, Sweden, May 1972.

2. Halpin, D. W. and W. W. Happ, "Digital Simulation of Equipment Allocation for Corps of Engineers Construction Planning," *Proceedings of the 17th Conference on Design of Experiments in Army Research*, Washington, D. C., October, 1971.

Chapter 9

1. McDaniel, Herman, *An Introduction to Decision Logic Tables*, New York, Wiley, 1968.

2. McDaniel, Herman, *Applications of Decision Tables*, Princeton, N. J., Brandon/Systems Press, 1970.

3. Rainer, R. K. "Predicting Productivity of One or Two Elevators for Construction of High-Rise Buildings," Ph.D. Dissertation, Auburn, Ala.; Auburn University, 1968.

Chapter 10

1. Gordon, G., *System Simulation*, Englewood Cliffs, N. J., Prentice-Hall, Inc., 1969.

2. Pritsker, A. A. B. and R. R. Burgess, "The GERT Simulation Programs: GERTS III, GERTS III Q, GERTS III C, and GERTS III R, NASA/ERC

Contract NAS-12-2113," Departmental Report, Virginia Polytechnic Institute, Department of Industrial Engineering, 1970.

3. Pritsker, A. A. B. and W. W. Happ, "GERTS: Part I—Fundamentals," *Journal of Industrial Engineering*, 17 (5), 1966, pp. 267-274.

4. Pritsker, A. A. B. and G. W. Whitehouse, "GERT: Part II—Probabilistic and Industrial Engineering Applications," *Journal of Industrial Engineering*, 17 (6), 1966, pp. 293-301.

5. Pritsker, A. A. B. and P. J. Kiviat, *Simulation with GASP II*, Englewood Cliffs, Prentice-Hall, Inc., 1969.

6. Pritsker, A. A. B., *User's Manual for GERT Simulation Program, NASA/ ERC NGR 03-011-034*, Arizona State University, July 1968.

Chapter 11

1. Gordon, Geoffrey, *System Simulation*, Englewood Cliffs, N. J., Prentice-Hall, Inc., 1969.

2. Naylor, T. H. et al., *Computer Simulation Techniques*, New York, Wiley, 1966.

Chapter 12

1. Gaarslev, Axel, "Stochastic Models to Estimate the Production of Material Handling Systems in the Construction Industry," *Technical Report No. 111*, The Construction Institute, Stanford University, Palo Alto, California, August 1969.

2. Halpin, D. W. and R. W. Woodhead, *CONSTRUCTO—A Heuristic Game for Construction Management*, University of Illinois Press, Urbana, Ill., March 1973.

3. Halpin, D. W. and R. W. Woodhead, "CONSTRUCTO—A Computerized Construction Management Game," *Construction Research Series, No. 14*, Department of Civil Engineering, University of Illinois, Urbana, Ill., December 1970.

4. Halpin, D. W., K. O. Reiff, and R. W. Woodhead, "Ein Baustellen-bezogenes Planspiel," *Bau & Bauindustrie*, Werner Verlag, Dusseldorf, Germany, November 1971 (in German).

5. Halpin, D. W. "An Investigation of the Use of Simulation Networks for Modeling Construction Operations," Ph.D. Dissertation, University of Illinois, Urbana, Ill., 1973.

6. Pritsker, A. A. B. and P. J. Kiviat, *Simulation with Gasp II*, Englewood Cliffs, N. J., Prentice-Hall, Inc., 1969.

7. Pritsker, A. A. B., *User's Manual for GERT Simulation Program, NASA/ ERC NGR 03-011-034*, Arizona State University, July 1968.

Appendices

Queueing Systems

A.1 GENERAL CONCEPTS

In many disciplines, situations arise in which discrete units arrive at a processor for servicing. The arriving units may require a common specific service from a standard performance processor but, in many cases, each arriving unit demands unique service from one or more different processors.

If the processing station is occupied or busy with another unit, the arriving unit is forced to wait; this results in a waiting line or queue. If, on the other hand, additional processing stations are provided as a means of eliminating the queueing of arriving units, the actual pattern or time distribution of arrival units may be such as to lead to inefficient use of the processing units with consequential excessive idleness.

Queueing situations are common to all industrial processes, particularly to construction processes and projects. The arrival of trucks at a shovel to be loaded with earth is a classical example of a queueing situation in construction. In this case, the processor (the loader or shovel) maintains its position and the trucks or haulers cycle in and out of the system. Another similar example is the use of dozers to push load scrapers. More common examples are ready-mix concrete trucks servicing hoppers, buggies, crane buckets, or wheelbarrows and the servicing function performed by material and man hoists on building sites.

Many questions that are of interest to the manager arise in connection with queueing problems.

1.   How long will units be delayed in the queue?

2.   How long will the queue be?

3.   How many units can be processed by the processor, considering delays caused by queueing?

4.   How can lack of service to arriving units be related to the idleness and inefficient use of processors?

5.   How many processors should be provided?

These are typical of some of the problems and the types of information that the manager can utilize in designing service facilities and in determining the production and resource utilization associated with such systems.

Queueing systems were first considered in a mathematical format by the Danish mathematician A. K. Erlang. In the course of studying the processing of telephone calls in Copenhagen early in this century, he was able to develop certain relationships that provide mathematically correct answers to the questions posed above. His studies provide the basis for what is now referred to as *queueing theory*. Extensions of Erlang's work have been applied to many industrial situations; more recently, certain applications to construction processes have been attempted.

Simple queueing systems can be represented schematically, as shown in Figure 2.23. They consist essentially of a processor station through which system flow units must pass. Implicit to this station is a processor unit that does the processing. This unit cycles as required back and forth from an idle to a busy state. A typical processor unit in a construction situation is a front loader, which cycles continuously from an idle state awaiting trucks to a busy state loading trucks. Obviously, if a processor (e.g., a loader) is busy with one unit, other units must wait. Therefore a queueing model also consists of a queue or waiting line. The units that have arrived for processing, but that cannot be processed immediately, take up a waiting position in this location. This is analogous in a loader system to the truck backup or queue. Finally, the system has a border or boundary that defines when units have entered and when they have departed. Upon completion of processing, units are considered to have exited the system. Arrivals in the system can pass directly through the waiting position and begin processing immediately if the processor is not busy. If the processor is busy, they enter the system and are held in the queue, pending availability of the processor. The boundary of the system for

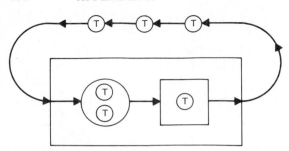

FIGURE A.1   A finite system.

incoming units is marked as the point at which they enter the queue and are either delayed or pass on directly to processing.

Units processing the system (e.g., trucks) are often referred to as calling units.* The processor units (e.g., the loaders) are referred to as servers. The number of servers indicates the number of channels or routes by means of which the calling units can be processed passing from the queue through a processing station and out of the system. If there is, for instance, only one loader, the system is called a single-server system. If there are several loaders, then the system becomes a multichannel or multiserver system. Units entering the system are said to enter from populations that can be either infinite or finite. In the case of a telephone system, the calls enter from a very large population and an infinite population is assumed. In an earth-loading situation, the number of trucks on the haul is a finite number. The number of trucks that can enter the system at any given time is known, since it is simply the number of trucks hauling minus the number in the system. For instance, in the earth-hauling model shown in Figure A.1, six trucks hauling are involved.

As shown, three trucks (indicated as (T)) are within the system boundaries. One truck is being loaded (i.e., processed) and two trucks are delayed in the queue. The other three trucks are outside of the system on the "back cycle." They are not within the system boundaries as defined. Therefore, the number of units that could arrive at any point in time, t, is simply the number hauling (fleet size = 6) minus the number in the system (3), or three trucks. Situations with both infinite and finite populations can be identified in construction processes.

The response of a queueing model is tied to the assumptions made regarding the rates of unit arrival, the processor rate, the type of population, and the discipline of units as they pass through the queue.

*This is obviously a throwback to Erlang's original work on telephone systems.

A.2 SYSTEM STATES

The system variables that are used to define or describe a system state (or status) at any time are called state variables. A particular system state is then identified by a set of instantaneous values of the state variables. If the value of one or more of the state variables is changed so that a new configuration can be recognized, then a new state exists.

Queueing theory problems can be readily described in terms of states defined as the number of units delayed in the queue, whether the processor is active or idle, and so forth. Based on the assumption made regarding the queueing problem model, a set of equations can be written to describe the queueing system under investigation. The concept of "states" is used in writing equations to describe a queueing system, and these equations are called equations of state. A system changes its state as time passes. In a queueing model, the configuration of the system at any given point in time can be described in terms of the number of units within the system. For instance, the finite queueing model shown in Figure A.1 can be said to be in state S_3, since the number of units within the system boundaries is three. In this example, since the truck fleet size is six trucks, seven possible system states can be identified (i.e., when 0, 1, . . . 6 trucks are in the system). The system can then only move between seven states as shown:

In general, the number of states in which a finite system can find itself is $M + 1$, where M is the number of calling units. For the model of Figure A.1, the system states are S_i ($i = 0, 6$). The number of states characteristic of a given system is a function of the number of parameters used to define the system and the range of values associated with each parameter. In this example, one parameter (trucks in the system) was used with a range encompassing the integer values from zero to six. Had two parameters been used, the states would be given as S_{ij}. The number of states would be $N_i \times M_j$, where N is the number of values of i and M is the number of value j can assume.

The queueing formulation of certain construction processes has been extended to include the concept of a storage or "hopper" in the system. This extension allows the server (i.e., loader in an earthmoving system) to store up productive effort in a buffer or storage during periods when units are not available for processing. In the context of the shovel-truck production system, this allows the loader to load into a storage hopper during periods when no trucks are available to be serviced directly.

Two parameters can be used to define hopper or storage-type systems. The i parameter, for example, could indicate the number of trucks in the system, and the j parameter could indicate the number of loads in the hopper or storage device. Therefore, if $i = 1, 2$, or 3 and $j = 1$ or 2, the system has six states, as follows:

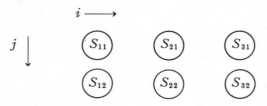

The utilization of states to define a queueing model greatly simplifies the writing of the equations that describe a particular situation.

Given a particular set of states, there is a probability P_i of being in one of those states S_i at time t and a set of probabilities T_{ij} of a transiting out of one state S_i to another state S_j. Therefore, a set of transition probabilities, T_{ij}, exists describing the chance of moving from S_i to S_j. The graphical equivalent of this is shown in Figure A.2. This shows the seven states of the finite system in Figure A.1 with the states shown as circles and the transition probabilities shown as links. Only those transition probabilities from state $S_n \to S_{n+1}$ or $S_n \to S_{n-1}$ are considered. In other words, consider only transitions moving from $S_3 \to S_4$ or $S_3 \to S_2$ for state S_3. In the context of the earth-hauling situation, this means that the system can only move from containing three trucks (S_3) to containing four trucks $(S_4$, if a truck arrives) or two trucks $(S_2$, if a truck departs). In fact, it might be possible to move from $S_3 \to S_0$, if three trucks leave the system simultaneously. Similarly, the system transits from $S_1 \to S_3$ if two trucks arrive simultaneously. Simultaneous arrivals or departures are called "*bulk*" arrivals or "*bulk*" departures. For mathematical simplicity, the assumption is made that Δt is defined small enough so that no more than one arrival or departure can occur in a given Δt. This assumption results in the diagram of Figure A.2. State diagrams including the transition probabilities as arcs are referred to as Markovian models.

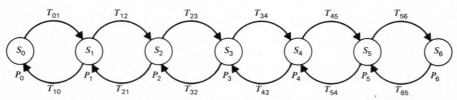

FIGURE A.2   State diagram with state (P_i) and transition $T_{(i,j)}$ probabilities.

A.3 MARKOVIAN MODELS

Markovian models are helpful in representing various situations in which a system moves from state to state based on a set of transition probabilities. Howard (1960) has presented a very clear characterization of the action of a Markovian process.

A graphic example of a Markov process is presented by a frog in a lily pond. As time goes by, the frog jumps from one lily pad to another according to his whim of the moment. The state of the system is the number of the pad currently occupied by the frog; the state transition is of course his leap.

In particular, Markovian concepts are helpful in analyzing queueing situations.

When the graphical Markovian model is properly defined, the process of writing the equations of state for the corresponding queueing model reduces to balancing the incoming and outgoing links. At any time t, the probability of being in S_i is specified as P_i. Therefore, at t there is a probability P_n of being in state n (i.e., S_n). There are several ways that the system can be in S_n following a short interval Δt (i.e., at $t + \Delta t$) as follows.

1. The system is in S_n at t and no arrivals occur and no departures occur during Δt.

2. The system is in S_n at t and one arrival occurs and one departure occurs.

3. The system is in S_{n+1} at t and one departure occurs during Δt.

4. The system is in S_{n-1} at t and one arrival occurs during Δt.

The first two cases cover the possibility of starting in a state n (S_n) and ending in that state (S_n). That is, the transition probability of starting in S_n and remaining in S_n is

$$T_{nn} = Pr \text{ (No arrivals)} \times Pr \text{ (No services)} \tag{1}$$

$$+ Pr \text{ (One arrival)} \times Pr \text{ (One service)}$$

In case 3, the transition probability of starting in a higher state ($n + 1$) and dropping down to S_n during Δt is given as

$$T_{(n+1)n} = Pr \text{ (No arrivals)} \times Pr \text{ (One service)} \tag{2}$$

Case 4 specifies the transition probability of starting in a lower state ($n - 1$) and moving up to S_n during Δt as

$$T_{(n-1)n} = Pr \text{ (One arrival)} \times Pr \text{ (No service)} \tag{3}$$

In general, the probability of being in state P_n at $t + \Delta t$ is given as

$$P_n(t + \Delta t) = P_n T_{nn} + P_{n+1} T_{(n+1)n} + P_{(n-1)} T_{(n-1)n} \tag{4}$$

If $n = 3$, for instance,

$$P_3(t + \Delta t) = P_3 T_{33} + P_4 T_{43} + P_2 T_{23} \tag{5}$$

A.4 EXPONENTIALLY DISTRIBUTED ARRIVAL AND SERVICE RATES

At this point, it is necessary to discuss the probabilities of arrivals and services in greater detail. As noted previously the assumption of exponentially distributed arrival and service times greatly simplifies the mathematical definition of a queueing system and the determination of the state probabilities $P_i(i = 0, M + 1)$.

The mathematical expectation of an arrival in the interval Δt is

$$\frac{1}{\lambda} = \int_0^\infty tf(t)\, dt$$

For exponentially distributed arrival times,

$$f(t) = \lambda e^{-\lambda t} \tag{6}$$

The probability that no arrival occurs in the interval $(0, t)$ is the same as the probability that the first arrival occurs after T.

$$P[t_a \geq T] = \int_T^\infty \lambda e^{-\lambda t}\, dt = e^{-\lambda t} \tag{7}$$

Therefore,

$$P\begin{bmatrix} \text{No arrival} \\ \text{in interval} \\ \Delta t \end{bmatrix} = P[t \geq \Delta t] = e^{-\lambda \Delta t} \tag{8}$$

The expression obtained by expanding this expression $(e^{-\lambda \, \Delta t})$ as a Taylor series is:

$$P\begin{bmatrix} \text{No arrival} \\ \text{in interval} \\ \Delta t \end{bmatrix} = e^{-\lambda \, \Delta t} = 1 - \lambda \, \Delta t + \frac{(-\lambda \, \Delta t)^2}{2!} + \frac{(-\lambda \, \Delta t)^3}{3!} + \cdots \tag{9}$$

If the higher-ordered terms are neglected (i.e., Δt very small) the expression reduces to

$$P\begin{bmatrix} \text{No arrival} \\ \text{in interval} \\ \Delta t \end{bmatrix} = 1 - \lambda \, \Delta t \tag{10}$$

Then the probability of an arrival in Δt is

$$P\begin{bmatrix} \text{One arrival} \\ \text{in interval} \\ \Delta t \end{bmatrix} = \lambda \, \Delta t \tag{11}$$

Since the assumption is made that the server times are exponentially distributed, the probability of no service completing in the interval $(0, T)$ is the probability that the service completes after T:

$$P[t, \geq T] = \int_T^\infty \mu e^{-\mu t}\, dt = e^{-\mu t} \tag{12}$$

where $1/\mu = $ the mathematical expectation of a service completion. Using the Taylor series expansion as above, the probability of no service in interval Δt becomes

$$P\begin{bmatrix} \text{No service} \\ \text{in interval} \\ \Delta t \end{bmatrix} = 1 - \mu\, \Delta t \tag{13}$$

and the probability of one service in t is

$$P\begin{bmatrix} \text{One service in} \\ \text{interval } \Delta t \end{bmatrix} = \mu\, \Delta t \tag{14}$$

A.5 INFINITE POPULATION QUEUEING MODELS

For systems with infinite populations and exponential arrival and server distributions, the transition probability defined in Equation 1 becomes

$$T_{nn} = \{(1 - \lambda\, \Delta t) \times (1 - \mu\, \Delta t)\} + \{(\lambda\, \Delta t) \times \mu\, (\Delta t)\}$$
$$= 1 - (\mu + \lambda)\, \Delta t + \lambda\mu\, (\Delta t)^2 + \lambda\mu\, (\Delta t)^2 \tag{15}$$

Again, neglecting the higher-order terms, since Δt is taken as very small, this reduces to

$$T_{nn} = 1 - (\mu + \lambda)\, \Delta t \tag{16}$$

The transition probabilities $T_{(n+1)n}$ and $T_{(n-1)n}$ become [substituting in (2) and (3)]

$$T_{(n+1)n} = \{(1 - \lambda\, \Delta t) \times (\mu\, \Delta t)\} \cong \mu\, \Delta t \tag{17}$$

$$T_{(n-1)n} = \{(\lambda\, \Delta t) \times (1 - \mu\, \Delta t)\} \cong \lambda\, \Delta t \tag{18}$$

Therefore, the probability of being in state n, $P_n(t + \Delta t)$, at $t + \Delta t$ becomes

$$P_n(t + \Delta t) = [1 - (\mu + \lambda)\, \Delta t]P_n + \mu\, \Delta t[P_{n+1}] + \lambda\, \Delta t[P_{n-1}]$$

or

$$P_n(t + \Delta t) = P_n(t) - (\mu + \lambda)P_n(t)\, \Delta t + \mu P_{n+1}(t)\, \Delta t + \lambda P_{n-1}(t)\, \Delta t \tag{19}$$

Transferring $P_n(t)$ to the left side of the equation and dividing by Δt, the expression is:

$$\frac{P_n(t + \Delta t) - P_n(t)}{\Delta t} = -(\mu + \lambda)P_n(t) + \mu P_{n+1}(t) + \lambda P_{n-1}(t) \tag{20}$$

Let $\Delta t \to 0$ (become infinitesimally small); then

$$\frac{d\,P_n(t)}{dt} = -(\mu + \lambda)P_n(t) + \mu P_{n+1}(t) + \lambda P_{n-1}(t) \tag{21}$$

If the system is assumed to be operating under steady-state conditions, then the probability of being in state n (i.e., P_n) does not change with time and

$$\frac{d\,P_n(t)}{dt} = 0$$

Therefore,

$$-(\mu + \lambda)P_n(t) + \mu P_{n+1}(t) + \lambda P_{n-1}(t) = 0 \tag{22}$$

Considering this situation as a Markovian chain, the model appears as shown in Figure A.3. The state probabilities have been associated with the "lily pads" in this diagram, and the frog leaps are set equal to λ for leaps going up one state and μ for leaps carrying the system down one state. Then Equation 22 can be written directly from the diagram by summing the flows into the node representing state n and setting them equal to the flows out. Flows into S_n are contingent on being in the state from which the flow originates (e.g., S_{n-1} or S_{n+1}). Therefore they are multiplied by the state probability of the state from which they originate. Flows out of S_n are contingent on being in S_n, and therefore are multiplied by P_n. Therefore, the balance becomes

$$\text{Flow out} = \text{Flow in}$$

$$\mu P_n + \lambda P_n = \mu P_{n+1} + \lambda P_{n-1}$$

or

$$-(\mu + \lambda)P_n + \mu P_{n+1} + \lambda P_{n-1} = 0$$

Using this concept, the diagram of state 0, S_0, is given by Figure A.4. The equation for P_1 in terms of P_0 is

$$\text{Flow out} = \text{Flow in}$$

$$\lambda P_0 = \mu P_1$$

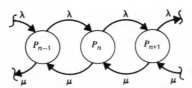

FIGURE A.3   State S_n in Markov chain.

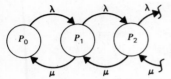

FIGURE A.4   State S_o in Markov chain.

or

$$P_1 = \frac{\lambda}{\mu} P_0 \tag{23}$$

This is also written $P_1 = \rho P_0$ where $\rho = \lambda/\mu$ is called the utilization factor. The general solution of the set of equations

$$\mu P_1 - \lambda P_0 = 0$$

and

$$\mu P_{n+1} + \lambda P_{n-1} - (\lambda + \mu) P_n = 0 \tag{24}$$

defining a queueing model with infinite population and exponentially distributed arrival and service times when solved in terms of P_0 becomes

$$P_n = \left(\frac{\lambda}{\mu}\right)^n P_0 \tag{25}$$

Since $1/\lambda = T_{\text{arrival}} =$ the mathematically expected or average arrival time, and $1/\mu = T_{\text{service}} =$ the mathematically expected or average service time, then

$$P_n = \left(\frac{\lambda}{\mu}\right)^n P_0 = \left[\frac{T \text{ service}}{T \text{ arrival}}\right]^n P_0$$

If, for instance, the average arrival time is 20 minutes and the average service time is 5 minutes, then

$$P_n = \left(\frac{5}{20}\right)^n P_0 = \left(\frac{1}{4}\right)^n P_0$$

It must be kept in mind that the equations and solution developed here are for a single channel infinite population system with exponentially distributed arrival and server times. It should also be kept in mind that the solution was obtained by setting $dP/dT = 0$ and therefore only applies to systems that are operating in a steady state. Solution of the equations for transient conditions is more mathematically complex and requires definition of the number of units in the system at time zero.

A.6 FINITE POPULATION QUEUEING MODELS

Finite population queueing models are of interest in construction, since in many situations a finite number of resources (a fleet of trucks, a crew of masons, etc.) are served by one or more resources in a cyclic fashion. This recycling of served units leads to a finite population model. For finite population systems with exponentially distributed arrival and service times, the Markovian graphical model can also be used.

The arrival transition probabilities must be modified, since the arrival rate in such systems is a function of the number of units outside of the system. That is, the rate of arrivals is proportional to the number of units that are external to the system and therefore in a mode that allows them to arrive. Returning again to the system shown in Figure A.1, three units are outside of the system and three are inside the system. The three units within the system cannot affect the arrival rate, since they are not in a mode in which they can arrive.

In this case, the arrival rate λ, which is the reciprocal of the average arrival time for one truck, $1/T_{\text{arrival (one truck)}}$*, must be multiplied by three to indicate the proportionality dependent on the number of trucks outside of the system. Therefore, the arrival rate of the system as shown is 3λ. If five trucks are on the back cycle, the arrival rate is 5λ. In general, the arrival rate of units entering the system is:

$$(M - i)\lambda$$

where

M = number of units in the finite population

i = number of units within the system

$\lambda = \dfrac{1}{T_{\text{av}}}$ where T_{av} if the average time a unit stays outside of the system

A Markovian model of the six-truck system shown in Figure A.2 is shown in Figure A.5. Again the state probabilities, P_i, have been associated with the state circles and the arcs represent the transition probabilities between states. The arrival rates have been modified to indicate the effect of units outside the system of any state. Therefore, the probability of a unit arrival within Δt when the system is in S_0 is 6λ. The comparable probability of a unit arrival when in S_5 is λ. The transit probability from $S_{n+1} \rightarrow S_n$ remains equal to μ. Using the method of equating inflows and outflows at each state node, $M + 1$

---

*The arrival time, T_{arrival}, in this situation is equal to the average "back cycle" time of a single truck. This is the average time a unit stays outside of the system.

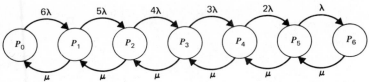

FIGURE A.5   Markov model for finite system ($M = 6$).

or 7 (seven) equations can be written. The equations written at each node in the model are as follows:

Node	Flow Out	=	Flow In
0 (S_0)	$6\lambda P_0$	=	μP_1
1 (S_1)	$(5\lambda + \mu) P_1$	=	$6\lambda P_0 + \mu P_2$
2 (S_2)	$(4\lambda + \mu) P_2$	=	$5\lambda P_1 + \mu P_3$
3 (S_3)	$(3\lambda + \mu) P_3$	=	$4\lambda P_2 + \mu P_4$
4 (S_4)	$(2\lambda + \mu) P_4$	=	$3\lambda P_3 + \mu P_5$
5 (S_5)	$(\lambda + \mu) P_5$	=	$2\lambda P_4 + \mu P_6$
6 (S_6)	μP_6	=	λP_5

It is possible to solve for the productivity of a finite queueing model such as the shovel-truck system by determining the probability that no units are in the system, P_0. Having determined P_0, the probability that units are in the system is $(1 - P_0)$, and this establishes the expected percent of the time the system is busy (i.e., productive). The production of the system is defined as:

$$\text{Prod} = L(1 - P_0)\mu C = L(\text{P.I.})\mu C \qquad (26)$$

where

μ = the processor rate (e.g., loads per hour)
C = capacity of the unit loaded
L = period of time considered
P.I. = Productivity index (i.e., the percent of the time the system contains units that are loading)

If, for instance, the P.I. is 0.65, the μ value is 30 loads per hour, the L value is 1.5 hours, and the hauler capacity is 15 cubic yards, the production value is

$$\text{Prod} = 1.5\ (0.65)\ 30\ (15) = 438.75 \text{ cubic yards}$$

The value of P_0 can be determined by writing the equations of state for the system and solving for the values of P_i ($i = 0, M$).

In addition to these equations, all state probabilities must sum to 1.0 and, therefore, the additional equation

$$\sum_{i=0}^{M} P_i = 1.0 \qquad (27)$$

is available. Since one of the node equations is redundant, this equation is substituted, providing the seventh equation required for solution of the P_i values $(i = 0, 6)$. Solving these equations in terms of P_0, the following values of the state probabilities result.

$$P_0 = 0.0121$$
$$P_1 = 0.0363$$
$$P_2 = 0.0906$$
$$P_3 = 0.1813$$
$$P_4 = 0.2719$$
$$P_5 = 0.2719$$
$$P_6 = 0.1359$$

The production of the system can be calculated using Equation 26. Assuming $\lambda = 6$ and $\mu = 12$, the production becomes

$$\text{Prod} = L(1 - P_0)\mu C \qquad (L = 1.5)$$

$$= 1.5(1 - 0.0121)12(15) = 266.73 \text{ cubic yards} \qquad (C = 15)$$

The general form of the solution for a finite system consisting of M units with exponentially distributed arrival and service times is:

$$P_0 = \left[\sum_{i=0}^{M} \frac{M!}{(M-i)!} \left(\frac{\lambda}{\mu}\right)^i \right]^{-1} \qquad (28)$$

$$P_i = \frac{M!}{(M-i)!} \left(\frac{\lambda}{\mu}\right)^i P_0$$

The verification of the values just calculated (with the node equation) using the general form Equations 28 is left as an exercise for the reader.

The value of P_0 can be reduced to nomograph format so that, given the values of λ and μ, the production index (P.I. $= 1 - P_0$) can be read directly from the chart. Such a nomograph for a system containing a single processor (e.g., loader) serving from 3 to 12 ($M = 3; 12$) arriving units is shown in Figure A.6. The assumed arrival and processing times are exponentially distributed. If $\mu = 12$, $\lambda = 4$, and $M = 4$, the P.I. value can be read on the x axis as 0.79. These nomographs for some typical queueing systems are given in Figures A.6, A.9, and A.12 (after Brooks and Shaffer, ref. 1).

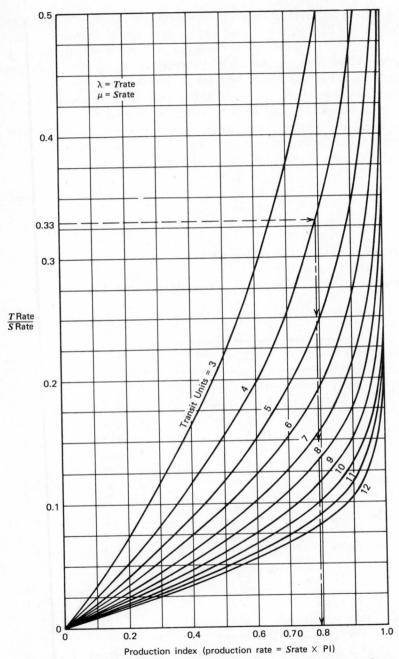

$\lambda = T\text{rate}$
$\mu = S\text{rate}$

$\dfrac{T\,\text{Rate}}{S\,\text{Rate}}$

Transit Units = 3

Production index (production rate = $S\text{rate} \times$ PI)

FIGURE A.6 Production forecast factors (Simple-Single-Server).

Queueing solutions to production problems involving random arrival and processing rates are of interest, since they allow evaluation of the productivity loss caused by bunching of units as they arrive at the processor. Analysis of this effect is not possible using deterministic methods, since they consider only the interference between units because of imbalances in the rates of interacting resources (e.g., truck and loader). This loss of productivity because of random transit times and its impact is discussed in Chapter Six.

In addition to the production value, it may be of interest to be able to estimate the mean number of trucks in the system. This can be easily developed with the calculated state probabilities, using the formulation:

$$\bar{N} = \sum_{i=0}^{M} P_i X_i \qquad (29)$$

where

$\bar{N}$ = mean number of units in the system
P_i = the probability of state i
X_i = number of units in the system state associated with P_i

Based on the information calculated above, the mean number of truck is

X_i	P_i	$P_i X_i$
0	0.0121	0
1	0.0363	0.0363
2	0.0906	0.1812
3	0.1813	0.5439
4	0.2719	1.0876
5	0.2719	1.3596
6	0.1359	0.8154
		$\bar{N} = 4.024$

Similarly, the average queue length can be calculated using the expression

$$\bar{Q} = \sum_{i=1}^{M} P_i(X_i - 1) \qquad (30)$$

where $\bar{Q}$ = mean number of trucks in the queue.

This expression reflects that in a single-server system, if one or more units are in the system at a given time, one of them is being processed. Therefore, the number of units in the queue is $(X_i - 1)$. Using this formulation, the mean number of trucks in the queue for the six-truck system is:

X_i	P_i	$(X_i - 1)$	$P_i(X_i - 1)$
1	0.0363	0	0
2	0.0906	1	0.0906
3	0.1813	2	0.3626
4	0.2719	3	0.8157
5	0.2719	4	1.0876
6	0.1359	5	0.6795
			$\bar{Q} = 3.036$

Information regarding the percent of the time that more than n units $(n < M)$ are in the system or the queue can be calculated by summing the appropriate subcomponents of the above expressions.

A.7 MULTISERVER FINITE POPULATION MODELS

In the finite system just considered, only a single-server channel was defined. If the system of Figure A.1 is modified slightly to have two loaders, a multiserver system (number of channels = 2) is defined. A slight modification of the Markov representation allows the writing of the steady-state equations for this system. The revised model is shown in Figure A.7.

The modification of the model relates to the transition probabilities associated with shifts from a higher to a lower state (i.e., $S_{n+1} \rightarrow S_n$). Since two loaders are available, the probability of transiting down from states containing two or more units $(S_n \geq S_2)$ is 2μ instead of μ. This indicates that the probability of a service completion in Δt when two loaders are defined is twice that when using only one loader. Similarly, if three loaders had been defined, the probability of downshifts for states containing three or more units would be 3μ. The model for a three-server system is shown in Figure A.8. It should be noted that the multiplier associated with μ in specifying downshifts cannot

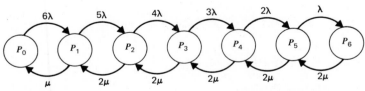

FIGURE A.7   Markov model for two-server system.

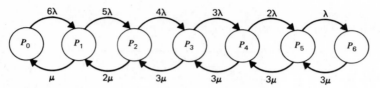

FIGURE A.8   Markov model for three-server system.

exceed the number of units in the state from which the shift occurs. In other words, the transition probability of downshifting from P_2 cannot be 3μ, since this implies that all three loaders are active with only two units in the system. Obviously, only two loaders can be active and, therefore, the transition probability is 2μ.

Returning to the two-server model, the equations of state for this system become:

Node	Out Flow	=	In Flow
0 (S_i)	$6\lambda P_0$	=	μP_1
1 (S_1)	$(5\lambda + \mu)P_1$	=	$6\lambda P_0 + 2\mu P_2$
2 (S_2)	$(4\lambda + 2\mu)P_2$	=	$5\lambda P_1 + 2\mu P_3$
3 (S_3)	$(3\lambda + 2\mu)P_3$	=	$4\lambda P_2 + 2\mu P_4$
4 (S_4)	$(2\lambda + 2\mu)P_4$	=	$3\lambda P_3 + 2\mu P_5$
5 (S_5)	$(\lambda + 2\mu)P_5$	=	$2\lambda P_4 + 2\mu P_6$
6 (S_6)	$2\mu P_6$	=	λP_5

Again, since one equation is redundant, the equation $\sum_{i=0}^{M} P_i = 1.0$ is used to provide the $(M + 1)$th equation required to solve for all state probabilities. The state probabilities calculated for this system by solving the set of linear equations with $\lambda = 6$ and $\mu = 12$ are:

$$P_0 = 0.0622$$

$$P_1 = 6\left(\frac{\lambda}{\mu}\right)P_0 = 0.1867$$

$$P_2 = 15\left(\frac{\lambda}{\mu}\right)^2 P_0 = 0.2333$$

$$P_3 = 30\left(\frac{\lambda}{\mu}\right)^3 P_0 = 0.2333$$

$$P_4 = 45 \left(\frac{\lambda}{\mu}\right)^4 P_0 = 0.1750$$

$$P_5 = 45 \left(\frac{\lambda}{\mu}\right)^5 P_0 = 0.0875$$

$$P_6 = 22.5 \left(\frac{\lambda}{\mu}\right)^6 P_0 = 0.0219$$

The calculation of productivity for this system is also modified, since there are two loaders. During the periods that two more units are in the system, the productivity is a function of 2μ. During the periods when only one unit is in the system (i.e., S_1), the production is constrained to μ.

Therefore, the expression for the production of this system is:

$$\text{Production} = \{\mu P_1 + 2\mu[P_2 + P_3 + P_4 + P_5 + P_6]\} CL$$

Using the same assumptions used previously regarding C and L leads to a productivity value for the two-server system of

$$\text{Prod} = \{12(0.1867) + 2(12)(0.7511)\}(15)(1.5)$$
$$\cong 456 \text{ cubic yards}$$

The productivity expression for three-server and multiserver systems generally must reflect the fact that the weighing factor used for μ in production calculations cannot exceed the n value of the state probability in multiples. In other words, a state that contains M units cannot yield a production of $N\mu$ if $M < N$. Therefore, the productivity expression for the three-server system becomes:

$$\text{Prod} = \{\mu P_1 + 2\mu P_2 + 3\mu(P_3 + P_4 + P_5 + P_6)\} CL$$

As an exercise, the reader should write the equations of state and solve for the production of the three-server system shown in Figure A.8.

As with the single server, nomographs that give the productivity index as a function of the number of servers, number of transit units, and utilization factor (λ/μ) are available. The appropriate chart for the two-server system is shown in Figure A.9. Entering this chart with the utilization factor (TRATE/SRATE) of 6/12 (0.5), the productivity index (P.I.) can be read as 1.70. System productivity is given as:

$$\text{Production} = (\text{P.I.}) \, \mu CL = (1.7)(12)(15)(1.5) = 459 \text{ cubic yards}$$

Methods similar to those for the single-server model can be used to calculate the mean number of units in the system and in the queue.

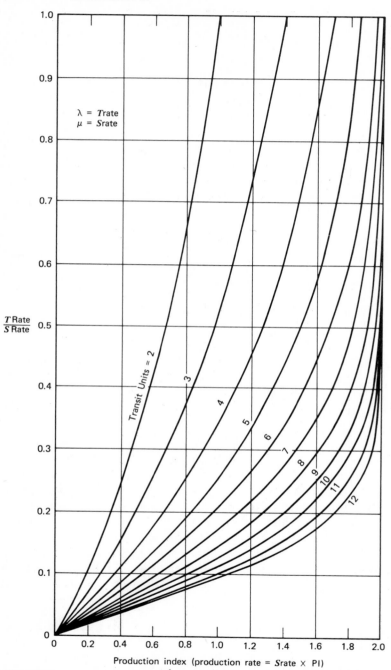

FIGURE A.9   Production forecast factors (Simple-Two-Server).

A.8 FINITE MODELS WITH STORAGE

Often in construction processes it is advantageous to store the productivity of the server, so that the server need not be idle while transit units are not available for service. This storage of effort allows a faster service rate when transit units do become available. The classical example of this in construction is the hopper used in earthmoving operations.* In fact, most storage towers or hoppers can be viewed as queueing models in which the server's productivity is stored. In the earthmoving situation, a loader lifts material into a storage tower until the capacity of the tower is reached. Upon the arrival of a truck unit, the material is released from the tower at an accelerated rate. Similar situations arise in concrete batching and transportation as well as in masonry operation. Moreover, any construction process in which the effort of one resource is stored temporarily pending its transfer can potentially be modeled as a hopper-type situation.

Again, it is helpful, in analyzing a system with storage, to develop the system states. As mentioned previously, two parameters are used to define the state of a system that includes a hopper. The number of units in the system is still of interest and must be defined. In addition, the number of loads or services in the hopper must also be considered. Therefore, the state is defined as S_{ij} where

$$i = \text{number of transit units in the system}$$

$$j = \text{number of service loads in the storage}$$

The parameter i has the range of integer values from 0 to M, where M is the number of transit units defined. The parameter j has the range of integer values from 0 to H, where H is the capacity of the storage in service loads. Some typical hopper situations are shown in Figure A.10.

In the first instance a tower with a capacity of four truck loads is used ($H = 3$). Therefore, the range of j is $(0, 3)$. In the second example, brick packets are stored on a scaffold. The space on the scaffold for stacking brick acts as a hopper in storing the effort of the labor resupplying bricks. In this case (as shown), the space available allows storage of six packets and, therefore, the hopper capacity is six ($H = 6$).

Consider a system in which the six trucks originally defined in Figure A.1 are serviced by one loader. This loader can either serve them directly (i.e., drop material directly through the hopper into the truck bay), or it can store loads in a hopper tower. Assume the capacity of the tower is two loads ($H = 2$). The symbols for arrival and server rates are λ and μ. The accelerated rate of loading afforded by the hopper is γ (gamma). The distribution of all times are described by the parameters $1/\lambda$, $1/\mu$, and $1/\gamma$, and are all assumed

---

*Because of this fact, storage systems in construction are commonly referred to as hopper systems.

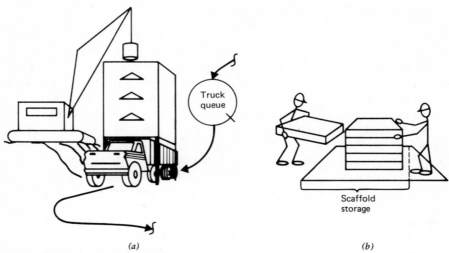

(a)

(b)

FIGURE A.10 Typical storage system.

to be exponential. The number of states or "lily pads" between which this system transits is defined as $(H + 1) \times (M + 1) = 21$. Again, a Markovian model is handy in developing the state equations for this hopper system (see Figure A.11).

The addition of the j parameter makes this model more complex than those previously developed. The highest row of the model is identical to the single-server model shown in Figure A.5 since, when the hopper is empty ($j = 0$), the system is actually a single-server system. However, when there are no trucks in the system ($i = 0$), the loader loads the hopper (rate = μ). There-

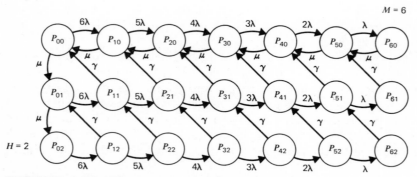

FIGURE A.11 Hopper system Markovian model.

fore, the arc representing loading of the hopper at rate μ, links P_{00} and P_{01}. Additionally, truck arrivals when the situation is in state P_{01} cause jumps to the right down the second row of states (S_{i1}, $i = 0$, 6). Should no truck arrive in a small Δt, the system can transit up to P_{02} following an additional load that fills the hopper to capacity. Once the system has arrived in any of the states in the middle or bottom rows, the loading of trucks is accomplished using the hopper. Therefore, the rate of loading is γ and the system transits from S_i, to $S_{i-1,j-1}$, in other words, up and to the left, as shown in Figure A.11.

The equations of state are written again by equating inflows and outflows at each node. This leads to 21 equations, one of which is redundant. The 21 state probabilities are calculated using 20 node equations and the equation summing probabilities to 1.0. The expression for production of the hopper system is given as

$$\text{Production} = \left\{ \mu \left(\sum_{i=1}^{M} P_{i0} \right) + \gamma \left(\sum_{j=1}^{H} \sum_{i=1}^{M} P_{ij} \right) \right\} CL$$

This expression reflects the fact that production occurs by loading out of states, causing a downshift in i. Downshifts occur in the top row at rate μ. In the other rows the rate of downshift and, therefore, of production is γ.

Fortunately, nomographs are available that allow determination of the productivity index for hopper systems. These charts use the factor λ/γ and the μ/γ factor to determine the P.I. The expression for production using these charts is:

$$\text{Production} = (\text{P.I.})\gamma\,CL$$

The nomograph for a system with $H = 2$ and $M = 6$ is shown in Figure A.12. Assuming $\gamma = 24$, and the same λ and μ values used previously ($\mu = 12$, $\lambda = 6$), Figure A.12 gives P.I. = 0.5. Therefore, system production is calculated as:

$$\text{Production} = (0.5)(24)(15)(1.5) = 270 \text{ cubic yards}$$

This indicates that the hopper does add a small amount of production to the single server without hopper. However, since the utilization factor of the loader is already high, the production increase using the hopper is not great.

Although limited in scope and in the range of field problems it can handle, queueing theory does provide a good vehicle for introducing some concepts basic to modeling construction operations. The concepts of unit flows and storage, system states, delays, and processing are fundamental both to queueing systems and to the modeling of the more complex dynamics of construction processes. These concepts can be introduced by considering some basic types of queueing models presented within the content of construction processes.

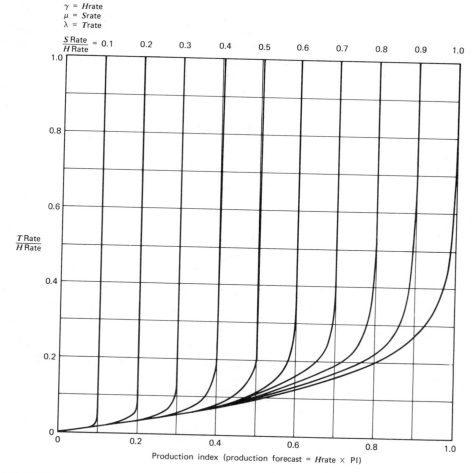

FIGURE A.12 Hopper-Single-Server production forecast factors. Hopper capacity = 2, transit units = 6.

Having developed a framework for considering processing and delay situations, the basic elements of a more relevant construction modeling environment can be defined.

A.9 FIELD APPLICATION

To calculate production using field data, it is necessary to arrive at values for λ and μ. If a hopper system is used, a value for γ is also required. These parameters must be developed from field observations. Typical data for a single server system in which five* trucks haul to a dump location from a single loader are given in Table A.1. This data indicates the sequence in which trucks arrived at the loader, the time of arrival, the time at which loading commenced, and the time of departure from the loader. As given, it is in minutes and seconds. Therefore, from the data, truck 1 arrives back in the system following its first cycle at 14 minutes 6 seconds.

For a finite queueing situation such as this, λ is the "back cycle" time. This is simply the time between departure from the system and reentry. Therefore the calculation reduces to simply:

$$\bar{A} = \frac{\sum_{j=1}^{M} \sum_{i=1}^{N_j} (AQT_{ij} - ELT_{(i-1)j})}{\sum N_j}$$

where

$\bar{A}$ = average arrival time
AQT_{ij} = the arrive-in-queue time for unit j on cycle i
$ELT_{(i-1)j}$ = the end-of-load time for unit j on the $(i-1)$ cycle
M = the number of units $(j = 1, M)$
N_j = the number of cycles for unit j $(i = 1, N_j)$

That is, the average arrival time for a single unit is the summation for all trucks of all back cycle times divided by the total cycles for all trucks. The back cycle for an individual truck is simply the difference between the entry time for the ith cycle and the departure time for the previous cycle $(i - 1)$.

To maintain the assumptions of the model, the calculation of μ must be made most carefully. Two situations are possible. If the queue is empty at the time of a unit arrival, there is no delay, and the time for commencement of loading and arrival in the system must be the same so as to maintain the assumptions of the model. The data do not always reflect this. The record of the arrival of truck 1 on its 10th cycle is the first entry in row 7 of the data. The queue at the time of arrival is empty. This can be established from the fact that the previous truck(s) departed at 150 minutes 21 seconds, and there is no loading until 1 arrives at 156:30.† According to the assumption of the model, loading should begin immediately on arrival. However, the actual data indicates a maneuver delay such that loading does not commence until 157:04. Therefore, the processing time in this case is:

$$S_{ij} = ELT_{ij} - AQT_{ij}$$

where

S_{ij} = loading time of unit j on cycle i
ELT_{ij} = the end of load time for unit j on cycle i
AQT_{ij} = the arrival in queue time of unit j on cycle i

*Although six trucks are indicated, truck 6 is deleted, since it cycles only twice.
†M : S. M = minutes, S = seconds.

Table A.1 FIELD DATA FOR EARTH HAULING PROBLEM
(Data from O'Shea *et. al.*, 1964)

Truck number	1	2	3	4	5	6	1	2
Arrive	0.00	0.00	0.00	9.08	9.20	9.30	14.06	16.17
Load	0.00	2.20	4.15	9.34	11.28	13.44	16.29	18.27
End of load	1.45	3.47	5.49	11.01	13.29	16.04	18.02	20.08

Truck number	3	4	5	1	3	2	6	5
Arrive	16.42	21.38	24.29	30.48	32.58	33.31	34.20	36.55
Load	20.35	22.45	26.04	31.23	34.40	37.01	39.20	42.04
End of load	22.10	25.14	28.45	33.46	36.28	38.51	41.31	44.27

Truck number	4	3	1	2	4	3	1	5
Arrive	38.47	45.55	46.28	50.07	59.00	59.37	60.55	62.43
Load	44.57	47.02	49.16	51.20	59.30	61.43	63.50	65.50
End of load	46.32	48.44	50.45	53.00	61.06	63.20	65.18	68.05

Truck number	2	4	3	1	5	2	4	3
Arrive	64.28	73.40	74.08	75.52	76.58	80.40	88.30	91.42
Load	69.21	74.19	76.54	79.20	81.34	84.40	89.02	92.50
End of load	70.56	76.22	78.44	81.03	84.07	86.45	90.27	95.18

Truck number	1	5	2	4	3	1	5	2
Arrive	92.24	93.45	96.58	106.26	107.55	110.11	110.45	115.11
Load	95.50	98.40	101.04	107.04	109.45	112.58	115.45	119.00
End of load	97.48	100.29	103.02	109.11	112.17	115.00	118.29	120.42

Truck number	4	3	5	1	2	3	4	5
Arrive	123.50	125.00	128.41	141.34	142.14	144.24	145.27	145.35
Load	125.05	128.38	132.01	141.47	143.12	145.07	146.52	148.39
End of load	127.54	131.20	135.38	142.36	144.32	146.23	148.10	150.21

Truck number	1	2	3	5	4	1	2	3
Arrive	156.30	156.43	158.52	159.41	159.57	170.18	172.36	174.56
Load	157.04	159.37	161.29	163.11	165.30	171.17	173.06	175.25
End of load	158.25	161.04	162.44	165.04	166.49	172.38	174.28	176.52

Truck number	4	1	2	3	4	1	2	3
Arrive	172.55	185.35	186.09	189.34	200.06	201.04	202.18	203.47
Load	184.25	186.17	188.10	190.00	200.52	204.00	206.07	209.57
End of load	185.49	187.37	189.29	191.19	203.36	205.27	208.30	210.47

Truck number	4	1	2	5	3
Arrive	217.38	218.16	220.24	221.40	222.50
Load	218.14	220.16	222.49	224.50	226.54
End of load	219.44	221.54	224.21	226.27	228.13

If the queue is not empty, the processing time becomes

$$S_{ij} = ELT_{ij} - ELT_{\text{(preceding unit)}}$$

Again, the controlling factor is not the begin load time as given in the data, since the model assumes that the begin load time of a waiting unit is assumed to occur simultaneously with the departure of its preceding unit.

To illustrate this, consider the following observations that have been converted from minutes and seconds to decimal values.

Truck Number	1	2	3
AQT	50.25	60.20	62.61
Begin load	50.75	60.55	63.62
ELT	57.80	63.15	65.35

All observations are for the sixth cycle of the truck listed. The correct model values are:

$$S_{62} = 63.15 - 60.20 = 2.95 \text{ minutes}$$

$$S_{63} = 65.35 - 63.15 = 2.20 \text{ minutes}$$

This amounts to a modification of the raw field data to accommodate the assumptions of the model.

One other modification of data is required to maintain the assumptions of the model. As noted previously, the model assumes a FIFO queue discipline. However, in the field, "breaks" in discipline sometimes occur. This happens when a unit arrives in the system but fails to process in the sequence of arrival. In effect, a later-arriving unit passes it and begins loading out of sequence, violating the FIFO criterion.

To illustrate this, consider the cycles of trucks 3 and 4 at the end of row 7 and beginning of row 8. Truck 3 arrives at 174:56 versus a 172:55 arrival time for truck 4. However, truck 3 loads at 174:56, while truck 4 does not load until 184:25. Obviously, a queue discipline break has occurred, and some assumption compensates for this. One approach is to transfer the unexplained delay from the in system delay to the back cycle time. The assumption is made that truck 4 arrives after truck 3. This is accomplished by modifying the AQT for truck 4 to a value greater than the AQT for truck 3. In this case, the $AQT(4)$ is modified to $AQT(3) + 1$ second or 174:57.

The implementation of these modifications is simplified using the worksheet shown in Table A.2. The form allows running account of the empty-occupied status of the queue, thereby indicating which basis to use in calculating μ_{ij}. The worksheet shows observations for truck 1. Similar sheets are

Table A.2 WORKSHEET FOR FINITE QUEUE ANALYSIS

Truck number (1)	Trip Number (2)	Truck Enters System (3)	Service Completed on Truck Entering Immediately Before (4)	Queue—Yes	Queue—No	Truck Exits System (5)	Queue—Yes Service Time (6) = (5) − (4)	Queue—No Service Time (7) = (5) − (3)	Truck Reenters System (8)	Back Cycle Time (9) = (8) − (5)
1	1	0	0		X	1.75	—	1.75	14.10	12.35
1	2	14.10	16.07	X		18.03	1.96	—	30.80	12.77
1	3	30.80	28.75		X	33.77	—	2.97	46.47	12.70
1	4	46.47	48.73	X		50.75	2.02	—	60.92	10.17
1	5	60.92	63.33	X		65.30	1.97	—	75.87	10.57
1	6	75.87	78.73	X		81.05	2.32	—	92.40	11.35
1	7	92.40	95.30	X		97.80	2.50	—	110.18	12.38
1	8	110.18	112.28	X		115.00	2.72	—	141.57	26.57
1	9	141.57	135.63		X	142.60	—	1.03	156.50	13.90
1	10	156.50	150.35		X	158.42	—	1.92	170.30	11.88
1	11	170.30	166.82		X	172.63	—	2.33	185.58	12.95
1	12	185.58	185.82	X		187.62	1.80	—	201.07	13.45
1	13	201.07	203.60	X		205.45	1.85	—	218.27	12.82
1	14	218.27	219.73	X		221.90	2.17	—	Never	N/A

End data

$$\sum = 19.31 \quad \sum = 10.00$$

$$\sum = 29.31$$

$$\sum = 173.86$$

$$T_s = \sum_{i=1}^{n} t_{si} \div n = \frac{29.31}{14}$$

$$T_a = \sum_{i=1}^{n} t_{bc_i} \div n = \frac{173.86}{13}$$

$$T_s = 2.0936 \quad \mu = \frac{1}{T_s} = 0.47764 \text{ lds/min}$$

$$T_a = 13.3738 \quad \lambda = \frac{1}{T_a} = .07477 \text{ arr/min}$$

maintained for all trucks in the system. A summary of the calculations of λ and μ is shown in Table A.3. Based on the calculated values of $\bar{A}$ and $\bar{S}$,

$$\rho = \frac{\bar{S}}{\bar{A}} = \frac{2.5}{13.1} = 0.19$$

Entering the nomograph with this value, the productivity index is 0.71 yielding a production value of:

$$\text{Production} = \mu(\text{P.I.})LC = \frac{60}{2.5}(0.71)(1)(1)$$

$$= 17.04 \text{ truck loads per hour}$$

This is close to the actual production as reported in O'Shea et al., 1964.

A.10 SHORTCOMINGS OF THE QUEUEING MODEL

This illustration of the field application of the queueing model reveals several difficulties encountered in its utilization as a precise method of forecasting production. The datum used in this example is one of 15 sets used to examine the application of these methods to forecasting shovel-truck fleet production. Two difficulties requiring correcting assumptions have already been noted.

1.   Breaks in queue discipline.

2.   The difference between observed begin load times and the assumption of the model.

Furthermore, the assumption of exponentially distributed arrival and server times did not obtain in the field. This is indicated clearly in Figures A.13 and A.14, which show scaled comparisons of representative sets of observed data versus the assumed exponential distributions. Figure A.13 shows three sets of arrival time data plotted versus the assumed exponential distribution. Similarly, two sets of load time data are compared to the exponential plot in Figure A.14. The method of scaling using λt and μt as scalers is illustrated by Table A.4. The data as observed approximate a lognormal distribution. This is consistent with findings reported by Teicholz.

In addition to these disparities between observed data and model assumptions, it is doubtful if the assumption of steady-state operation is justified. The sets of data from which the example was taken were collected during half shift periods of 4 to 5 hour durations. Simulation studies of queueing systems similar to the example indicate that it takes a longer period to "settle down" and reach a steady-state level of operation.

Modification of the model to correct for some of the differences in field and theoretical performance can be achieved. It is possible to utilize lognormal

Table A.3 A SUMMARY OF μ AND λ CALCULATIONS

Truck	Trips	T_a	λ	Loadings	T_s	μ	Sequence
1	13	13.37	4.486	14	2.09	28.659	N/A*
2	13	12.49	4.803	14	2.34	25.618	N/A
3	14	11.86	5.056	15	2.33	25.773	$T_a = 11.723$ $T_s = 2.359$
4	12	12.97	4.624	13	2.98	20.161	$T_a = 13.145$
5	10	15.23	3.939	11	2.89	20.729	N/A
6	1†	34.33	N/A	2	2.63	22.817	N/A
	$\sum = 63$	$\overline{T}_a = 13.07$	$\overline{\lambda} = 4.59$	$\sum = 69$	$\overline{T}_s = 2.50$	$\bar{\mu} = 24.0$	

Notes. 1. Trips = Round trips. 2. λ is calculated in arrivals per hour. 3. T_s = mean service time in minutes; T_a = mean arrival time in minutes. 4. The sequence column indicates the corrected values of T_a and T_s when breaks in the FIFO discipline occurred in the observed data.

*N/A = Not applicable.

†This row (truck 6) not included in $\overline{T}_a$ or $\overline{\lambda}$ calculations.

478

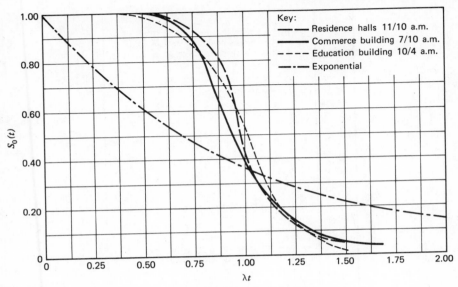

FIGURE A.13   Typical actual arrival distributions.

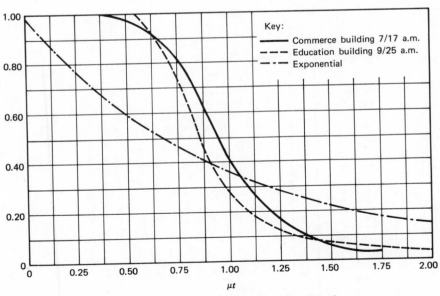

FIGURE A.14   Typical actual loading time distributions.

Table A.4 TYPICAL ACTUAL ARRIVAL DISTRIBUTIONS after O'Shea *et. al.*, 1964.

$\lambda = 3.13$ Arrivals per Hour Commerce Building 7/17 A.M.				$\lambda = 2.27$ Arrivals per Hour Residence Halls 11/10 A.M.				$\lambda = 4.31$ Arrivals per Hour Education Building 10/4 A.M.			
t_{min}	λt^*	N	$S_0(t)$	t_{min}	λt^*	N	$S_0(t)$	t_{min}	λt^*	N	$S_0(t)$
30	1.566	2	0.048	35	1.325	3	0.094	20	1.436	1	0.024
25	1.305	5	0.119	33	1.249	6	0.190	18	1.293	5	0.122
22.5	1.175	9	0.214	29	1.098	8	0.250	17	1.220	7	0.171
20.0	1.044	12	0.286	28	1.06	10	0.313	16	1.150	10	0.244
19.0	0.992	16	0.381	27	1.022	13	0.406	15.5	1.113	12	0.293
18.0	0.940	22	0.524	26	0.984	18	0.563	15.0	1.08	16	0.390
17.0	0.890	28	0.666	25	0.946	21	0.656	14.0	1.01	20	0.488
16.0	0.835	30	0.714	24	0.909	25	0.787	13.5	0.970	23	0.561
15.0	0.783	36	0.857	22	0.833	28	0.875	13	0.934	27	0.659
14.0	0.731	39	0.928	20	0.757	29	0.906	12	0.862	32	0.780
13.0	0.679	40	0.952	14	0.530	32	1.000	10	0.718	35	0.854
11.0	0.574	41	0.976					8	0.575	39	0.951
9.0	0.470	42	1.000					5	0.359	41	1.000

$^*\lambda t$ = arrivals per hour.

arrival and server processes more closely approximating field observations.*
It is also possible to solve the queueing model for the start-up or transient
phase of operation. In such solutions, $dP/dt \neq 0$ but is a function of t. However, solution of the modified model is mathematically complex and not amenable to simple nomographic presentation. It is also doubtful that the increase
in precision afforded by such techniques is warranted. The relatively simple
two-cycle (dual link) interacting systems that queueing theory addresses can,
in most cases, be adequately handled by deterministic methods, as described
in Chapter Six. A familiarity with the basic concepts of queueing theory is,
however, helpful in understanding the dynamics of more complex systems in
which chains of queues interact.

In particular, the concepts of idleness and delay, flow and activity are
required to develop viable models for construction management. The discrete
unit flow nature of queueing systems provides a springboard for better understanding of unit processes as they occur on the job site.

REFERENCES

1. Brooks, A. C. and L. R. Shaffer, "Queueing Model for Production Fore-
casts of Construction Operations," Unpublished report, Department of Civil
Engineering, University of Illinois, Urbana, Ill., undated.

2. Gaarslev, Axel, "Stochastic Models to Estimate the Production of Material Handling Systems in the Construction Industry," *Technical Report No.
111*, The Construction Institute, Stanford University, Palo Alto, Calif., August
1969.

3. Gordon, G., *System Simulation*, Englewood Cliffs, N.J., Prentice-Hall,
Inc., 1969.

4. Howard, Ronald A. *Dynamic Programming and Markov Processes*,
Cambridge, Technology Press of Massachusetts Institute of Technology, 1960.

5. Jackson, J. R., "Networks of Waiting Lines," *Operations Research*, 5 (4),
August 1957, pp. 518–521.

6. Kiviat, P. J., "GASP—A General Simulation Program," Applied Research Laboratory, United State Steel Corporation, Monroeville, Pennsyl-
vania, July, 1963.

7. Mangelsdorf, T. M., "Waiting-Line Theory Applied to Manufacturing
Problems," Master's Thesis, Massachusetts Institute of Technology, Cambridge, 1955.

*For a more detailed discussion, see Gaarslev, 1969.

8. Martin, G. C., "Hopper-Truck Queueing Theory Model," Unpublished report, Department of Civil Engineering, University of Illinois, Urbana, Ill., undated.

9. Morse, P. M., *Queues, Inventories and Maintenance*, New York, Wiley, 1958.

10. Naylor, T. H., et al., *Computer Simulation Techniques*, New York, Wiley, 1966.

11. O'Shea, J. B., G. N. Slutkin, and L. R. Shaffer, "An Application of the Theory of Queues to the Forecasting of Shovel-Truck Fleet Productions," *Construction Research Series No. 3*, Department of Civil Engineering, University of Illinois, Urbana, Ill., June 1964.

12. O'Neill, R. R., "An Engineering Analysis of Cargo Handling vs. Simulation of Cargo Handling Systems," *Report 56-37*, Department of Engineering, University of California, Los Angeles, September 1956.

13. Palm, C., "The Distribution of Repairmen in Servicing Automatic Machines," *Industritidningen Norden*, 75, 1947, pp. 75–80.

14. Rainer, R. K., "Predicting Productivity of One or Two Elevators for Construction of High-Rise Buildings," Ph.D. Dissertation, Auburn University, Auburn, Ala., 1968.

15. Spaugh, J. M., "The Use of the Theory of Queues in Optimal Design of Certain Construction Operations," *Technical Report of National Science Foundation (NSF G 15933)*, Department of Civil Engineering, University of Illinois, Urbana, Ill., 1962.

16. Teicholz, P., "A Simulation Approach to the Selection of Construction Equipment," *Technical Report No. 26*, Construction Institute, Stanford University, June 1963.

PROBLEMS

1. Uncle Fudd has a loader and three trucks. He wants to know the production of this system of haulers. From information he provides, you determine the loader rate, μ, to be 21 loads per hour and the truck arrival rate, λ, to be 6 units per hour. Draw the Markovian model for this system, write the equations of state, and solve for the state probabilities. What is the productivity of the system? Check your findings using the solution nomographs (Appendix D).

2. The loader in problem 1 is too expensive, and Uncle Fudd gets a smaller one. The smaller loader has a rate of 12 loads per hour. Using the savings from the sale of the more expensive loader, he buys another truck, expanding his haul fleet to four trucks. To offset the decrease in loader rate, he installs a hopper to increase production. The hopper rate is 30 loads per hour and the truck arrival rate remains six trucks per hour. Draw the Markovian model for this situation, write the equations, and solve for the state probabilities. What is the system production?

3. Ripp Co., Inc. must move a crane to unload precast concrete wall panels and wants to estimate how long it will tie up the crane. The panels will be transported from the casting yard using the company's four trucks, which can haul two panels per trip. Mean times are as follows.
   (a) Twenty-four minutes to unload two panels.
   (b) Nineteen minutes to travel to casting yard (empty).
   (c) Ten minutes to load two panels.
   (d) Thirty-one minutes to travel (unloaded) to the job site.
   Using a queueing formulation, determine how many 8-hour shifts with the crane will be required to unload 100 panels.

4. (a) Five masons are supported by two laborers. The laborers carry bricks to the masons at a rate of one packet containing 10 bricks every 2.2 minutes on the average. The placement rate by the masons averages 10 bricks every 6.2 minutes. There is no space on the scaffold for temporary storage. Solve the system of state equations and determine the productivity of the system.
   (b) Using the situation as described above, assume that one laborer is now on vacation and the scaffold has been modified to allow stacking of brick packets on it. Calculate the hourly production of this modified system and hence determine the stack storage required (and therefore the laborer overtime) to ensure that the single laborer can support the masons during their 8-hour work shift.

5. The CONJOB Construction Co. is completing a job in which pouring of the basement slab was delayed until the building was nearly complete. The pour will be made by wheelbarrows filled from a bin that is fed from a chute (elephant trunk) through one of the windows. The bin holds a maximum of 6 cubic units of concrete. The mixer on site is capable of producing 72 cubic units of concrete per hour, and the wheelbarrows will be filled with 2 cubic units per trip. The rate of loading the wheelbarrows from the bin is one every 20 seconds. The floor pour requires 360 cubic units of concrete. There are five wheelbarrows available and the travel cycle is 8 minutes. Construct the Markovian model for this system and write the equations of state. Calculate the productivity of the system using the appropriate nomograph (Appendix D).

6. A lake-front coffer dam is to be constructed to expedite the construction of a pier. The coffer will be constructed using a double row of sheet pile filled with granular material to form a coffer wall. The lateral stability for the sheet pile walls will be developed by tying them back to one

another. The tie installation is to be accomplished using two cranes and eight installation crews. The mean time for initial installation of the rods is 10 minutes. During this time the rods are spot welded in place and the crane is required to hold them in position. Following this, final welding of the rods is completed, requiring the crew for an additional 25 minutes, on the average. The crane, however, is free to work on other rods during this period. How many rods can be installed per hour? (Problem as described in Brooks and Shaffer, Ref. 1.)

7. The field data given below is similar to that shown in Table A.1. Using a form similar to the one shown in Table A.3, calculate mean μ and λ values for each truck. For the system described, determine the system productivity in truck loads per hour. The values given are in the format M. S where M is minutes and S is seconds. A single loader is utilized. What is the mean length of the truck waiting line?

Truck number	1	2	3	4	5	6	7	8
Arrive	0.00	0.00	0.00	6.10	7.05	8.45	14.01	14.52
Load	0.18	5.28	3.16	9.39	7.55	11.31	14.36	16.31
End of load	2.03	7.28	4.50	11.01	9.21	13.17	16.05	17.55

Truck number	1	3	9	5	2	4	6	7
Arrive	15.40	20.29	22.12	24.38	23.20	27.56	33.10	34.27
Load	18.24	21.38	24.18	29.20	26.51	31.50	33.50	36.33
End of load	19.43	23.50	26.33	30.46	28.50	33.49	36.15	38.02

Truck number	8	1	3	4	2	5	4	6
Arrive	35.00	36.04	41.40	41.30	45.46	46.35	51.41	55.10
Load	38.28	40.19	41.59	45.15	48.33	50.35	55.24	57.25
End of load	40.01	41.40	44.00	47.11	49.44	54.55	56.53	58.50

Truck number	7	8	1	3	2	5	4	6
Arrive	57.10	57.30	57.40	62.51	63.24	69.13	73.00	73.40
Load	59.18	61.22	63.35	65.53	68.39	71.25	73.38	75.43
End of load	60.53	63.08	65.20	68.11	70.49	73.10	75.18	77.50

Truck number	8	3	2	7	5	4	6	
Arrive	76.36	85.51	86.30	87.01	88.08	92.42	93.44	
Load	97.32	100.21	104.45	106.27	108.09	0.00	109.58	
End of load	99.51	102.22	106.02	107.39	109.28	0.00	111.42	

Truck number	8	3	7	5	6	8	2	
Arrive	112.12	121.00	121.20	124.33	126.15	130.05	134.06	
Load	115.27	123.23	126.35	128.42	131.54	133.44	135.46	
End of load	116.11	125.58	128.04	130.23	133.01	135.12	137.11	

Truck number	3	5	8					
Arrive	142.10	147.08	148.03					
Load	142.38	147.49	149.50					
End of load	144.41	149.25	151.05					

APPENDIX B

A REVIEW OF THE MODELING PROCEDURE

This appendix presents a detailed discussion of the steps used in the modeling procedure and uses the tunneling model development of Section 5.7 as an illustration. Flowcharts describing the steps followed in developing the tunneling systems model are given in Figures B.1 to B.4. A discussion of each block in the flow diagram of Figure B.1 will reveal some of the basic principles involved in modeling systems in general.

Blocks 1 and 2

These blocks define the candidate set of flow units. A flow diagram describing this procedure is given as Figure B.2. In identifying these units, certain questions are helpful.

1. What units are processed by the system?
2. What support units are required to accomplish this processing?
3. What support units delay the processing for significant periods of time?
4. What units are to be monitored regardless of their criticality to processing?

In developing the tunneling model, the flow unit candidates were listed first. From this listing the unit was chosen that represented the processed unit (i.e., the pipe sections). This is the unit that enters the system, is operated on (processed), and then exits the system. In many cases its flow path forms the basic structure building component of the final system model. In some cases, however, as was seen in the mason example, this unit may be implicit in the cycle of one of the support units and therefore may be superfluous to the system as a separate cycle.

Block 3

Having determined what units are processed and which are required for this processing, the next step is to define the work tasks required for processing. These are the work tasks required for processing of the mainline flow unit and involve resources which are to be monitored. Definition of these work tasks may lead to the addition of some new units to the candidate list and deletion of others. It is an important step in the modeling procedure, since it determines the level of detail that the model will reflect.

In the case of the tunneling problem, the pipe sections are operated on by four major resources. This led to three major work tasks of interest in the

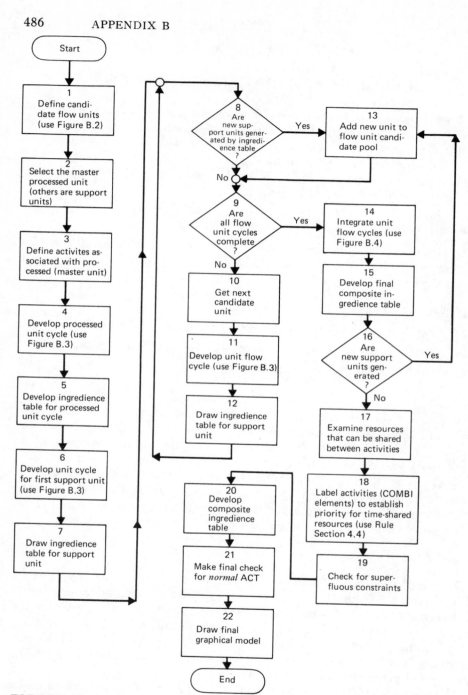

FIGURE B.1   Model-building algorithm.

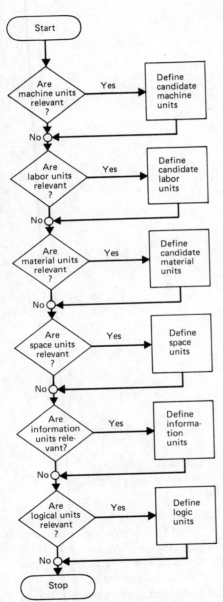

FIGURE B.2   Flow unit identification chart.

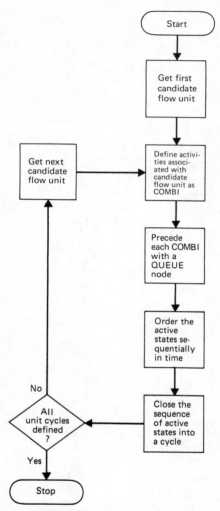

FIGURE B.3   Flow diagram for development of flow unit cycle.

processing sequence. The four resources of interest that are themselves flow unit candidates are (1) the crane, (2) the jack set, (3) the jacking collar, and (4) the tunneling machine. Since the jack set and the jacking collar operate on the pipe sections virtually simultaneously, the two have been considered as acting together in a single work task. If greater detail is desired regarding the system, the REPLACE work task (3) could be divided into two work tasks,

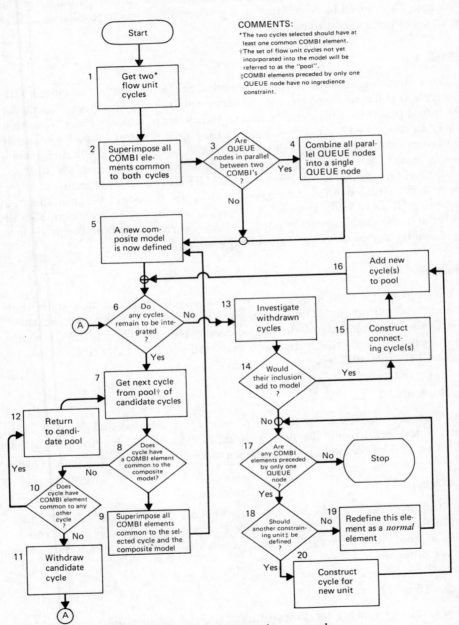

FIGURE B.4 Flow diagram for integration procedure.

one for the replacement of the collar and the other for the repositioning of the jack set. This would have resulted in a processed unit cycle consisting of four work tasks instead of three.

Block 4

When modeling individual resource cycles, the active states associated with the cycle are all considered to be ingredience constrained. Therefore, in drawing the initial processed unit cycle as well as the cycles of support units, a waiting state always precedes an active state. During flow integration and when final composite ingredience table is developed, it will be determined whether a work task is, in fact, resource constrained. If it is not it will be modeled as aNORMAL work task. Otherwise, it will remain a COMBI processor. This means that for every active state defined for a given unit cycle, there will be an associated preceding QUEUE node as well. Therefore, in any initial cycle, the number of active and idle states should be equal. A flow diagram describing the procedure for defining flow unit cycles is given in Figure B.3.

Block 5

The ingredience table displays the relationship between the flow units that interact with the processed unit and the work tasks in the mainline or processed unit cycle. Development of the ingredience table may yield further insights into the operation of the system and may indicate flow units to be added to original list of candidates.

Blocks 6 and 7

One of the support units defined previously is selected and its flow unit cycle is developed using the flow diagram in Figure B.3. Having completed definition of the flow cycle, an ingredience table is developed for the new cycle.

Blocks 8 to 13

In the process of developing the ingredience table of newly defined cycle, certain flow units not specified in Block 1 may suggest themselves. If this is the case, the new candidate flow units are added to the list of support units not yet proposed.

Blocks 10 to 12

These blocks repeat, for flows other than the master unit and the first support unit selected, the process of defining flow cycles and ingredience tables. Should construction of the ingredience table suggest new units, the Block 8 to 13 sequence causes addition of these units to the candidate unit pool. Consideration, when contemplating adding another flow unit to the system, must be given to the trade-off between increase in complexity versus impact on system operation. If a unit provides important information or its constraining effect on the system is significant, then it can be added. In general, however, the number of flow unit types should be held to the absolute minimum.

Blocks 14 and 15

When flow cycles for all of the candidate units have been defined, the cycles are integrated using the procedure specified in the integration flow diagram (Figure B.4). At this point, a tentative ingredience table for the entire model is developed to establish the ingredience constraints for each activity in the total model. The composite ingredience table indicates for each work task the preceding QUEUE nodes. By examining the columns of the composite ingredience table, establish which work tasks are common to two or more flow cycles. The number of checks in each column indicates the number of ingredients required for commencement of the work task. The columns allow direct development of the graphical segments for each resource constrained work task. At Block 16, a final check is made to see if any new flow unit candidates have been generated by the construction of the composite ingredience table. If new units are suggested, they are added, and cycles for them are developed. This action is initiated by returning to Block 13.

Blocks 17 and 18

Flow units that are shared between several work tasks must be given a routing priority. Flow units that require a routing priority can be determined by examining the QUEUE nodes of the composite model. If a QUEUE node is followed by more than one COMBI element, a routing priority for each of the follower COMBI elements must be established. In accordance with the labeling rule of Section 4.4, a flow unit is always routed to the following work task whose resource requirements are first satisfied. If there is a tie, the following work task with the lowest numerical label is selected. All idle states that are followed by two or more active states should be checked to see if the routing priority imposed by the labeling of the following work tasks is consistent with the desires of the modeler.

Block 19

In some cases, integration of the work tasks to produce the final model structure will lead to nested feedback loops. A nested feedback loop situation is shown in Figure B.5. In this case, the flow unit "B" cycle is nested within that of flow unit A. Assume that only one A unit and one B flow unit are

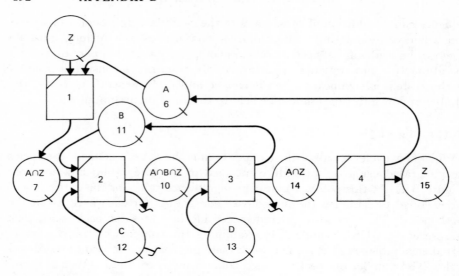

FIGURE B.5   Nested feedback loop.

defined. The system then implies that work task 2 cannot start until the units previously passing 2 have cleared work task 3. However, unit A travels on to 4 and then returns to 1. This constrains the flow of units through 1 such that this flow cannot commence until the previous Z unit has cleared 4. This effectively screens units from arriving at 1 and 2 until A has recycled from 4. Therefore, the return of B from 3 is superfluous, since it is sure to be at idle state 11 before A returns to idle state 6. Since no units can enter the segment 1, 2, 3, 4 until A returns, the constraining effect of B is preempted. In the example presented, it can be said that the cycling of B has no constraining effect on the system. The modeler may wish to retain idle state 11 in the model, however, to monitor the delays experienced by B because of the nesting effect as well as constraint caused by flow unit C.

Blocks 20 to 22

Having eliminated superfluous flow cycles and labeled work tasks to ensure the desired routing priorities, the final ingredience table for the total system can be developed. The final composite ingredience table will indicate the work tasks that, in the final model, are preceded by only one idle state, and, there-fore, are not constrained other than by the completion of the preceding work task. Such work tasks can be defined as NORMAL work tasks. The final graphical model to include final labeling and designation of NORMAL work tasks is now completed.

APPENDIX C

PROBLEM SPECIFICATION LANGUAGE

C.1 THE COMPUTER INTERFACE

The CYCLONE system user feeds information to the computer using an input vocabulary or work set that provides the vehicle for

1. Professional statement of the problem.
2. User-oriented access to the power of the computer.

The CYCLONE program is written in FORTRAN and can be adapted without too much difficulty to systems having FORTRAN compilation capability. In the program, the formal logic is a network representation of the system of interest. A description of the number of work tasks involved, their attributes and interrelationships provides information that is sufficient to drive rigorous simulation algorithms.

This appendix introduces and discusses the problem specification language used in the CYCLONE system.

The CYCLONE system utilizes a free-form input reader and a set of relatively simple problem specification words that allow the user to specify his problem using either batch mode card or on-line input. When specifying input to the CYCLONE program, each item in the input stream should be spaced apart from all preceding and subsequent items. That is, a blank should be used between each name, label, number, and the like. Because of this free-formatting feature, exact spacing (other than as noted) or placement of the input entries on the input cards or on the type carriage when working on-line is not critical.

The set of problem specification words used in the CYCLONE system allows the user to develop the data set required for problem solution directly from precedence diagrams of the situation to be modeled. The CYCLONE vocabulary and the hierarchical structure that describes the order of problem description as shown previously in Chapter Twelve is given in Figure C.1.

The actual specification of a problem in the problem-oriented language can be divided into five major sections.

1. General system information.
2. Network input.
3. Flow unit initialization.
4. Report information.
5. Work task duration information.

The problem specification words appropriate to each of these sections are shown in Figure C.1. Those words that are not required and that may be deleted when specifying a given system are shown in brackets (e.g., $< \cdot \cdot >$). The optional words are either:

1. Used to make the input more readable and thus give the user a feeling of conversational interaction with the computer.

2. Used to define parameters that have default values in the program and therefore need not be specified by the user unless desired.

The section headers associated with input sections are as follows.

Input Section	Header
1. General system information	There is no header word used in conjunction with this section. The two required procedural words, NAME and RUNS, may be thought of as header words.
2. Network input	NETWORK INPUT
3. Flow unit initialization	FLOW UNIT INPUT or EQUIPMENT INPUT
4. Report information	REPORT HISTOGRAMS
5. Work task duration information	DURATIONS

C.2 GENERAL SYSTEM INFORMATION

Certain items of information pertaining to the system as a whole must be specified to the computer. These items of information pertain to the labeling of reports and the specification of program parameters. The report labels that may be specified are:

1. The date of the problem run.

2. A project number to be associated with output generated by the system.

The date consists of three integer values indicating the month, the day, and up to a four-digit representation for the year. An integer value of not more than four digits may be specified by the user as the project number. If the project number is submitted as a negative value, a diagnostic trace of the system

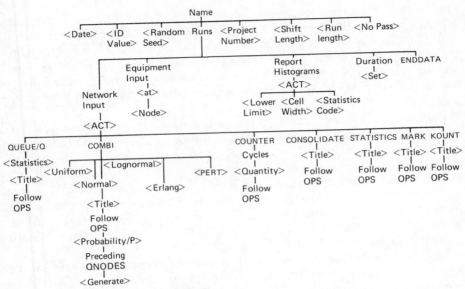

FIGURE C.1 CYCLONE POL hierarchy.

entity flows is printed as output (see Figure 12.14). The labels that must be submitted are the project name and the RUNS header. The project name is submitted in quotation marks following the procedural word NAME. Input items specified in quotation marks will be referred to as input *strings*.* The following input segment illustrates the specification of the data, project number, and name in the CYCLONE word set.

DATE 4 8 1975
PROJECT NUMBER 113
NAME "HI RISE LIFT PROBLEM"

The other major category or procedural words used in the general system information section pertains to the definition of program parameters. The only parameter that must be specified when submitting a job is the number of runs or replications of the simulation experiment that are to be performed before the simulation is terminated. The number of replications is defined as an integer value that is submitted following the procedural word RUNS.

The definition of an ID value (see Table C.1) is optional. This value establishes the length of the list that keeps permanent information regarding the

---

*Apostrophes are used in lieu of quotation marks when using the IBM system.

attributes of the system elements and information regarding the location and scheduled movement of system flow entities. The ID value is proportionally related to the size of the CYCLONE system model to be analyzed. The default value for this parameter is 100. This value is economical for most systems of interest found in construction. The user overrides this value by submitting the words ID VALUE in the input stream followed by the desired integer value.

During the simulation, stochastic variables as required for work task durations are generated. This requires the generation of a stream of random numbers that draw random values from the appropriate probability distribution. Associated with each random number stream is a random number seed. The use of a unique seed generates a uniquely defined set of stream of random numbers. The user has the option of defining a unique random number seed for each simulation experiment. The procedural expression used in this instance is RANDOM SEED, or simply SEED, followed by a decimal number not to exceed 10 digits.

The user can also specify a time limit that controls the total duration of each replication of the experiment. This parameter establishes a value that, when exceeded by the simulated clock value, TNOW, causes the replication to terminate and summary statistics for the replication to be taken. This parameter works in conjunction with the CYCLES parameter (defined for the COUNTER ACCUMULATOR element in the section on NETWORK INPUT) to cause the replication to terminate. Whichever value, RUN LENGTH or CYCLES, is exceeded first governs. The desired time duration is specified using the procedural words RUN LENGTH followed by a decimal value. The decimal number represents the time cutoff in minutes.

The system production achieved is printed following each replication of the simulation experiment. The production based on system hourly and shift output is given. The user may designate a shift length to be used in this calculation by using the procedural words SHIFT LENGTH followed by the decimal value in hours to be used for the shift duration.

In some systems the modeler wishes to preclude the possibility of one flow unit passing another in the transit of elements within the network. That is, it is desired that no transit time be generated that would cause the unit being processed to complete a work task prior to or ahead of another unit that is already in transit. A practical example of this might be the modeling of a haul road where no passing is possible. Therefore, all trucks on the road must maintain their sequence (i.e., no passing), and faster trucks cannot finish the travel work task ahead of slower trucks already in transit.

To implement this function, the construct NO PASS is specified in the general information segment. This function insures that all entities transiting a given work task maintain the transit discipline and that units that begin transit first finish transit first.

The procedural words used in the general system information section with their functions and default values are summarized in Table C.1. A typical input segment specifying general information might appear as follows.

> **NAME "HOPPER SIMULATION PROBLEM"**
> **DATE 7 24 1972**
> **ID VALUE 50**
> **RANDOM SEED 1431**
> **PROJECT NUMBER 1**
> **SHIFT LENGTH 8.0 HRS***
> **RUNS 2**

Using default values for all but required input items, the above segment can be reduced to

> **NAME "HOPPER SIMULATION PROBLEM"**
> **RUNS 2**

C.3 NETWORK DEFINITION

In this section, words used for the specification of the simulation network topology will be defined. These words define the structure of the system through which system entities must flow. The procedural words used in the NETWORK INPUT segment are given with their associated functions and default values in Table C.2.

In CYCLONE, the definition of a modeling element establishes a file in the program storage arrays, which contains the attribute information relating to the element. The first file location in the storage arrays is reserved for the storage of work tasks scheduled for completion. This is a dynamic file that is updated at the end of each simulation time step. The information relating to the work tasks that make up the network (distribution type, parameter set, etc.) remains the same throughout the simulation. However, files must be allocated for the storage of this information. These files are numbered, starting with the integer value 2. *Therefore, no element should be designated as node 1.*

In assigning parameter set numbers to the various work tasks, use of parameter set 1 calls an intrinsic in the program that sets the work task duration to a constant value of 0.0. Therefore, user-specified parameter sets should be defined starting with the parameter set numerical label of 2 and numbering in ascending order. Parameter set 1 should be used only when a constant value of 0.0 is desired.

---

*The use of the time units HRS and MIN is optional, since hours and minutes are the assumed units for the shift length and run length respectively.

Table C.1   PROCEDURAL WORDS USED IN GENERAL
INFORMATION SECTION

Pro Word(s)	Function	Default
NAME	Used to specify the name to be printed on reports relating to the specified system.	None—specification required
DATE	Specification of the month, day, and year to be printed on reports pertaining to this system (e.g., 3 10 1973).	0/0/0
ID VALUE	Used to specify the size of the array used to store information on arcs and flow entities in the system.	100
RUNS	Specification of the number of times the system is to be simulated.	None—specification required
RANDOM SEED	Used to specify the number that initializes or starts the stream of random numbers used in the simulation.	1267
PROJECT NUMBER	Specifies the number to be printed on reports for project numerical label. If the value is specified as a negative number, a trace of the entity flow will be printed.	1
SHIFT LENGTH	The length in hours of each shift. This value is used for the shift production figure printed in the production report.	4.5
RUN LENGTH	This value specifies the time cutoff in minutes to be used in each simulation run. In the network input section (below), a simulation termination based on number of cycles may also be specified. If both are speci-	9999

(Continued)

Table C.1   (*Continued*)

Pro Word(s)	Function	Default
	fied, the criterion that is realized first is utilized to terminate the simulation.	
NO PASS	This function, when specified, insures that transit discipline is maintained by units transiting network elements and those units that begin transit first complete transit first (i.e., FIFO discipline).	OFF

The use of the network defining word set for specification of typical network elements is illustrated in Chapter Twelve. The symbolic notation is the same as that used in Chapter Twelve.

( ) This symbol indicates that an item is optional.

{ } This symbol indicates that one or a list of items is required.

| This vertical line indicates "or" in graphical notation.

Generally, the input line for a COMBI processor element is as follows:

(ACT) {numerical label} {COMBI} $\begin{pmatrix} \textbf{CONSTANT} \\ \textbf{NORMAL} \\ \textbf{UNIFORM} \\ \textbf{ERLANG} \\ \textbf{LOGNORMAL} \\ \textbf{PERT} \end{pmatrix}$

(PARAMETER SET set number) ("title")

(FOLLOW OPS) $\left\{ \text{label(s) of following element(s)} \right.$

$\left(\begin{Bmatrix} \textbf{PROBABILITY} \\ \textbf{|P} \end{Bmatrix} \begin{pmatrix} \textbf{decimal value} \\ \textbf{from 0.0 to 1.0} \end{pmatrix} \right) \Big\}$

{PRECEDING QNODES ⟨label(s) of queue nodes⟩
    (GENERATE ⟨value of number generated⟩)}

In the event that the GENERATE function is used in conjunction with a preceding queue node, the procedural word GENERATE is added to the input stream.

Table C.2   NETWORK INPUT PROCEDURAL WORDS

Procedural Word(s)	Function	Default Values
ACT N	Used to designate the numerical label associated with the modeling element, where N is the integer value of the label.	Use is optional—not required (N is required)
COMBI or C	Indicates the associated element is ingredience constrained and performs a combination function.	None
CONSOLIDATE N	Indicates the associated element acts to consolidate or group N incoming units into one output unit, where N is an integer value specified in decimal format(e.g., 3.0, 8.).	None
KOUNT N	Indicates that the associated element is used as a counter maintaining a register of unit passages where N is the number of the register.	None
COUNTER	Used to define an accumulator element that counts the number of entity passages and accumulates system production.	None—specification is required
CYCLES N	Required with accumulator elements and indicates the number of system cycles, N, to be simulated.	None—specification is required
FOLLOW OPS	Used to specify the logic between node elements by indicating the nodes that follow the element being defined.	None—specification is required
GENERATE N	Indicates that the associated element generates N output units for each input unit, where N is an integer value specified in decimal format (e.g., 2.0, 16.0).	None
MARK	Indicates that the associated element performs a marking function.	None

(Continued)

Table C.2   (*Continued*)

Procedural Word(s)	Function	Default Values
NETWORK INPUT	Alerts program that input defining network elements and network topology is to follow.	Specification required
PARAMETER SET N	Used to specify the number of the parameter set to be used in generating transit times for the associated element (N is an integer value in decimal format).	Parameter set = 1
PRECEDING QNODES	Indicates that specification of the queue nodes that precede the associated COMBI processor.	None—specification required.
PROBABILITY or P	Used following FOLLOW OPS values to specify the probability of transit along the arcs defined by the following elements.	1.0 (i.e., 100%)
QUEUE or Q	Used to label an element that precedes a COMBI element and in which units are delayed subject to realization of ingredience requirements.	None—specification required
QUANTITY	Specifies the multiplier to be used in counting system production each time a unit passes the accumulator.	1.0
STATISTICS	Indicates that the associated element is used as a statistics collection element.	None
TITLE	Indicates that a descriptive title for the element being defined will follow.	
UNIFORM NORMAL LOGNORMAL ERLANG PERT	This set of words is used to indicate the type of probability distribution associated with the element being defined.	Specifies a CONSTANT time parameter associated with work task.

The general defining input for a QUEUE node is:

(ACT) {numerical label} {Q|QUEUE} (STATISTICS) ("Title")
(FOLLOW OPS)

{numerical label(s) for following element(s)}

The single letter Q can be substituted for QUEUE in the above input. Statistics regarding unit flows at queue nodes are defined by using the procedural word STATISTICS in the input stream.

The general defining line input for a NORMAL work task is:

$$\textbf{(ACT) \{numerical label\}} \begin{pmatrix} \textbf{CONSTANT} \\ \textbf{NORMAL} \\ \textbf{UNIFORM} \\ \textbf{ERLANG} \\ \textbf{LOGNORMAL} \\ \textbf{PERT} \end{pmatrix}$$

(PARAMETER SET set number) ("title")

(FOLLOW OPS) $\Big\{$**label(s) of following element(s)**

$$\left(\left\{ \begin{matrix} \textbf{PROBABILITY} \\ \textbf{|P} \end{matrix} \right\} \left\langle \begin{matrix} \textbf{decimal value} \\ \textbf{from 0.0 to 1.0} \end{matrix} \right\rangle \right) \Big\}$$

The input required to define a function node in the general case is:

$$\textbf{(ACT) \{integer numerical label\}} \begin{cases} \textbf{STATISTICS} \\ \textbf{MARK} \\ \textbf{CONSOLIDATE} \langle \textbf{GROUP SIZE} \rangle \\ \textbf{KOUNT} \quad \langle \textbf{Register Number} \rangle \end{cases}$$

("title") (FOLLOW OPS) {decimal labels of following elements

({PROBABILITY|P} ⟨decimal value⟩)}

The functions are not mutually exclusive, and it is possible to have multi-function nodes (e.g., STATISTICS CONSOLIDATE). The group size associated with the procedural word CONSOLIDATE is the value N, which indicates the number of flow units that must arrive to cause one unit exit (i.e., the number to be consolidated). This is specified as a decimal (e.g., 5.0). The register number following the procedural work KOUNT is also specified as a decimal number.

The general format for the ACCUMULATOR COUNTER element is:

(ACT) {numerical label} COUNTER (CYCLES ⟨number of cycles⟩)

(QUANTITY ⟨quantity multiplier⟩) (FOLLOW OPS) {label(s) of
following element(s)}

If no QUANTITY parameter has been specified, the default value is used. The default value for the QUANTITY parameter is one (1).

It is possible to associate a probability of transit with arcs emanating from a CYCLONE element. This is specified using the procedural word PROBA-BILITY or the abbreviation P in conjunction with the follow operation attribute for elements other than the QUEUE node.

C.4 UNIT INITIALIZATION

In the network input segment the paths and elements through which various entities flow were defined. In this section of the input, the entities are initialized into the system. The two questions that must be answered relative to the initializing of units are:

1. At what point in the system are entities to be initialized?

2. What are the type and number of entities to be initialized?

The CYCLONE POL allows simple specification of these attributes. The following words are used for entity initialization:

Word(s)	Function	Default
UNIT INPUT or EQUIPMENT INPUT	These words alert the program to entity initialization information to follow.	None—must be specified.
UNIT(S)	Used to introduce the discrete number of a given type of resource that is to be initialized in into the system.	Use optional
AT, NODE	The words are used to introduce the number of the QUEUE node at which the specified entities are to be initialized.	Use optional

In this input segment, titles for each of the entity types can also be inserted between quotation marks (i.e., as a string). This entity title will be reprinted in the data reprint. The section header and a typical entity initialization appears as follows.

EQUIPMENT INPUT

3 "TRUCKS" AT NODE 5.

or

3 "TRUCKS" AT 5.

or

3 5.

C.5 WORK TASK DURATION INPUT

In the NETWORK INPUT section, values indicating the parameter set number associated with each modeling element are specified. For instance, PARAMETER SET 2 was associated with COMBI work task 3 in Figure 12.5. At this point in the specification of the problem, the parameters included in these sets are specified. Four fields are associated with each set, and the parameters specified in each field are given in Table C.3. The control words used for submitting these parameter values to the computer follow.

Table C.3 PARAMETER SET VALUES

Distribution Type	Parameter Defined in			
	Field 1	Field 2	Field 3	Field 4
CONSTANT	Constant time value	Not used	Not used	Not used
NORMAL LOGNORMAL	Mean value	Minimum value	Maximum value	Standard deviation
UNIFORM	Not used	Minimum value	Maximum value	Not used
ERLANG	Mean divided by value in field 4	Minimum value	Maximum value	Number of exponential deviates
PERT	Most likely value, M	Optimistic value, A	Pessimistic value, B	Not used

Word	Function	Default
DURATIONS	This control word alerts the program that input pertaining to the definition of parameter sets will follow.	None-required
SET	This word is used to specify the numerical label or pointer used to access and store the parameter value sequence.	Use optional

Following the use of the control word SET, the integer representing the set number is submitted. This integer is then followed by a set of real-valued numbers in sequence, starting with the field one value and continuing through field four. The DURATION specification segment for a typical network is:

DURATIONS
SET 2   2.185   0.728   5.12   0.546
SET 3   6.2   2.07   10.84   0.403

C.6 REPORT SPECIFICATION

The last segment of input data required by the CYCLONE system specifies the nature and type of defined statistics presented in reports generated at the end of each CYCLONE simulation. The CYCLONE program provides automatic reports on the COUNTER element and the QUEUE nodes preceding each COMBI processor. In addition, statistics are generated on those elements designated as STATISTICS elements in the NETWORK INPUT section of the input (see above). In this section, values are submitted that establish the cell width and lower limiting values of the frequency histograms printed in the final report. The types of statistic to be collected on each element are also specified in this input segment. The statistics codes used for requesting these statistic types are shown below.

Number Code	Description
1	The time of *first* arrival at a state or node.
2	The time of *all* arrivals at a state or node.
3	The time *between* arrivals at a state or node.
4	The time *interval* required between arrival at one state and arrival at a second state.
5	The time for one consolidation of M arrivals on nodes designated as CONSOLIDATION nodes.

In addition to these requested statistics, the CYCLONE program automatically prints statistics at the end of each run indicating:

1. The average of units delayed in Q nodes preceding combination states.

2. The percent of the time units were delayed in the Q nodes preceding combination states.

3. The average hourly production rate for the run.

4. The average shift production based on the shift duration (SHIFT LENGTH) specified in the general information input section.

The procedural word ENDDATA is used to delimit the problem defining input.

APPENDIX D

QUEUEING NOMOGRAPHS

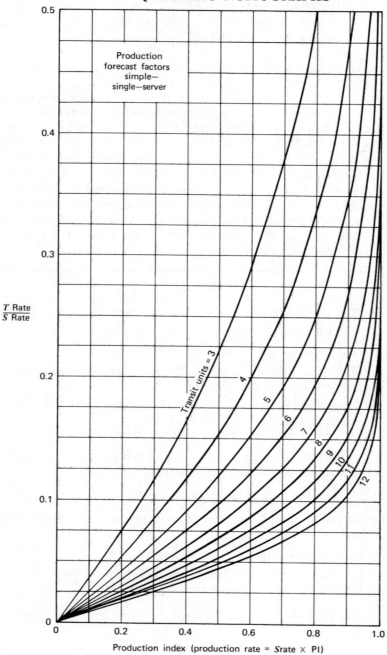

Production
forecast factors
simple—
single—server

$\dfrac{T \text{ Rate}}{S \text{ Rate}}$

Transit units = 3

4

5

6

7

8

9

10

11

12

Production index (production rate = Srate × PI)

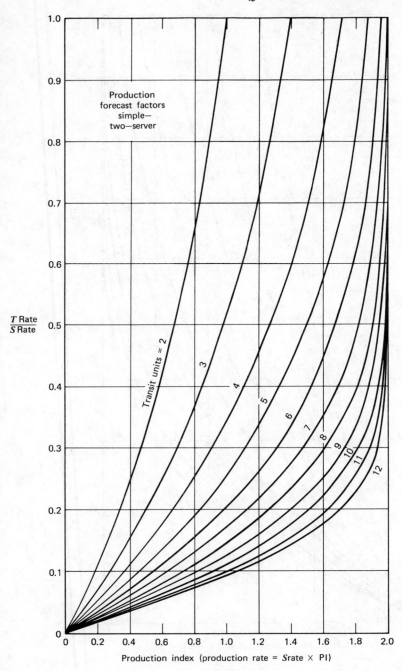

Production
forecast factors
simple—
two—server

$\dfrac{T\,\text{Rate}}{S\,\text{Rate}}$

Transit units = 2

3

4

5

6

7

8

9

10

11

12

Production index (production rate = Srate × PI)

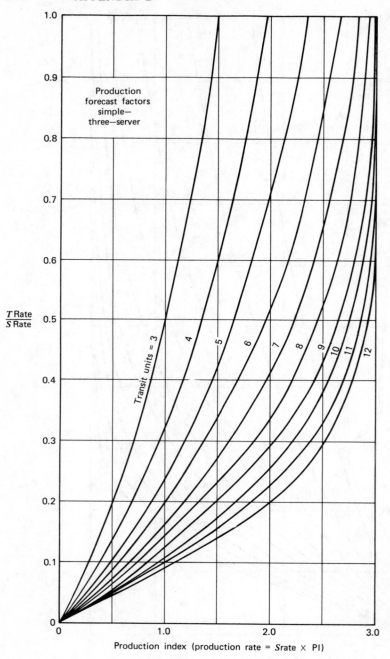

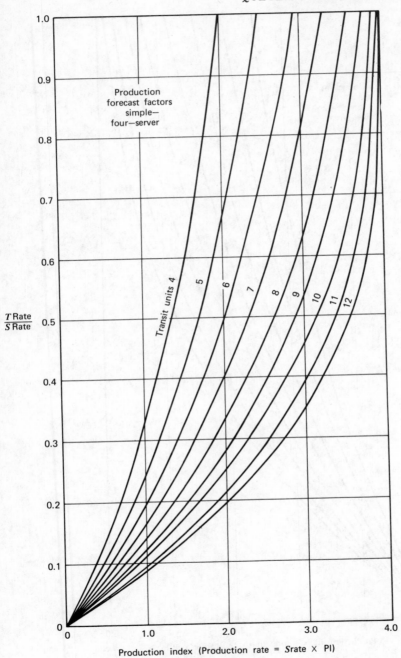

Production index (Production rate = Srate × PI)

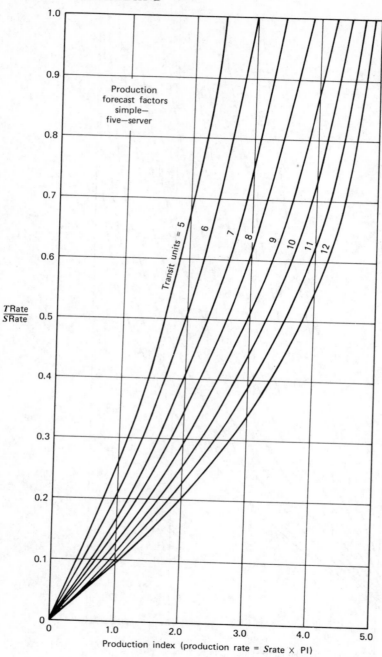

Production forecast factors simple— five—server

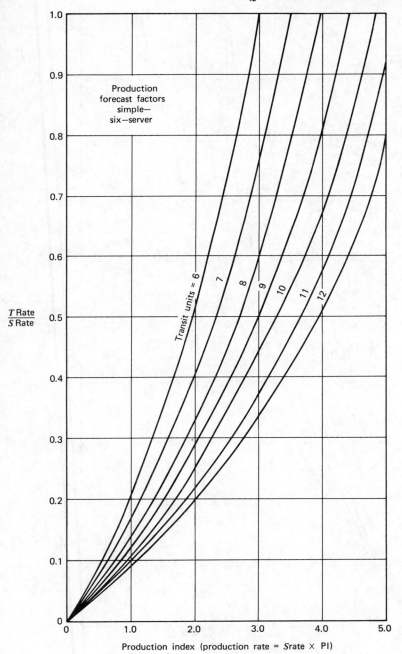

Production index (production rate = Srate × PI)

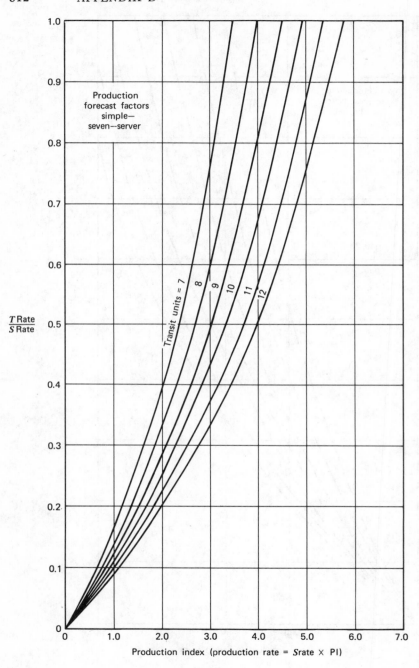

Production forecast factors
simple—
seven—server

$\dfrac{T\,\text{Rate}}{S\,\text{Rate}}$

Transit units = 7

8 9 10 11 12

Production index (production rate = Srate $\times$ PI)

Hopper—Single—Server
Production forecast factors Hopper Capacity = 1 Transit units = 2

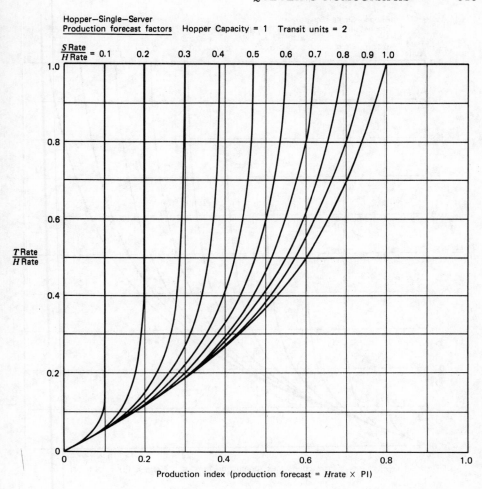

$\frac{S\,\text{Rate}}{H\,\text{Rate}}$ = 0.1 0.2 0.3 0.4 0.5 0.6 0.7 0.8 0.9 1.0

$\frac{T\,\text{Rate}}{H\,\text{Rate}}$

Production index (production forecast = Hrate × PI)

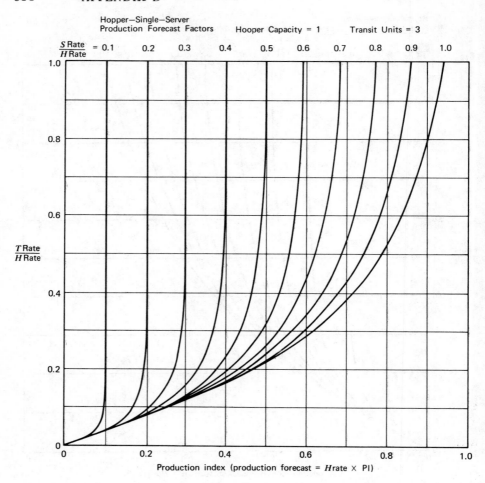

Hopper—Single—Server
Production Forecast Factors   Hooper Capacity = 1   Transit Units = 3

$\frac{S \text{ Rate}}{H \text{ Rate}}$ = 0.1   0.2   0.3   0.4   0.5   0.6   0.7   0.8   0.9   1.0

$\frac{T \text{ Rate}}{H \text{ Rate}}$

Production index (production forecast = Hrate × PI)

Hopper—Single—Server
Production Forecast Factors   Hopper Capacity = 1   Transit Units = 4

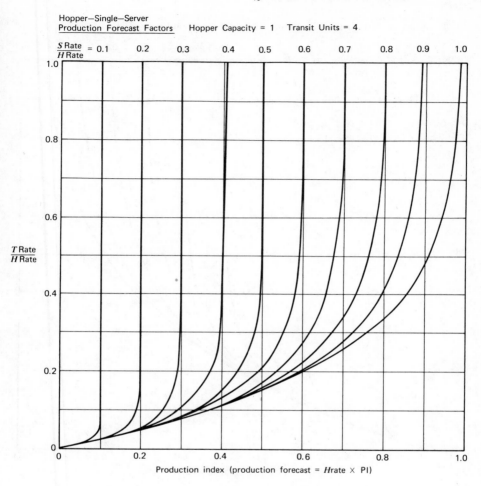

Production index (production forecast = Hrate × PI)

Hopper—Single—Server
Production Forecast Factors   Hopper Capacity = 2   Transit Units = 3

$\dfrac{S \text{ Rate}}{H \text{ Rate}}$ = 0.1   0.2   0.3   0.4   0.5   0.6   0.7   0.8   0.9   1.0

$\dfrac{T \text{ Rate}}{H \text{ Rate}}$

Production index (production forecast = Hrate × PI)

Hopper—Single—Server
Production Forecast Factors   Hopper Capacity = 2   Transit Units = 4

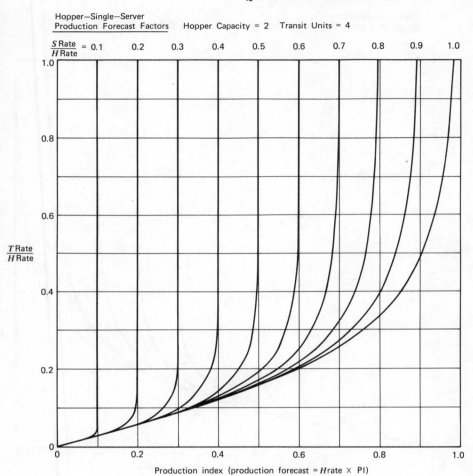

Production index (production forecast = Hrate × PI)

Hopper—Single—Server
Production Forecast Factors   Hopper Capacity = 2   Transit Units = 5

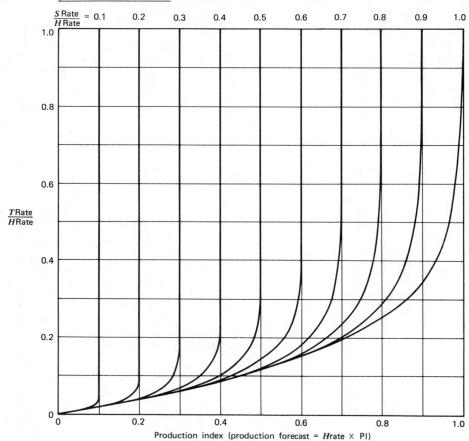

$\dfrac{S\,\text{Rate}}{H\,\text{Rate}} = 0.1$ 0.2   0.3   0.4   0.5   0.6   0.7   0.8   0.9   1.0

$\dfrac{T\,\text{Rate}}{H\,\text{Rate}}$

Production index (production forecast = Hrate $\times$ PI)

Hopper—Single—Server
Production Forecast Factors   Hopper Capacity = 3   Transit Units = 4

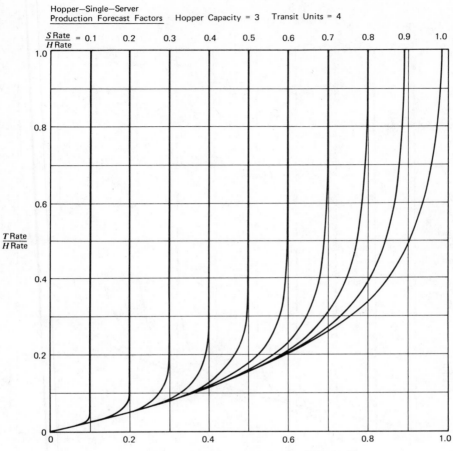

Production index (production forecast = Hrate × PI)

Hopper—Single—Server
Production Forecast Factors   Hopper Capacity = 3   Transit Units = 5

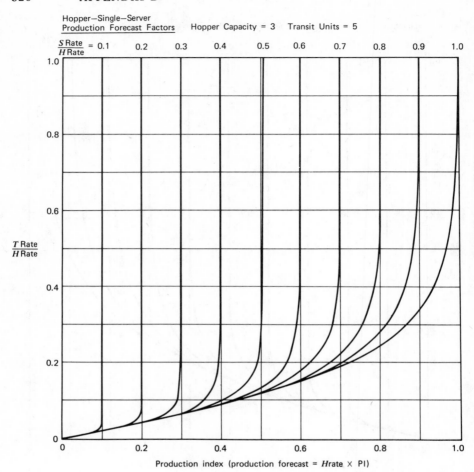

Production index (production forecast = *H*rate × PI)

Hopper—Single—Server
Production Forecast Factors   Hopper Capacity = 3   Transit Units = 6

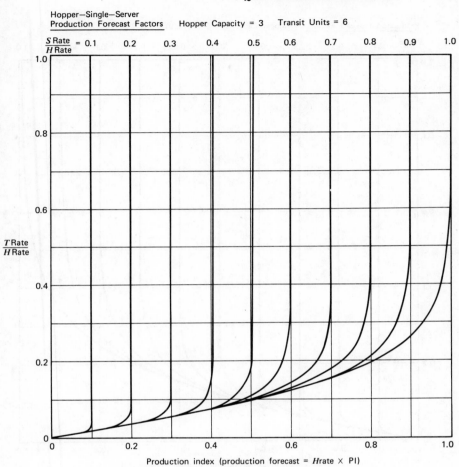

Production index (production forecast = Hrate × PI)

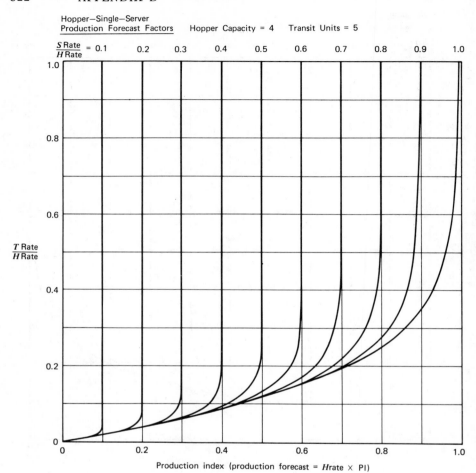

Hopper—Single—Server
Production Forecast Factors   Hopper Capacity = 4   Transit Units = 5

$\frac{S\,\text{Rate}}{H\,\text{Rate}}$ = 0.1   0.2   0.3   0.4   0.5   0.6   0.7   0.8   0.9   1.0

$\frac{T\,\text{Rate}}{H\,\text{Rate}}$

Production index (production forecast = Hrate × PI)

Hopper—Single—Server
Production Forecast Factors   Hopper Capacity = 4   Transit Units = 6

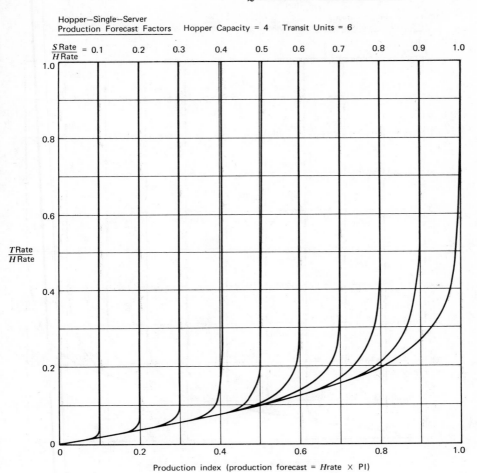

Production index (production forecast = Hrate $\times$ PI)

Hopper—Single—Server
Production Forecast Factors   Hopper Capacity = 4   Transit Units = 7

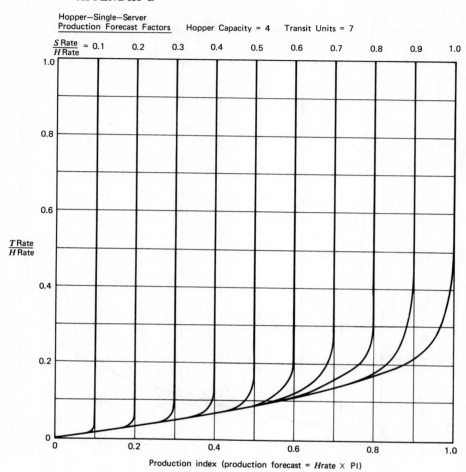

Production index (production forecast = Hrate × PI)

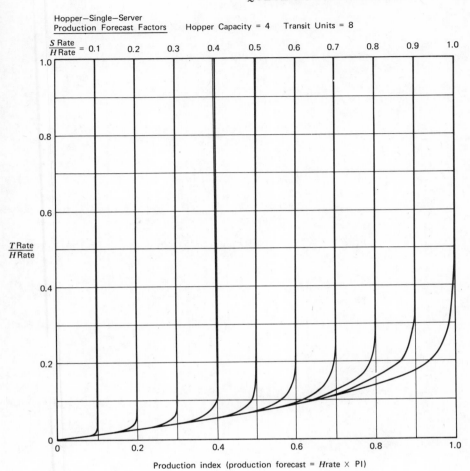

Hopper—Single—Server
Production Forecast Factors   Hopper Capacity = 4   Transit Units = 8

$\frac{S\ \text{Rate}}{H\ \text{Rate}}$ = 0.1   0.2   0.3   0.4   0.5   0.6   0.7   0.8   0.9   1.0

$\frac{T\ \text{Rate}}{H\ \text{Rate}}$

Production index (production forecast = Hrate × PI)

INDEX